INNOVATION, PATH DEPENDENCY, AND POLICY

Innovation, Path Dependency and Policy

The Norwegian Case

Edited by
JAN FAGERBERG
DAVID C. MOWERY
and
BART VERSPAGEN

OXFORD
UNIVERSITY PRESS

Great Clarendon Street, Oxford OX2 6DP

Oxford University Press is a department of the University of Oxford.
It furthers the University's objective of excellence in research, scholarship,
and education by publishing worldwide in

Oxford New York

Auckland Cape Town Dar es Salaam Hong Kong Karachi
Kuala Lumpur Madrid Melbourne Mexico City Nairobi
New Delhi Shanghai Taipei Toronto

With offices in

Argentina Austria Brazil Chile Czech Republic France Greece
Guatemala Hungary Italy Japan Poland Portugal Singapore
South Korea Switzerland Thailand Turkey Ukraine Vietnam

Oxford is a registered trade mark of Oxford University Press
in the UK and in certain other countries

Published in the United States
by Oxford University Press Inc., New York

First published 2009

British Library Cataloguing in Publication Data
Data available

Library of Congress Cataloging in Publication Data
Data available

Typeset by SPI Publisher Services Ltd., Pondicherry, India
Printed in Great Britain
on acid-free paper by the
MPG Books Group, Bodmin and King's Lynn

ISBN 978–0–19–955155–2

5 7 9 10 8 6 4

Acknowledgments

This book is based on a research project ("Innovation, Path-Dependency, and Policy") financed by the Norwegian Research Council (KUNI program, project no. 154877). The support of the Research Council was indispensable to the creation and drafting of the book. Professor Mowery's participation was supported in part by a sabbatical leave from the University of California, Berkeley and by a grant from the US National Science Foundation (Cooperative Agreement 0531184). Research Council funding supported a number of workshops at which project participants presented and discussed their work with one another and scholars from Norway and abroad. These workshops greatly enhanced the quality of the work and we would like to thank all of the participants for their advice and encouragement. Equally important for the final drafting and editing of this volume was the support of the Centre for Advanced Study at the Norwegian Academy of Science and Letters in Oslo, which sponsored the "Understanding Innovation Group" that included the editors and some of the contributors to the volume during the academic year 2007–2008. The editors would like to express their gratitude to Director Willy Østreng and his staff for providing a supportive and stimulating research environment for the final stages of the work on this volume. We also are indebted to Koson Sapprasert for his excellent assistance in the preparation of the manuscript. Finally, the volume benefited from the editorial and administrative talents of Susan Beer, Matthew Derbyshire, and David Musson of Oxford University Press.

Jan Fagerberg
David C. Mowery
Bart Verspagen

Contents

List of Contributors

Heidi Wiig Aslesen, Norwegian School of Management (BI) and Norwegian Institute for Studies in Innovation, Research, and Education (NIFU STEP), Oslo

Fulvio Castellacci, Norwegian Institute of International Affairs (NUPI), Oslo

Tommy H. Clausen, Nordland Research Institute, Bodø and Centre for Technology, Innovation, and Culture (TIK), University of Oslo

Ole Andreas Engen, University of Stavanger

Jan Fagerberg, Centre for Technology, Innovation, and Culture (TIK), University of Oslo

Terje Grønning, Institute for Educational Research, University of Oslo

Magnus Gulbrandsen, Norwegian Institute for Studies in Innovation, Research, and Education (NIFU STEP), Oslo

Svein Erik Moen, Institute for Labour and Social Research (FAFO), Oslo

David C. Mowery, Haas School of Business, University of California—Berkeley

Svein Olav Nås, Norwegian Institute for Studies in Innovation, Research, and Education (NIFU STEP), Oslo

Lars Nerdrum, Department of Policy Analysis, Lifelong Learning, and International Affairs, Norwegian Ministry for Education and Research, Oslo

Knut Sogner, Norwegian School of Management (BI), Oslo

Bart Verspagen, Faculty of Economics and UNU-MERIT, Maastricht University, Netherlands and Centre for Technology, Innovation, and Culture (TIK), University of Oslo

Olav Wicken, Centre for Technology, Innovation, and Culture (TIK), University of Oslo

List of Figures

List of Tables

List of Abbreviations

AFGC	AKVAFORSK Genetics Center
AIAG	Aluminium Industrie Aktien-Gesellschaft
AIDS	Acquired immunodeficiency syndrome
AKVAFORSK	Institutt for akvakulturforskning (Institute of Aquaculture Research)
ALCAN	Aluminum Limited of Canada
ALCOA	Aluminum Company of America
AME	Advanced MicroElectronics
ANOVA	Analysis of Variance
ÅSV	Årdal og Sunndal Verk
BACO	British Aluminium Company
BI	Norwegian School of Management
CGE&Y	Cap Gemini Ernst & Young
CIS	Community Innovation Survey
CMI	Chr. Michelson Institute
COMPADEC	Compagnie pour l'Etude et le Development des Echanges Commerciaux
CONDEEP	Concrete Deep Water Structure
CRINE	Cost Reductions in a New Era
DBF	Dedicated Biotechnology Firm
DNN	Det Norske Nitridaktieselskap
EB	Elektrisk Bureau
ECA	Economic Cooperation Administration
ESF	European Science Foundation
EU	European Union
FFI	Forsvarets forsknings institutt (Norwegian Defence Research Establishment: NDRE)
FUGE	Research Council of Norway Functional Genomics program
FUNN (SKATTEFUNN)	Forskning og Utvikling i et Nyskapende Næringsliv (Public R&D subsidy: tax credit scheme)

GDP	Gross Domestic Product
GGDC	Groningen Growth and Development Centre
GSM	Global System for Mobile Communications
GWA	Goodwill Agreement
HEI	Higher education institution
HIV	Human immunodeficiency virus
IAI	International Aluminium Institute
ICT	Information and Communication Technology
IFA/IFE	Institute for (Nuclear) Energy Research
KV	Kongsberg våpenfabrikk (Kongsberg Weapons Factory)
LME	London Metal Exchange
LNG	Liquid Natural Gas
MabCent	Centre for Marine Bioactivities and Drug Discovery
MARINTEK	Norwegian Marine Technology Research Institute
MCER	Ministry of Church, Education, and Research
MER	Ministry of Education and Research
MTI	Ministry of Trade and Industry
NACO	Norsk Aluminum Company (Norwegian Aluminium Company)
NATO	North Atlantic Treaty Organization
NAVC	Norwegian Association of Venture Capital
NAVF	Norges Allmennvitenskapelige Forskningståd (Norwegian Research Council for Science and Humanities)
ND	Norsk Data
NDRE	National Defence Research Establishment (also FFI)
NFFR	Norwegian Research Council for Fisheries
NIFU STEP	Norsk institutt for studien av innovasjon, forskning og utdanning (Norwegian Institute for Studies in Innovation, Research, and Education)
NIS	National Innovation System
NLH	Norges Landbrukshøgskole (Norwegian College of Agriculture)
NLTH	Norsk Leverandårer til Havbruksnæringen (Norwegian Equipment Producers Association)
NMT	Nordisk Mobil Telefoni (Nordic Mobile Telephone)
NOFIMA AS	Norwegian Fisheries and Food Research Inc.
NOK	Norwegian Krone (currency)
NORSOK	Norsk Sokkels Konkurranseposisjon

NOU	Norges offentlige utredninger (Norwegian Official Report)
NR	Norsk Regnesentral (Norwegian Computing Centre)
NRC	Norwegian Research Council (=RCN)
NRF	Norsk Rødt Fe (Norwegian red cattle)
NTH	Norges Tekniske Høgskole (Norwegian Institute of Technology)
NTNF	Norges Teknisk-Naturvitenskapelige Forskningsråd (Norwegian Research Council for Scientific and Industrial Research)
NTNU	Norges Teknisk-Naturvitenskaplige Universitet (Norwegian University of Science and Technology)
NYSE	New York Stock Exchange
OD	Oljedirektoratet (Oil Directorate)
OECD	Organization for Economic Cooperation and Development
OSE	Oslo Stock Exchange
PIL	Prosessindustriens Landsforening (Federation of Norwegian Process Industries)
PPP	Purchasing Power Parity
PRO	Public Research Organization
PSR	Public Sector Research
R&D	Research and Development
RA	Raufoss Ammunition Factory
RCN	Research Council of Norway (=NRC)
RF	Rogalandsforskning (Rogaland Research)
SDFI	State's Direct Financial Interest
SGN	Société Générale des Nitrures
SI	Sentralinstitutt for industriell forskning (Central Institute for Industrial Research)
SIMRAD	Simonsen Radio
SINTEF	Stiftelsen for industriell og teknisk forskning (Foundation for Industrial and Technological Research)
SIS	Sectoral Innovation System
SITC	Standard International Trade Classification
SME	Small and Medium-sized Enterprise
STAN	Structural Analysis
STEP	Senter for innovasjonforskning (Centre for Innovation Research)
STK	Standard Telofon og Kabelfabrik A/S (Standard Telephone and Cable Company)

TEKES	Teknologian jo innovotioiden kehittämiskeskus (Finnish Funding Agency for Technology and Innovation)
TF	Televerkets Forsknings institutt (Telephone Agency's Research Institute)
TIK	Senter for tecknologi, innovasjon og kultur (Centre for Technology, Innovation, and Culture)
TNO	Toegepast Natuurwetenschappelijk Onderzoek (Organization for Applied Scientific Research)
TTA	Technology Targeted Area
UiB	Universitetet i Bergen (University of Bergen)
UiO	Universitetet i Oslo (University of Oslo)
UNCTAD	United Nations Conference on Trade and Development
UNU-MERIT	Maastricht Economic and Social Research and Training Centre on Innovation and Technology, United Nations University and Maastricht University
WFE	Whole Fish Equivalent
WTO	World Trade Organization
VAW	Vereinigte Aluminiumwerke AG
VC	Venture capital
VTT	Valtion Teknillinen Tutkimuskeskus (Technical Research Centre of Finland)

1

Introduction: Innovation in Norway

Jan Fagerberg, David C. Mowery, and Bart Verspagen

Norway was once one of the poorer countries in Europe. According to Maddison (2003), in 1870 Norway's GDP per capita was only three quarters of the Western European average. By 1973, however, Norway had caught up with most Western European countries, and by 2001, Norway's GDP per capita was one quarter higher than the Western European average.[1] Hence by the beginning of the twenty-first century, Norway had become one of the richest countries in the world.

How can such a remarkable episode of economic "catchup" be explained? The explanation of international differences in economic performance has been a central theme for economists since Adam Smith first raised the question in his study of *The Wealth of Nations* (1776). Until recently, however, most economists' thinking about the subject focused on such factors as natural-resource endowments, labor supply, and capital accumulation. More recently there has been a shift of focus towards intangibles such as knowledge or innovation, and several new theoretical frameworks have emerged. These include evolutionary economics (Nelson and Winter 1982), new growth theory (Romer 1990; Aghion and Howitt 1992) and the literature on "national systems of innovation" (Lundvall 1992; Nelson 1993; Edquist 2004). This book employs these new theoretical and empirical approaches to examine the contributions of knowledge and innovation to Norway's economic development.

Our emphasis on knowledge and innovation as essential factors for economic prosperity uses a broad definition of these concepts. Thus we consider more than the organizations such as universities and research institutes that develop and transmit knowledge, or organizational units within firms, such as R&D departments, that seek to develop and exploit knowledge. This broader perspective is essential for several reasons. First, economic growth benefits less from the creation of knowledge per se than from its application to the production of new and existing goods and services. An exclusive focus on the creation of new technologies that ignores their exploitation risks overlooking essential cross-national differences in the translation of new knowledge into economic gains. The effective exploitation of new knowledge or technology is especially important for small countries such as Norway, whose contribution to the global pool of new knowledge necessarily will be dwarfed by the potential contributions to Norway's economic growth from exploitation of this pool. Second, in Norway as well as elsewhere, considerable learning and innovation occur beyond the boundaries of organizations created specifically to support innovation (Lundvall 1992, 2007). Ignoring the contributions to economic prosperity from these "non-formal" innovation-related activities may create a biased account of the sources of economic growth that in turn yields misleading policy guidance. Third, since sectors and industries differ in the ways in which learning and innovation occur within their boundaries (Pavitt 1984; Malerba 2004), a broad perspective toward the understanding of innovation is especially important in examining nations such as Norway, with a pattern of specialization that differs significantly from that of most other high-income economies.

These arguments strongly suggest that a few quantitative indicators of innovative performance are insufficient for understanding national innovation systems and providing policy advice. Instead, it is necessary to explore the innovation dynamics in leading economic sectors, investigate the interaction between the firms in these sectors and the "knowledge infrastructure" of the country, and assess the role of policy in this context. We adopt such an approach in this book. The first part of the volume (Chapters 2–4) examines the historical development of Norway's national innovation system, emphasizing the role of the Norwegian knowledge infrastructure and government policies toward innovation. The second part (Chapters 6–10) provides detailed studies of innovation within important sectors of the Norwegian economy, including aluminum, aquaculture, and the ICT sector. The third and final part (Chapters 11–13) analyzes the current structure and performance of Norway's knowledge infrastructure (public research institutes and universities) and policies for financial support of innovation-activities in industry. The remainder of this introductory chapter introduces the theoretical perspective applied in this book, provides some basic descriptive and comparative information on the modern Norwegian innovation system, summarizes the findings of the subsequent chapters and considers the lessons from this research.

INNOVATION, PATH DEPENDENCY, AND POLICY

The "national innovation system" (NIS) concept first appeared in work by Christopher Freeman (Freeman 1987), Bengt Åke Lundvall (Lundvall 1992), and Richard Nelson (Nelson 1993), and this analytic framework has since been extensively discussed in both scholarly and policy-analytic work. The NIS approach posits that innovation is an interactive activity that involves different various actors and organizations, and further argues that these patterns of interaction tend to be relatively stable over time and develop distinctive national features. Despite the popularity of the concept, however, very few studies have analyzed the development of individual national innovation systems in depth. Moreover, as Edquist (2004) points out, scholars disagree on how best to apply the innovation system concept to individual nations. Hence, some clarification on our use of the NIS approach is in order.

The term "innovation" is central to the NIS approach, highlighting its focus on the economic exploitation of knowledge, rather than only its creation. Schumpeter (1934) argued that innovation should be distinguished from invention, the creation of new ideas. New ideas may occur anywhere in the economic system, but attempts to carry these out into practice, e.g. commercialization, require specialized organizations suited to that purpose. In advanced economies such as Norway, private firms normally undertake this role, although other types of organizations, such as those in the public sector, can and do innovate. The bulk of the discussion in this volume, however, focuses on innovation within firms.

Schumpeter also provided a definition of innovation as a "new combination" of existing sources of knowledge and resources (Schumpeter 1934). Nevertheless, innovation is a cumulative phenomenon. It builds on existing knowledge, including past inventions and innovations, while at the same time providing the basis for new innovative activity in the future. Hence, choices made in the past influence innovation today, and contemporary innovation activity will similarly leave its imprint on the opportunity set facing future entrepreneurs. Thus, "history matters"—innovation is "path-dependent." Schumpeter's term, "new combinations," also points to the fact that innovations commonly depend on many different types of new and existing knowledge, capabilities, and resources, not all of which may reside within a firm. In many cases, knowledge, capabilities, and the like must be acquired from other "external" actors, including firms, research laboratories, universities, or individuals. Thus, although innovation primarily occurs in firms, it is at the same time an interactive process, in which many different social agents within the public and private sectors, may be involved (Lundvall 1988, 1992; van de Ven 1999).

The importance and extent of path dependency within innovation processes have been emphasized and debated in a large literature (Arthur 1989, 1994; David 1986; North 1990; Grabher 1993; Pierson 2000; Martin and Sunley 2006; Liebowitz and Margolis 1994, 1995). Evolutionary approaches to the analysis of innovation, such as those utilized in this volume, emphasize variety creation, adaptation, selection and retention, all of which are linked to time and path dependence. At any point in time many new ideas emerge, but only those that (at the time) are sufficiently well adapted to the selection environment are likely to be applied and form the basis for continuing adaptation and improvement. The introduction of an innovation is associated with the Schumpeterian process of technological competition (Fagerberg 2003), characterized by entry (and exit) of firms, continuous innovation, gradual development of standards, the adaptation or creation of institutions, etc.

Although Schumpeter emphasized the broad similarities among these evolutionary processes across time and economic sectors, there are also important differences among industries or technological fields in the characteristics of their "sectoral" innovation systems (Pavitt 1984; Carlsson and Stankiewicz 1991; Malerba 2004). For example, in pharmaceuticals or biotechnology, codified knowledge, university research, and formal instruments for protection of intellectual property (e.g. patents) are very important, while in some other fields, such as for example the auto industry, ship-building and construction, these factors are less important than in-house learning, interaction with customers and suppliers, or secrecy (Malerba 2004; von Tunzelmann and Acha 2004).

Sectoral innovation systems are characterized by well-defined knowledge bases, as well as contrasting patterns of evolution and industrial dynamics. A national system of innovation consists of firms in many different sectors operating within by a common (national) "knowledge infrastructure" and a common institutional and political framework. Hence institutions, governance, as well as politics, are relevant to the analysis of national systems of innovation (Pierson 2000; Whitley 2002). Institutions or "rules of the game" (North 1990) are difficult and costly to establish but facilitate economic interactions once adopted. Hence, institutions may be an important source of path dependency in their own right (North 1990). This may also hold for policies, which often tend to be "remarkably durable" (Rose 1990). For example, the system of "national concessions" in Norway's development of its natural resources was important for both aluminum in the early twentieth century and oil and gas in the 1970s, as Chapters 6 and 7 point out.

Since national innovation systems include firms belonging to different sectoral systems, the sectoral composition of a given national economy influences the operation and structure of its national innovation system, even as the national innovation system affects the performance of its constituent sectoral systems. In small, open economies such Norway that are highly specialized in a small number of sectors, this relationship between the sectoral and the national level is powerful, and we devote considerable attention to the interaction between the sectoral and national innovation systems within Norway in this volume. The relationship between sectoral and national innovation systems is a coevolutionary one, in which sectoral characteristics (and the needs of firms in these sectors) influence the development of the knowledge infrastructure, institutions and policies at the national level, while at the same time the latter characteristics influence the subsequent evolution of the national economy (including its sectoral composition).

The national innovation system also is a selection environment for new entrepreneurial ventures, and path-dependency influences this selection environment. New ventures that have little in common with economically strong existing sectors may find that the national innovation system is poorly adapted to their needs. Narula (2002), for example, argued that Norway's innovation system for this reason has provided little support for new, knowledge-intensive sectors.[2] Although path dependency has been important in the evolution of the Norwegian and other national innovation systems, the development of these systems is affected by more than past developments alone. Innovation systems are open systems; new initiatives do appear within them, and the selection processes that winnow out these initiatives are complex and operate at multiple levels. Norwegian economic history contains a number of examples of successful new initiatives that relied for their creation on foreign entrepreneurs, capital or markets. The establishment of Norsk Hydro, for example, although spearheaded by Norwegians, succeeded only because of support from foreign investors, and foreign investment and technology have played important roles in other important new industries in Norwegian history (Lie 2005).

It is also unrealistic to portray the knowledge infrastructure, entrepreneurs, and the politicians within even a relatively small nation such as Norway as monolithic. Among other things, the democratic political system of Norway has supported the growth of different political groups with conflicting perceptions of the economic future that Norwegian entrepreneurs have exploited to gain political and financial support for new undertakings. Chapters 2–4 describe the rise of a group of "industrial modernizers" in postwar Norway who exercised considerable power within the Norwegian government and state-owned industry. Although some of their efforts, particularly the attempt to create national "high-tech" champions based on ICT, met with limited success, they enjoyed considerable political and economic support for decades. The "rise and fall" of Norway's postwar policies of industrial modernization provide a fascinating example of the complex interactions among major stakeholders in the Norwegian economy and knowledge infrastructure that gave rise to self-reinforcing processes and policy intervention that were also influenced (and sometimes checked) by external events.

AN OVERVIEW OF THE NORWEGIAN INNOVATION SYSTEM

In common parlance, innovation is often associated with high-technology industries, such as information and communication technologies, scientific research in large-scale facilities in firms or universities, and professionals working in urban environments. Norway, however, has no major international firms in high-tech industries, and no university that ranks among the top 50 worldwide. Moreover, Norway's population is small (currently 4.6 million) and the country is among the 50 countries with the lowest population density in the world (about 12 people per km^2). Its capital and largest city, Oslo, has just over half a million inhabitants. These characteristics are rarely associated with strong national innovative performance, especially in the industries typically defined as "high-technology."

Figure 1.1 compares Norwegian GDP per capita (measured in purchasing power parity) with regional GDP per capita in Western Europe.[3] The thin line

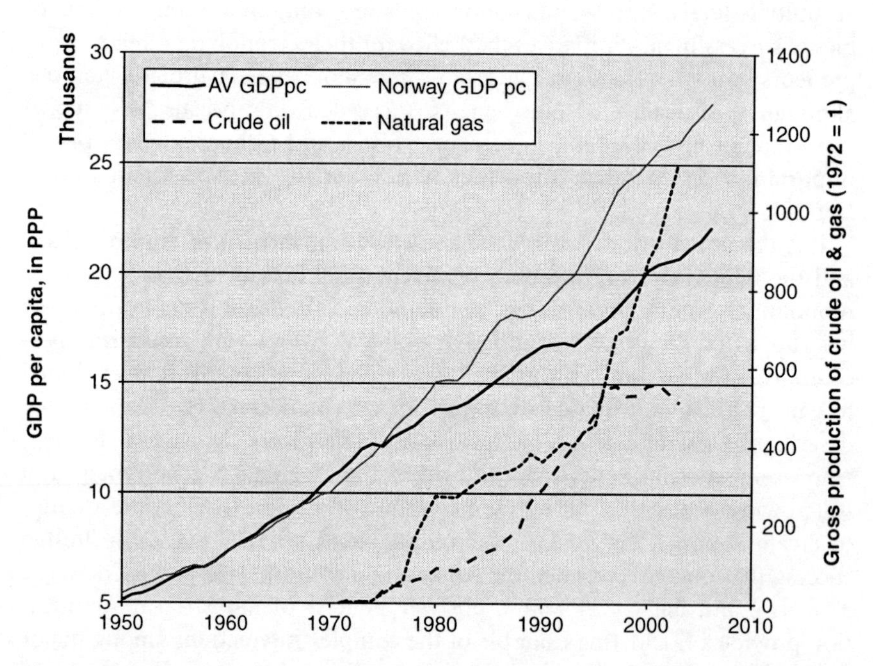

Figure 1.1. Norwegian economic growth and the rise of the oil and gas sector, 1950–2007.

Source: GGDC Total Economy Data Base (www.ggdc.net).

shows the Norwegian level, the thick black line indicates the Western European average. As we noted earlier, postwar Norwegian GDP per capita was roughly equal to the Western European average until the first oil crisis of the 1970s, which led to recession and lower growth elsewhere in Europe. Norway was much less seriously affected by the recession, and experienced more rapid growth than the other countries in Western Europe after the mid-1970s. This Norwegian "growth spurt" is related to the discovery of the offshore oil and gas fields in Norwegian waters that began production in the early 1970s (the two dotted lines in Figure 1.1 depict Norwegian oil and gas production). Although oil and gas production remained low in the first half of the 1970s, output subsequently grew rapidly, and this sector's importance within the Norwegian economy increased dramatically during from 1975 onwards. As a result, Norwegian GDP per capita soared.

Norway was not the only northwest European nation to discover and exploit offshore oil and gas deposits during the 1960s and 1970s—the United Kingdom, Denmark, and the Netherlands all benefited from similar discoveries. Nonetheless, the transformative effects of oil and gas appear to have been most significant in the Norwegian economy. Although Norway's oil and gas sector accounts for a small share of national employment, its development opened up a huge market that Norwegian manufacturing and services firms were well placed to exploit. Firms in sectors such as shipbuilding, engineering, ICT, and other business services expanded their sales in this rapidly expanding market, aided by supportive governmental policies (see Chapter 7). In the Netherlands, another small open economy, oil and gas production was associated with deindustrialization, the so-called "Dutch disease." In Norway, however, the growth of the oil and gas sector benefited domestic manufacturing industry, output from which grew more rapidly than otherwise might have been the case (Cappelen *et al.* 2000). The rapidly increasing tax income from the oil and gas sector also enabled Norway's government to pursue a more expansionary fiscal and monetary policy than the more austere policies elsewhere in Western Europe during the 1980s and 1990s. As a consequence, Norwegian rates of labor force participation and economic growth were consistently higher—and unemployment markedly lower—than in Western Europe as a whole.

After a quarter-century of rapid growth, Norwegian GDP per capita was approximately one quarter higher than the West European average. However, only about one half of this difference can be explained as rents from oil and gas production.[4] Thus Norwegian GDP per capita is above the average for Western Europe even when the direct effects of oil and gas are removed from the data. Assessing the exact contribution of oil and gas to Norwegian economic growth is beyond the scope of this book. But its impacts on the Norwegian innovation system were substantial, and receive considerable attention in subsequent chapters.

Although oil and gas now appears to be the most economically important Norwegian resource-related industry, Norway's economic development historically has relied on the exploitation of a number of rich natural resources. Most of these were related to the geography of the country, such as activities related to the sea (fishing, shipping, and related industries), and the opportunities created by Norway's mountainous terrain for mining and production of hydroelectric power, which provided the basis for the nation's electrometallurgical and electrochemical industries. Although these sectors now account for a smaller share of Norwegian GDP than in previous periods, they remain important sources of income and employment in some regions of Norway and retain considerable influence in Norwegian domestic politics. They also contribute significantly to Norway's exports.

Figure 1.2[5] illustrates the Norwegian pattern of specialization in production (of tradable goods and services) in 2002 relative to the European average. The index has a zero mean and varies between unity (indicating products that are produced only in Norway) and minus one (not produced in Norway). It shows that in addition to its large oil and gas, sector, Norway remains highly specialized in its areas of traditional strength, particularly fisheries, shipping

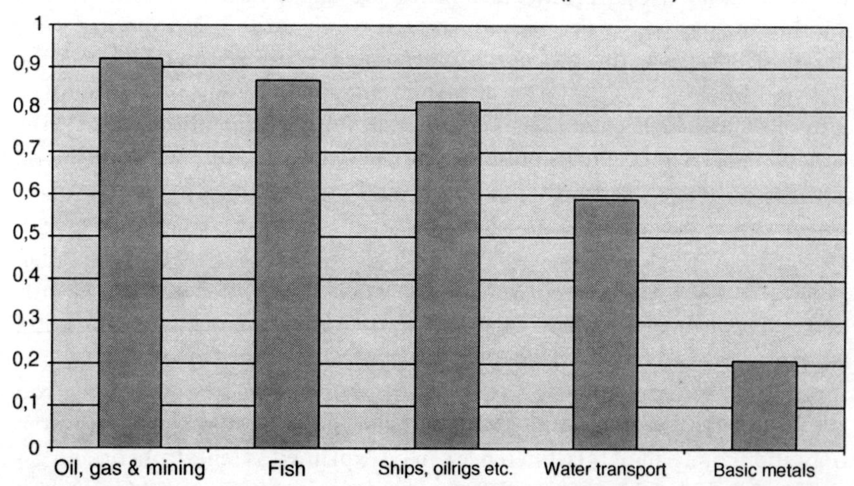

Figure 1.2. The five leading areas of Norwegian specialization, based on production of tradable goods and services in 2002.

Source: GGDC 60 Industries Data Base (www.ggdc.net).

and related industries. During the second half of the twentieth century, Norway also pioneered the development of fish-farming, and the nation remains among the global leaders in this industry. As we noted earlier, the shipbuilding industry that formerly supported Norway's fisheries and shipping sectors has retained its economic significance within Norway by diversifying into production of equipment for exploration and production of oil and gas. The basic metals sector, a large user of hydroelectric power, is another natural resource-based sector in which Norway remains specialized.

The relationship between Norway's pattern of economic specialization and its innovation system is a central theme of this book and the topic of long-running policy debates in Norway. One view of the role of technology in economic growth argues that a strong high-technology industrial base (consisting of ICT, biotech, new materials, pharmaceuticals, and selected other industries) is necessary for continuing prosperity. As several of the chapters in this volume show, however, Norway's resource-based sectors (aluminum, oil and gas, and fish farming) have for decades been highly innovative, drawing on domestic innovation, technology transfer from foreign sources (the success of which relied on substantial indigenous Norwegian "absorptive capacity"), and Norway's universities and research institutes.

One manifestation of the strong performance of Norway's economy during the past thirty years is its high rate of labor productivity growth, which has averaged more than 2.5 percent per year since 1975 (OECD 2007). Norway's strong economic performance, however, has been associated much lower levels of R&D investment and measured innovative activity than in most other high-income European economies. Figure 1.3 compares R&D spending as a share of GDP in Norway with that of other high-income industrial economies, and shows that Norway's R&D/GDP ratio of 1.6 percent is in the lower half of the reference group. Moreover, like most other countries with low R&D intensity, Norway's economy is characterized by a relatively large share of government-financed R&D, which consists mainly of R&D carried out in universities and institutes within the public sector. Most of the countries whose R&D intensity in Figure 1.3 exceeds that of Norway have a much higher industry-financed share.

Although R&D spending is a widely used indicator of innovation, it is only one of several important contributing factors in successful innovation; the importance of R&D investment relative to other factors varies substantially among economic sectors (Fagerberg *et al.* 2004). Does the unusual (relative to other European economies) Norwegian pattern of specialization explain its lower levels of R&D investment? For example, it is possible that the sectoral innovation systems in Norway's fields of specialization operate differently, or rely on sources of innovation that themselves require lower levels

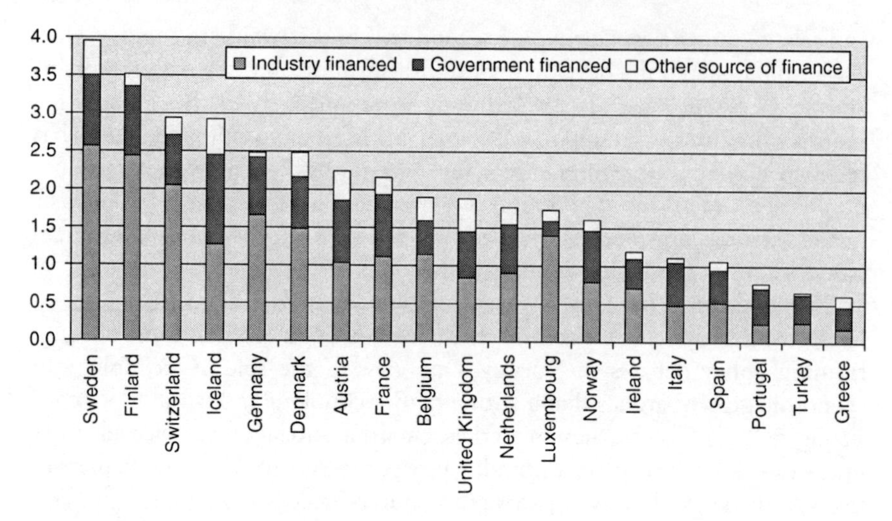

Figure 1.3. R&D as a percent of GDP: Norway and a reference group of European economies, 2004.
Source: OECD.

of R&D investment than in other European economies. One approach to examining this question controls for cross-national differences in economic specialization patterns when comparing R&D investment levels across countries. Figure 1.4 compares the share of value added accounted for by Norway's business R&D (R&D performed within industry) with similar figures for other Western European countries as reported by the OECD ("actual") and weighted by the industrial structure of the country with which Norway is compared ("adjusted").[6] If Norwegian firms on average invest more in R&D than firms in the same sectors in the other country, the ratio will be above one and vice versa.

The results reported in Figure 1.4 indicate that Norway's economic structure does influence its low economy-wide R&D/GDP ratio. In five out of the six comparisons (the exception being Sweden, a nation with one of the highest R&D/GDP ratios in the world), Norwegian firms perform as much business-enterprise R&D as do foreign firms in the same sectors. The finding that the low level of Norwegian R&D is influenced by the pattern of specialization is corroborated by the results of other studies (OECD 2007), and by evidence from other high-income economies that are specialized in natural resource production.[7] Nonetheless, as was pointed out earlier, R&D is only one factor in innovation, and R&D investment data may not capture other important aspects of sectoral or national innovation-related activity, including the

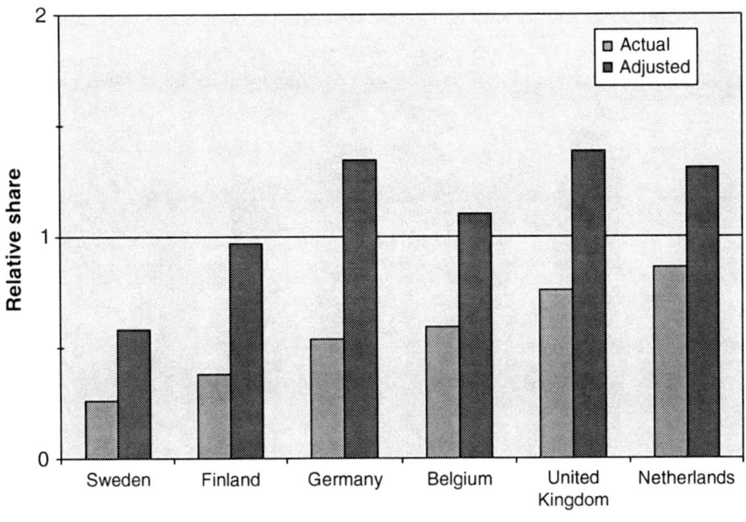

Figure 1.4. Norway's share of business R&D in GDP relative to those of other countries, actual and adjusted for structural differences, 2001/2002.

Source: Authors' calculations based on OECD and Eurostat data.

adoption and modification, as well as the creation, of technology. One source of data on a broader set of innovation-related activities that we use extensively in subsequent chapters is the Community Innovation Survey (CIS), covering firms throughout Europe. Innovation in this survey is a broad concept that includes the introduction of production and processes that are new to the firm, not necessarily new to the market.

Figure 1.5 compares the share of innovative firms in Norway with that of other European countries (as reported by the fourth version of the CIS survey undertaken in 2004, "CIS4").[8] The measure "share of innovative firms" is the number of firms that report having undertaken successful product or process innovation divided by the total number of reporting firms for the country in question. As in the previous figure the Norwegian share is compared with those for other economies on an "actual" and "adjusted" basis, the latter comparison being adjusted for cross-national differences in industrial structure. Thus, if Norwegian firms are more innovative than firms in the other country, the share will be above one or vice versa. The comparative data in Figure 1.5 suggest that the share of innovative firms in Norway is comparable to that of a number of other Western European countries but significantly lower than Sweden and Germany. Interestingly, and in contrast to R&D (Figure 1.4), the

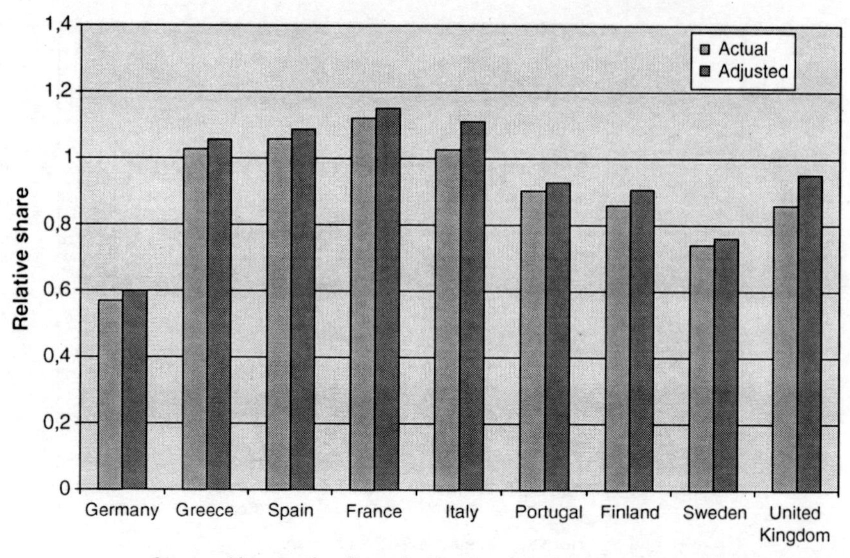

Figure 1.5. Innovative firms as a share of all Norwegian firms relative to other European economies, actual and adjusted for structural differences, 2004.
Source: Eurostat (CIS 4).

result does not appear to be very sensitive to cross-national differences in specialization patterns. Although Norway and Sweden are sometimes categorized as having similar economic policies and structures (Katzenstein, 1985), the evidence in Figure 1.5 suggests some important differences in their national innovation systems.

The Community Innovation Survey also reveals important information about other qualitative features of the Norwegian innovation system that are discussed in more detail in Chapter 5. The interactive nature of innovation means that success in innovation depends on the ability of firms to engage in innovation cooperation and interact with customers (Lundvall 1988; von Hippel 1988). The evidence indicates that Norwegian firms resemble those in other Nordic countries in their relatively high levels of cooperation with other firms and organizations in innovation. Firms in Norway and other Nordic countries also tend to value the role of customers in innovation more highly than do firms from other European countries.

Innovation is not only—or mainly—about inventing new things, but depends as well on commercial exploitation of the opportunities created by new knowledge (Kline and Rosenberg 1986, Fagerberg 2004). One measure of a country's ability to identify, absorb and exploit new knowledge, often

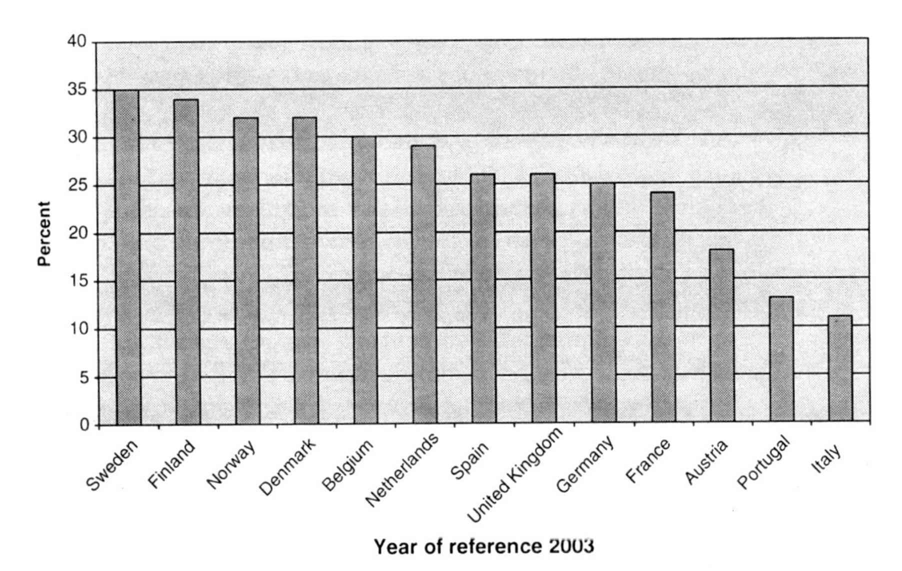

Figure 1.6. Percentage of Population with tertiary education (age 25–64), Norway and a reference group, 2000–2004.

Source: OECD (2006), Education at a Glance.

termed "absorptive capacity" (Cohen and Levinthal 1990), is the level of education among its population, particularly levels of higher education (Figure 1.6). Norway and other Nordic countries have substantially higher shares of tertiary-education degreeholders than is true of many other European economies. Another indicator of absorptive capacity is the level of adoption of important new technologies within an economy. Figure 1.7 compares the level of Norwegian adoption in 2005 of one such "general purpose technology," personal computers, with that of other European nations, revealing that the Nordic countries, including Norway, display the highest rates of adoption for PCs. These various indicators point to an important strength of the Norwegian innovation system, its strong performance in knowledge diffusion and cooperation in innovation. This performance characteristic is typically not captured in conventional indicators of innovation inputs or outputs.

The Norwegian economy has generated strong growth in productivity, employment and income since 1970, and this performance reflects more than the effects of oil and gas. At the same time, however, Norway has an unusually low share of R&D in GDP, particularly in the business sector, and the CIS data also suggest low levels of industrial innovation in Norway by comparison with some other high-income European economies. Other characteristics of industrial innovation in Norway, however, such as the level of collaboration

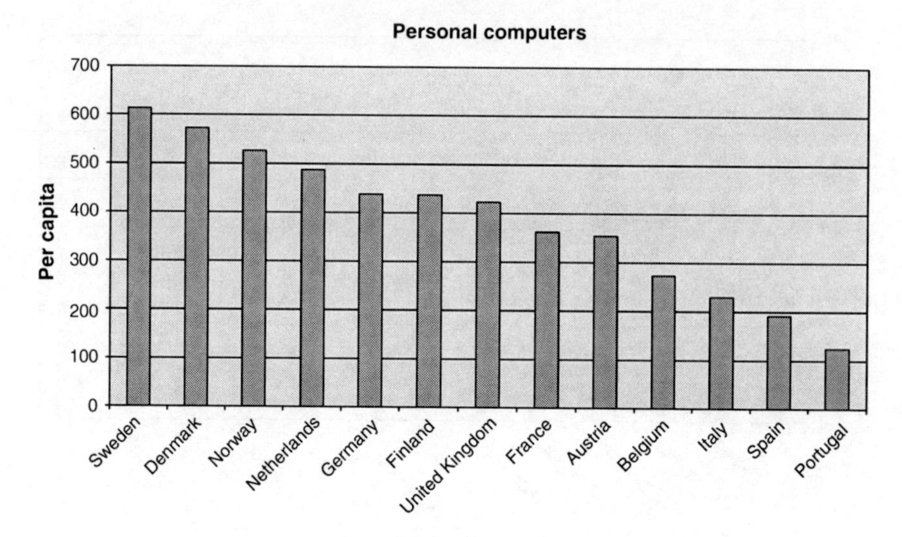

Figure 1.7. Penetration of PCs within the population, Norway, and a reference group, 2005.

in innovation, the importance of customers as sources of information for innovation, the high qualifications of the labor force and the limited indicators suggesting good performance in technology adoption, are relatively strong by comparison with most other European economies. These apparently contradictory indicators and findings underscore the need for a more detailed analysis of sectors, institutions, and economic dynamics to provide greater insight into the characteristics of the Norwegian innovation system. The remainder of this volume presents a collection of such studies, and we summarize their key findings in the next section.

NORWAY'S INNOVATION SYSTEM: HISTORICAL DEVELOPMENT AND CURRENT STRUCTURE

Part 1 provides an overview of the Norwegian national innovation system and the chapters in this volume that discuss various aspects of the system's development and performance. Most of the chapters in this volume adopt a historical perspective. This approach is essential to an understanding of the current structure and performance of Norway's innovation system, since industries and firms co-evolve with the public research infrastructure, institutions and policy. The evolving industrial structure and firm characteristics help shape a nation's knowledge infrastructure and innovation policy, and the emergent policy framework and infrastructure in turn influence the survival of existing and new industries and firms. Chapters 2–5 provide an overview of the development of the Norwegian innovation activity, public research infrastructure, policies and support schemes. Part 2 (Chapters 6–10) presents detailed studies of the development of the innovation systems of important sectors within Norway's industrial structure, while the third part of the volume (Chapters 11–13) analyzes Norway's contemporary public R&D infrastructure and R&D subsidies.

The development of the Norwegian innovation system

Chapter 2 develops a broad conceptual framework for understanding the historical development of the Norwegian innovation system, arguing that Norway's economic development has been characterized by the emergence over time of industrial systems with different approaches to innovation. The *small-scale decentralized* development path is characterized by small firms that invest little of their own funds in innovation-related activities. In contrast, *large-scale, centralized enterprise* is dominated by organizations that seek to exploit economies of scale and scope in capital-intensive industries in which foreign investment has been prominent in Norway. As we note below, however, even the firms within these industries historically were slow to develop in-house R&D. Finally, the *knowledge-intensive, network-based* development path is characterized by R&D-intensive firms in "new" industrial sectors that are closely linked to the national R&D infrastructure of public laboratories and universities. In Norway, as in other high-income economies, these three development paths and corresponding sectoral innovation systems coexist, rather than one being succeeded historically by another. Norway thus is home to a diverse and complex "ecology" of innovation systems, illustrated by the contrasting examples of aquaculture, aluminum, and information technology, all three of which have played important roles in Norway through much of the twentieth century.

Complementing this historical overview, Chapters 3 and 4 respectively discuss the development of Norway's public research infrastructure and the changing structure of public policy toward innovation in Norway. Norway was a European "latecomer" in developing a national university system, particularly in the technical field. Norway instead developed a public research infrastructure during the late ninetieth and early twentieth centuries as a piecemeal response to changing industrial needs. Inevitably, this process of institutional development gave priority to supporting established (and politically influential) industries, such as mining, fisheries, and agriculture.

Specialized research and higher education facilities devoted to these industries were established in Norway before the foundation of the Norwegian Technical University in 1910. The latter, once established, became an important source of qualified personnel for industry, particularly the scale-intensive, resource-based enterprises that played an important role in the Norwegian economy beginning in the early twentieth century. Nevertheless, the firms within Norway's large-scale industries rarely performed much in-house R&D. This does not mean that they did not innovate, but to the extent they did so they relied primarily on experience, the qualifications of their personnel and the exploitation of external knowledge, rather than on R&D within the firm.

Chapter 5 provides a more detailed examination of the structure and innovative performance of Norwegian firms in industry and services, comparing these dimensions of the national innovation system with those of other European innovation systems. Using innovation indicators derived from the Community Innovation Survey, the chapter develops a sectoral taxonomy of Norwegian innovation, and examines the sectoral patterns of industrial dynamics. The portrait of innovative activity within Norwegian industry that emerges from the analysis suggests that the three "developmental paths" outlined in Chapter 2 retain some influence on the modern Norwegian innovation system, although these historical "fingerprints" are somewhat blurry. From an international perspective, the Norwegian "resource-based innovators" are the most salient part of the innovation system, showing a relatively strong focus on process innovation, and production-oriented innovation motives. Nevertheless, as in most other industrial economies, the "science-based innovators" form the sector that exhibits the highest levels of innovation effort, measured in terms of inputs and outputs. Perhaps surprisingly, the "science-based innovators" are also the group of industries that show the highest level of firm concentration, the lowest level of dynamism and structural change. Chapter 5 also notes that the most recent CIS data indicate that the innovative performance of both the "science-based innovators" and "resource-based innovators" in Norway declined somewhat during the 2001–2003 period. This finding suggests that the long-term challenges faced by Norway could be substantial and raises questions both about the results of Norwegian government policies to support innovation (see below for further discussion).

Sectoral systems of innovation in Norway

As we noted earlier, during the first half of the twentieth century, Norwegian firms did not invest heavily in in-house R&D but relied mainly on external sources for knowledge. After 1945, however, firms in the scale-intensive sector began to invest more heavily in the development of their technological capabilities. Chapter 6 illustrates how firms in the Norwegian aluminum industry, which initially were both economically and technologically dependent on foreign firms, gradually changed course, largely as a result of a changing international competitive environment, and began to rely on Norway's public research infrastructure, which in response improved its capabilities in this area. According to industry representatives cited in the chapter, Norwegian aluminum firms now are now among the most technically advanced producers in the global aluminum industry.

Norwegian oil and gas firms, another scale-intensive, capital-intensive and resource-based industry, initially did not invest extensively in developing their knowledge base through R&D, relying instead on recruitment of qualified personnel and the purchase of R&D and engineering services (see Chapter 7). In addition, the early development of the industry in Norway (like aluminum) relied heavily on foreign firms. But the sheer size of the sector, along with the technological complexity of its activities, created a large market for Norwegian engineering firms and knowledge-intensive business services of all kinds. Much of this consulting and engineering support was provided by the public research infrastructure, particularly the institute sector. Only after a severe external shock, in the form of a collapse in the price of oil during the first half of the 1980s, did Norwegian oil and gas firms begin to develop a more strategic approach to R&D. Among other things, this strategic approach led to a joint public–private research effort (NORSOK) aimed at substantial reductions in the costs of offshore oil exploration and production in Norway. Chapter 7 argues that NORSOK spurred private R&D investment by Norwegian oil and gas firms, but private R&D investment has begun to exceed public R&D spending in the sector only since 2001.

Chapter 8 examines aquaculture, an industry in which Norwegian entrepreneurs and relatively small firms, many of which were in the fisheries sector, have pioneered in the development of new technologies for raising fish and other sources of seafood. The industry's location in coastal and northern Norway, its roots in the politically salient fisheries sector, and the small-scale, decentralized and entrepreneur-driven character of the industry all contributed to the development of government policies that promoted the industry while attempting to protect its decentralized, small-scale characteristics. These policies, along with other characteristics of the Norwegian industry, have produced an aquaculture industry structure in Norway that contrasts with those of other countries, particularly Scotland and Chile, that entered the industry later. Persistent financial problems among the smaller firms in

Norwegian aquaculture nevertheless have led to a less restrictive regulatory framework, and the industry has been restructured through mergers and acquisitions. The modern Norwegian aquaculture industry now has a heterogeneous structure, characterized by a small number of large, increasingly global firms that coexist with a large group of small, family owned firms. The origins of the aquaculture industry in the "small-scale, decentralized" developmental path remain apparent in the characteristics of innovation within the sector. Norwegian aquaculture innovation remains important and continues to be dominated by a combination of trial and error methods and exploitation of external sources, among which the Norwegian public research infrastructure has played a central role.

Science-based industries, such as ICT (discussed in Chapter 10), pharmaceuticals, and more recently biotechnology, are widely regarded as cornerstones of advanced-economy national innovation systems. Chapter 9 discusses the development of Norwegian biotechnology, a science-based industry that has received much attention from policymakers throughout the industrial world. Nevertheless, Norwegian firms have yet to achieve major commercial success in this industry, and have not developed regional agglomerations of the type commonly regarded as a defining (and essential) characteristic of a strong biotechnology industry. As the chapter points out, Norway may develop an alternative approach to the development of biotechnology, based on the expertise of Norwegian firms in some areas of marine and animal biotechnology. Interestingly, one of the strengths of Norway's biotechnology industry appears to be in segments such as marine biotechnology, that are rooted in Norway's "small-scale decentralized" resource-based industries (e.g. fisheries, aquaculture, animal breeding).

Although neighboring Sweden and Finland are home to global leaders in ICT (Ericsson and Nokia), Norway is very weak in this industry, as Chapter 10 recounts, in spite of lengthy and costly government support for the development of Norwegian ICT from the 1950s through the 1980s. These promotional policies for ICT were supported by a politically powerful group of technocrats, the so-called "modernizers" (see Chapters 2–4), promoting the development of "national champions" in this area. This strategy relied on Norwegian firms' commercialization of applications of ICT products developed from Norwegian military R&D programs instituted during the 1950s and 1960s. Despite the considerable funding and political support for these policies, however, and the early success of Norwegian IT firms such as Norsk Data, by the 1990s, Norway's ICT industry had nearly collapsed.

The current absence of large, multiproduct Norwegian ICT firms is especially incongruous in light of the significant technological contributions in telephony and radio communications technologies from Norwegian scientists and engineers. Norway pioneered many advances in mobile communications, for example, because of the nation's unusual geography and needs for such innovations, and the GSM technology that now is a global standard for mobile telephony was invented in Norway. Nevertheless, in spite of its commercial failings and the eventual collapse of many of the erstwhile national champions during and after the 1990s, Norwegian ICT firms became important suppliers of knowledge-intensive services and specialized products to other parts of the Norwegian economy, especially the oil and gas industry.

Contemporary institutions and programs supporting innovation in Norway

The historical development of Norway's economy and the supporting innovation system has left "fingerprints" on the contemporary Norwegian innovation system, as Chapter 5 points out. The third part of this volume examines the contemporary institutions and programs supporting innovation in Norway. Chapter 11 and 12 respectively discuss the role of Norway's universities and public research institutes in today's national innovation system. Chapter 11 uses data on research funding, publications and patenting to suggest that in Norwegian universities, similarly to those elsewhere in Europe, university–industry research collaboration has expanded since 1980. During the 1980s, the share of industry funding of university R&D increased, and the 1990s saw growth in PhD students finding work in firms, particularly in the oil-and gas industry, which now is the single largest employer of PhDs in Norway's private sector. The "University Act Amendments," passed in 2003, seek to encourage university patenting and licensing of faculty inventions, but too little time has passed since their passage to assess their effects with confidence.

Both Chapters 11 and 12 emphasize the importance of interaction and collaboration between universities and industry, the research institutes and industry, and between the research institutes and Norwegian universities. Indeed, the relatively close links between Norway's research institutes and universities, as well as the tendency for Norwegian firms to pursue collaborative innovation strategies, distinguish Norway's innovation system. For historical reasons, Norway's public research institutes have long played a particularly important role in spurring such cooperation. In fact, 30–40 percent of the firms in many Norwegian manufacturing industries report that they collaborate with public research institutes, and user surveys indicate that the firms on average value such cooperation highly (Chapter 12). The discussion in

Chapter 12 concludes that the Norwegian research institutes contribute to the nation's overall technological performance, and that the frequent criticism of the institutes for diverting resources from firms and universities is largely unsubstantiated.

The final chapter in Part 3 (Chapter 13) examines the operation and effects of Norway's R&D subsidy programs for industrial R&D. The chapter shows that Norwegian subsidies for industrial R&D (including EU-supported subsidies) mostly flow to relatively large firms with evidence of prior innovation activity and a proven ability to penetrate export markets. Conversely, firms with no history of innovation and smaller firms are less likely to receive such support, although EU subsidies do appear to support smaller firms' R&D to a slightly greater extent.[9] Taken as whole, however, public subsidies for industrial R&D in Norway appear to be more conducive to existing national strengths and paths of technological development than more novel avenues. Moreover, the chapter also demonstrates that Norwegian "research" subsidies stimulate additional private-firm investment in R&D, while the reverse is true for "development" subsidies, which tend to crowd out firms' own development expenditures. Thus, government financial support for "riskier" innovation-related activities seems to be more effective in expanding firms' R&D investment than support to projects closer to the commercialization stage. This is consistent what we should expect from innovation theory (Fagerberg *et al.* 2004). As projects approach the commercialization phase uncertainty typically is reduced, and firms are more likely to make the necessary investments with or without public subsidies.

Overall, the evidence in the chapters in Part 3 support a mixed verdict on the effectiveness of contemporary Norwegian government policies in supporting innovation and the development of new, knowledge-intensive activities. Norway has a substantial public-sector research infrastructure that has developed incrementally over the years in response to firms' needs. The Norwegian Research Council and government ministries dominate the funding of this infrastructure, and during the past decade, the links between Norwegian universities and the public research institutes have become closer in some areas, while in other areas, competition between the institutes and universities has intensified. Users of the public research institutes, especially firms in the resource-based sectors that play such an important role in the Norwegian system, appear to be broadly satisfied with the quality and accessibility of their services. Nevertheless, Narula (2002) argued that the Norwegian public research infrastructure, particularly the research institutes, is used less extensively by smaller, science-based firms in pharmaceuticals and biotech and that many firms in these industries rely on foreign sources for contractual R&D

services. However, reliance on foreign R&D should not necessarily be regarded as a problem. In many fields of science and engineering, the most advanced expertise will be available only from foreign sources, and interaction with foreign centers of excellence can contribute no less to innovation in areas where domestic competence is strong than in those with weaker capabilities. A potentially more serious issue is the evidence in Chapter 13 indicating that Norway's R&D subsidy programs may be biased against smaller firms and firms without prior innovation activity. There may be a danger that this support may mainly help "insiders" (and the established industries in which these firms are operating) rather than "outsiders" seeking to develop new knowledge-intensive activities.

CONCLUSION

Norway's economic performance has been characterized as a "puzzle" or "paradox" (OECD 2007; Grønning *et al.* 2008). Productivity and income are among the highest in the world, even without the extra contribution of the nation's successful oil and gas sector. But Norwegian R&D investment accounts for a small share of GDP by comparison with other industrial economies, and other measures of Norwegian innovation activity, although imprecise, also are not very impressive. How can this be explained?

One explanation posits that the statistical measures of innovation-related activities are poor proxies for the underlying phenomenon that we seek to measure. However, the premise that a close statistical relationship should exist between aggregate measures of R&D, innovation output and economic prosperity may also be flawed, particularly in a national innovation system in which resource-intensive industries are prominent. In fact, this premise assumes the existence of the "linear model" of innovation that was critically assessed by Kline and Rosenberg (1986), among other scholars. Rather than being an exogenous factor leading to predictable economic results, innovation is an endogenous phenomenon that is shaped through interaction between firms and their environments. Arguably, the evolutionary reasoning underlying Kline and Rosenberg's argument suggests that innovation is best studied as a historical process. The emergence and evolution of an innovation system rests on a co-evolutionary process in which the development of firms and industries on the one hand interacts with and affects a national public research infrastructure, policies and institutions, on the other. Such co-evolutionary processes may also give rise to path dependencies of various sorts, e.g. processes that systematically favor some types of activities (or solutions or ideas) while constraining others.

This historical and evolutionary perspective is useful in understanding the Norwegian innovation system and the Norwegian paradox. At the beginning of the twentieth century, the Norwegian economy relied on external sources for new technologies. Technologies from foreign sources were adapted to Norwegian conditions by technically trained people who often had received their education abroad. A national public research infrastructure evolved slowly in response to the needs of Norwegian firms and industries. A mining college was founded under Danish rule during the eighteenth century and around the turn of the twentieth century, Norway's primary industries exercised sufficient political influence to lead to the formation of public research institutes in agriculture and fisheries. Only with the emergence of large-scale, capital-intensive industries based on the exploitation of natural resources in the early part of the twentieth century was Norway's technical university established, nearly a century after Sweden's technical university was founded, and, with time, contract research organizations adapted to these industries' needs. As Chapter 3 noted, Norwegian university scientists and engineers became active in industrial consultancy in the first half of the twentieth century, and during the following decades Norway's research institutes, of which many are public (or semi-public), expanded their operations in response to users' needs. Foreign sources of technology and capital also played an important role in the establishment or expansion of many of Norway's large-scale, resource-intensive industries.

By the mid twentieth century, Norway's national innovation system had acquired many of its current features. Norwegian firms were innovative in many respects and demanded highly educated labor. But they invested little in internal R&D. Instead they exemplified the predictions of evolutionary theory (Nelson and Winter 1982) by using "localized search" in problem-solving, seeking technical knowledge from other firms, public sources, academia etc. Only when the search for solutions from external sources was unsuccessful did Norwegian firms invest substantially in intrafirm R&D. This investment in in-house R&D became more significant as some Norwegian firms approached the international knowledge frontier. In a number of Norway's resource-based industries, such as the aluminum and oil and gas industries, this exhaustion of external sources of technical solutions has occurred relatively recently, during the last few decades. Thus, the historically dominant approach to innovation within much of Norwegian industry throughout the twentieth century relied on interaction with other actors in the system, in combination with modest levels of investment in intrafirm R&D. As the analysis in Chapter 5 shows, this approach continues to leave its fingerprints on the national innovation system.

The historically low level of investment by Norwegian firms in intrafirm R&D does not imply that they did not innovate. Arguably, the extensive structural changes that have occurred in the Norwegian economy during the last century have been accompanied by a stream of economically important innovations. For example, the rise of the large scale, capital-intensive path of economic development in the early twentieth century was based on the exploration of a new natural resource, hydroelectric energy, by entrepreneurs such as Sam Eyde, who in a classically Schumpeterian fashion, developed a "new combination" of knowledge, capabilities and resources. The Norwegian oil and gas industry faced daunting challenges in producing oil and gas under conditions of unprecedented complexity and hazardousness, and developed new technological and organizational solutions (CONDEEP for example, see Chapter 7). The Norwegian aquaculture industry also relied on a stream of important, incremental innovations in fish farming, processing, and disease control. But none of these major innovations depended on large-scale intrafirm R&D investment. Arguably, innovations of this type, which affect the entire production system of natural-resource industries, may not be classified as innovations by the CIS-type surveys that focus on technological (product and process) innovations (Smith 2004).

The contributions in this volume emphasize the contributions of institutions and politics to the path-dependent development of innovation systems. Indeed, the path-dependent character of the evolution of the Norwegian national innovation system is a product of politics, as well as institutional development. For example, the continued existence and extensive government support for the "small-scale, decentralized" path of industrial development in Norway was in part the outcome of intense political struggles that during the interwar period produced an institutional regime that endures to the present day (Chapter 2). Moreover, this set of political commitments and institutional supports shaped the creation and organization of the Norwegian aquaculture industry half a century later (Chapter 10).

Another example of institutional persistence that had far-reaching consequences is the "concession laws" from the early decades of the twentieth century. These laws were originally drafted to create a system for national, democratic control of natural resources, specifically, hydroelectric power, and influenced the early years of Norway's aluminum industry (Chapter 6). But Chapter 7 notes that the same regulatory system was also crucial to the development of Norway's offshore oil and gas sector more than half a century later.

The technological and organizational development of the Norwegian oil and gas industry might well have followed a very different path (as arguably was the case for the offshore oil and gas industries in Denmark and the United Kingdom that exploited the North Sea finds in their territorial waters) in the absence of the regulation system created during the early twentieth century for an entirely different sector. Hence, institutional innovation—and politics—exert a great influence over the development of national innovation systems, and the nature of this influence deserves more attention. Previous work on national systems of innovation has devoted little attention to these matters, possibly because much of it examines "snapshots" of various innovation systems at a specific point in time and lacks historical depth. Arguably, one of the advantages of the historical, evolutionary perspective applied in this study is that it advances our understanding of the roles played by institutions and politics in innovation.

The title of this volume, innovation, path-dependency and policy, points to three interrelated aspects of Norwegian economic development. First, applying a broad perspective to the study of innovation and long-run economic change, we recognize that innovation has been important factor in Norway's impressive economic performance, although the characteristics of Norway's industrial base and the processes of innovation that it supports mean that much of this innovation has eluded straightforward measurement. Perhaps the most important factor in Norway's innovative performance has been the ability of Norwegian entrepreneurs, firms, and public sector actors to recognize opportunities, mobilize resources, adapt existing capabilities and develop new ones, and develop appropriate institutions and policies. The system's adaptability thus appears to be one of the important factors contributing to Norway's successful technological and economic development., This adaptability reflects other social, cultural, institutional and/or political characteristics of the nation that we cannot pursue here, but which also present a promising line of research.

Second, the development of Norway's national innovation system is a historical process characterized by strong path dependency. The Norwegian innovation system has been dominated by resource-based innovation, in contrast to those of other Nordic economies (both the Swedish and Finnish innovation systems have included resource-based sectors, but these now are much less dominant in each nation). The development of new industries that are less closely linked to natural resources, in spite of considerable support from public policy, has been much less successful. The lack of new, knowledge-intensive industries in Norway is less a result of active resistance from established firms in politically powerful sectors than a reflection of the continued vitality of innovation-led growth and productivity in these established sectors. Norway's resource-based sectors (both the large- and the small-scale examples) have displayed considerable dynamism in developing knowledge and adapting to new challenges. Finally, as we pointed out above, political and institutional changes in Norway have fundamentally influenced the structure of the economy and its innovation-related activities. Thus, in the case of the Norwegian innovation system, path dependency is as much a political and institutional phenomenon as an economic one.

Norway's history of innovation and economic growth should not be viewed as a basis for complacency about the future, which poses significant challenges. Although the oil and gas sector will remain economically important, there can be no doubt that the period of rapid economic growth based on the exploitation of Norway's offshore oil and gas is close to its end and that future growth will rely on other sources. A second important change is the end of the century-long era of cheap hydroelectric energy, the abundant supply of which led to the establishment of electrometallurgical and electrochemical industries in Norway. These industries will have to compete on the basis of superior technology with foreign firms that benefit from lower energy costs. Hence, although natural resources may play an important role in Norway's future economic growth, maintaining the nation's strong performance will require an increase in the level and scope of innovative activity, and this goal should figure prominently on the policy agenda. Raising the share of the overall Norwegian firm population that is active in innovation, rather than focusing primarily on firms in "high tech" sectors, is an important target for innovation policies. In light of the importance of innovation for Norway's future economic growth, it is disquieting to observe that in contrast to most other European economies, the share of Norwegian firms reporting that they were active innovators has declined during the first years of the new millennium (Chapter 5), underscoring the seriousness of the challenges facing Norway.

NOTES

1. The Western European countries included in this comparison are Austria, Belgium, Denmark, Finland, France, Germany, Italy, Netherlands, Norway, Portugal, Spain, Sweden, Switzerland, and the United Kingdom.
2. Based on interviews in selected Norwegian firms that invest in R&D, Narula (2002) concluded that large established firms in areas of traditional economic strength in Norway expressed greater satisfaction with the performance of Norway's national knowledge infrastructure than did small entrepreneurial firms in science-based sectors, such as pharmaceuticals and biotech. Narula interpreted

these interview-based findings as indicative of a "lock-in" phenomenon operating at the level of the Norwegian innovation system, which he argued was more supportive of innovation in established industries than for innovation in new industrial sectors, such as biotechnology or pharmaceuticals.

3. All data on GDP per capita and productivity used in this section are drawn from the GGDC total economy database, version of August 2007 (www.ggdc.net). The countries included in the comparison are Austria, Belgium, Denmark, Finland, France, Germany, Italy, Netherlands, Norway, Portugal, Spain, Sweden, Switzerland and the United Kingdom, e.g. the same as in Maddison (2003) refered to earlier.
4. Rent in this context is defined as return to capital above what is "normal" in the economy as a whole. Oil rents represented 14 percent of GDP in 2001, and Norwegian GDP per capita was about 27 percent above the Western European average.
5. The source for the data in Figure 1.2 is the GGDC 60 Industries Database (http://www.ggdc.net). The index (I) in the figure is the so-called "revealed comparative advantage" measure, normalized to vary between 1 and −1, e.g. $I = (r - 1)/(r + 1)$, where r is the ratio between a sectors share of GDP in Norway and the same share for the World as a whole.
6. For example, the ICT industry is very R&D-intensive, and accounts for a large share of Swedish GDP. Norway's ICT industry, however, is small. This structural difference between the two economies contributes to the higher ratio of R&D performed in industry to GDP in Sweden (when compared to Norway). By using a common set of sectoral weights when comparing Norway and Sweden, we are able to control for the effects of such structural differences. The same methodology is used in Figure 1.5's depiction of firm-level indicators of innovative performance, based on the CIS4 data.
7. Norway's level of domestic R&D investment has more in common with other natural-resource based economies such as Australia and Canada than with its closest European neighbors. In 2004, overall R&D investment accounted for 1.6–2.0 percent of GDP in these three countries, with industry accounting for about one half of the domestic R&D investment. In Finland and Sweden, in contrast, R&D as a share of GDP was in the 3.5–4 percent range, with approximately two thirds financed by industry. Source: OECD.
8. The data in the figure are similar to those analyzed in Chapter 5, but are drawn from a more recent edition of the CIS, and are adjusted for cross-national differences in industrial structure i.e. weighted by the structure of the country with which Norway is compared.
9. Moreover, it should be noted that the continuation of the "FUNN" scheme, the so-called "SKATTEFUNN," has been designed to benefit small, entrepreneurial firms to a larger extent than FUNN. The requirement in FUNN of cooperation with external research providers has for example been abolished. A recent evaluation of SKATTEFUNN concludes that the new scheme is successful in its aims (Cappelen *et al.* 2008).

REFERENCES

Aghion, P. and P. Howitt (1992) "A Model of Growth through Creative Destruction," *Econometrica*, 60 (2), pp. 323–51.

Arthur, W. B. (1989) "Competing Technologies, Increasing Returns, and Lock-in by Historical Events," *The Economic Journal*, 99, pp. 116–31.

——(1994) *Increasing Returns and Path Dependency in the Economy*, Ann Arbor: University of Michigan Press.

Cappelen, Å., T. Eika, and I. Holm (2000) "Resource Booms: Curse or Blessing?," *Manuscript presented at the Annual Meeting of American Economic Association 2000*, Statistics Norway, Oslo.

Cappelen, Å., E. Fjærli, F. Foyn, T. Hægeland, J. Møen, A. Raknerud, and M. Rybalka (2008) Evaluering av SkatteFUNN—Sluttrapport, Rapporter 2008/2, Statistisk Sentralbyrå, Oslo.

Carlsson, B. and R. Stankiewicz (1991) "On the Nature, Function and Composition of Technological Systems," in Carlsson, B. (ed.) *Technological Systems and Economic Performance: The case of Factory Automation*, New York: Springer.

Cohen, W. and D. Levinthal (1990) "Absorptive Capacity: A New Perspective on Learning and Innovation," *Administrative Science Quarterly*, 35 (1), pp. 128–52.

David, P. A. (1986) "Understanding the Economics of QWERTY: The Necessity of History," in Parker, W. N. (ed.) *Economics History and the Modern Economist*, London: Basil Blackwell.

Edquist, C. (2004) "Systems of Innovation: Perspectives and Challenges," in Fagerberg, J., D. C. Mowery and R. R. Nelson (eds.) *Oxford Handbook of Innovation*, Oxford: Oxford University Press.

Fagerberg, J. (2003) "Schumpeter and the Revival of Evolutionary Economics: An Appraisal of the Literature," *Journal of Evolutionary Economics*, 13, pp. 125–59.

——(2004) "Innovation: A Guide to the Literature," in Fagerberg, J., D. C. Mowery, and R. R. Nelson (eds.) *Oxford Handbook of Innovation*, Oxford: Oxford University Press.

Fagerberg, J., D. C. Mowery, and R. R. Nelson (eds.) (2004) *Oxford Handbook of Innovation*, Oxford: Oxford University Press.

Freeman, C. (1987) *Technology Policy and Economic Performance Lessons from Japan*, London: Pinter.

Grønning, T., S. E. Moen, and D. S. Olsen (2008) "Low innovation intensity. High growth and specialized trajectories: Norway," in Edquist, C and L. Hommen (eds.) *Small-Country Innovation Systems: Globalisation, Change and Policy in Asia and Europe*, Edward Elgar, UK, pp. 281–318.

Grabher, G. (1993) "The Weakness of Strong Ties: the Lock-in of Regional Development in the Ruhr Area," in Grabher, G. (ed.) *The Embedded Firm*, London: Routledge.

Katzenstein, P. J. (1985) *Small States in World Markets: Industrial Policy in Europe*, New York: Cornell University Press.

Kline, S. J. and N. Rosenberg (1986) "An Overview of Innovation," in R. Landau and N. Rosenberg (eds.) *The Positive Sum Strategy: Harnessing Technology for Economic Growth*, Washington DC: National Academy Press, pp. 275–304.

Lie, E. (2005) Oljerikdommer og internasjonalisering. Hydro 1977–2005. Pax Forlag.

Liebowitz, S. J. and S. E. Margolis (1994) "Network Externality: An Uncommon Tragedy," *Journal of Economic Perspectives*, 8 (2), pp. 133–50.

—— —— (1995) "Path Dependence, Lock-in, and History," *Journal of Law, Economics and Organization*, 11 (1), pp. 205–26.

Lundvall, B.-Å. (1988) "Innovation as an Interactive Process: From User-Producer Interaction to the National System of Innovation," in Dosi, G., C. Freeman, R. R. Nelson, G. Silverberg, and L. Soete (eds.) *Technical Change and Economic Theory*, London: Pinter.

—— (1992) *National Systems of Innovation: Towards a Theory of Innovation and Interactive Learning*, London: Pinter.

—— (2007) "Innovation System Research and Policy: Where it came from and where it might go," Paper Presented at CAS seminar, December, Oslo: Centre for Advanced Study at the Norwegian Academy of Science and Letters.

Maddison, A. (2003) *The World Economy: Historical Statistics*, Paris: OECD.

Malerba, F. (2004) "Sectoral Systems: How and why innovation differs across sectors," in Fagerberg, J., D. C. Mowery, and R. R. Nelson (eds.) *Oxford Handbook of Innovation*, Oxford: Oxford University Press.

Martin, R. and P. Sunley (2006) "Path Dependence and Regional Economic Evolution," *Journal of Economic Geography*, pp. 1–43.

Narula, R. (2002) "Innovation Systems and 'inertia' in R&D Location: Norwegian Firms and the Role of Systemic Lock-in," *Research Policy*, 31, pp. 795–816.

Nelson, R. R. (ed.) (1993) *National Innovation Systems: A Comparative Study*, Oxford: Oxford University Press.

Nelson, R. R. and S. G. Winter (1982) *An Evolutionary Theory of Economic Change*, Cambridge, Mass.: Harvard University Press.

North, D. C. (1990) *Institutions, Institutional Change and Economic Performance*, Cambridge, Cambridge University Press.

OECD (2007) *Economic Surveys: Norway*, Paris: OECD.

Pavitt, K. (1984) "Sectoral Patterns of Technical Change: Towards a Taxonomy and Theory," *Research Policy*, 13, pp. 343–73.

Pierson, P. (2000) "Increasing Returns, Path Dependence, and the Study of Politics," *The American Political Science Review*, 94 (2), pp. 251–67.

Romer, P. M. (1990) "Endogenous Technological Change," *Journal of Political Economy*, 98 (5), pp. S21–S102.

Rose, R. (1990) "Inheritance Before Choice in Public Policy," *Journal of Theoretical Politics*, 2 (3), pp. 263–91.

Schumpeter, J. A. (1934) *The Theory of Economic Development*, Cambridge, Mass.: Harvard University Press.

Smith, A. (1776) *An Inquiry into the Nature and Causes of The Wealth of Nations*, London: Strahan and Cadell.

Smith, K. (2004) "Measuring Innovation," in Fagerberg, J., D. C. Mowery, and R. R. Nelson (eds.) *Oxford Handbook of Innovation*, Oxford: Oxford University Press.

van de Ven, A. H. (1999) *The Innovation Journey*, Oxford: Oxford University Press.

von Hippel, E. (1988) *The Sources of Innovation*, New York: Oxford University Press.

von Tunzelmann, N. and V. Acha (2004) "Innovation in low-tech industries," in Fagerberg, J., D. C. Mowery, and R. R. Nelson (eds.) *Oxford Handbook of Innovation*, Oxford: Oxford University Press.

Whitley, R. (2002) "Developing innovative competences: the role of institutional frameworks," *Industrial and Corporate Change*, 11, pp. 497–528.

Part I

The Historical Evolution and Current Structure of the Norwegian National Innovation System

2

The Layers of National Innovation Systems: The Historical Evolution of a National Innovation System in Norway

Olav Wicken

The national innovation system (NIS) of Norway is characterized by diversity. This chapter examines the multiple and heterogeneous historical processes, each defined as a *path*, that have given rise to such diversity. Each of the paths has involved specific types of social groups, organizations, knowledge bases, and institutional set-ups, and for each path a specific type of innovation structure has been developed.

We define three main historical paths emerging from three major industrial transformation processes in Western history defined as Industrial Revolutions (Bruland and Mowery 2004). Each of these transformations created a new industrial path. The Norwegian NIS therefore may be described as the historical outcome of three diverse paths that have created three distinct "layers" within the national system. The creation of a new path does not mean that the old paths of the economy remain static. Instead, each path historically has undergone radical transformations in response to changing environments. The main dynamics of the innovation system are therefore linked to *path transformation and path creation processes.*

PATH TRANSFORMATION: SMALL-SCALE DECENTRALIZED INDUSTRIALIZATION

The transformation of paths over time is illustrated by the development of the oldest path, "small-scale decentralized industrialization," during the early 1900s. This path is characterized by small-scale personal or family owned companies using informal knowledge, and has been a core aspect of the

Box 2.1. Small-scale decentralized path

Small-scale companies
Production processes normally characterized by low capital intensity
Local knowledge and open exchange of information and learning in local community
Open search for information abroad or in other regions of Norway
Institutional set-up defined by the local community (companies to accept "rules of the game" established by other firms and local institutions)
No or little internal R&D, but expanding use over time of science-based flows of knowledge and technologies

Norway's industrial, political, and economic history. In primary production (i.e. agriculture, fishing, fish farming), in manufacturing industry, as well as in most of the service sector, the characteristics of the first path are central to the characterization of Norway's economic history as "Democratic Capitalism" (Sejersted 1993).

This path originated in the first Industrial Revolution, when most developments in Europe took place in small workshops, using traditional knowledge, and forms of organization.[1] The resulting small-scale companies were closely held and regulated by local norms and rules. This "localism," with a very large number of independent farmers and land owners, fishermen, traders, forest owners, ship owners, etc. formed the political and cultural basis for much of Norway's economy during the nineteenth century, and many of its elements remain important. These small-scale, often rural economic units were closely integrated in local communities, and were operated by self-employed owner/managers.

The long-term survival of the small-scale decentralized path in the Norwegian economy has depended on multiple processes of transformation through which new forms of knowledge and organizations have been incorporated into the existing forms of production. Today, small-scale Norwegian companies exploit both informal knowledge and science-based information in innovation processes and collaborate with other companies or universities/research institutes to solve problems. Many small-scale companies have succeeded in remaining or becoming competitive knowledge-intensive production units and organizations.

PATH CREATION: LARGE-SCALE CENTRALIZED INDUSTRIALIZATION

The most important path creation process during the twentieth century is the evolution of a path dominated by large-scale economic organizations. The

Box 2.2. The large-scale centralized path

Large-scale of organization
Production often is capital-intensive
Firms are able to shape their own environment and influence the "rules of the game"
Systematic search for relevant knowledge—specialized expertise hired by company
Internal research processes organized in separate laboratories or development departments
Collaborative learning with other companies and/or research communities
Collective, individual, and hierarchical learning processes in the work place

large-scale company became the dominant form of economic organization beginning in the mid twentieth century in Norway, and this path has enjoyed strong political support throughout its evolution. Particularly during the "oil age" of Norway's recent history (since roughly 1980) a path characterized by large-scale companies that are able to influence and shape their own environment (or institutional set-up) became strong.

The emergence of a new form of industrialization with characteristics distinguishing it from the old path is called a *path creation process*.[2] The creation of the large-scale industrial enterprise path within Norway relied on both external actors (particularly foreign investors) and local actors and organizations. The Norwegian *NIS is open*, and foreign capital and knowledge were of crucial importance for this path creation process. The new layer in the economy involved new social groups (investors, engineers, scientists, managers, consultants, etc.) and organizations that exploited new forms of knowledge (science, engineering knowledge, management knowledge, finance knowledge, law, etc.). The emerging groups and forms of knowledge became important for learning and innovation processes in the new path.

The large-scale form of industrialization influenced Norwegian national politics and therefore, the institutional environment of these industries. Particularly after the Second World War, Norwegian political institutions have encouraged the development of the modern industrial enterprise (see Chapter 4, as well as the chapters on the oil and gas and aluminum industries). This path was a central focus of Norwegian industrial policy making, and this policy focus was reinforced by the emergence of the oil sector.

A NEW PATH AS ENABLING SECTOR: R&D-INTENSIVE NETWORK-BASED INDUSTRIALIZATION

The large-scale, centralized form of production became the basis for Norway's resource-based industrial development during the twentieth and early twenty-first centuries. A new path emerged during the last part of the twentieth century, reflecting a challenge from new forms of production organization to the old paths of Norwegian economic development. This path creation process, however, has not yet produced the basis for new paths of industrial development in Norway. In contrast to Sweden and Finland, the new path was incorporated into the older paths, as many R&D-intensive, smaller firms became technology producers and problem solvers for old industries (especially oil and gas) and the public sector.

This path creation process was closely linked to the emergence of electronics, computer production, telecommunications, and automation systems. Many of the new companies within these sectors may be defined as firms where the production unit has become a laboratory. Although most of the firms in Norway's large-scale centralized path of development conducted their R&D in specialized organizational units, the new emerging path was characterized by a blurring of the distinction between *R&D* and production. In the Norwegian context most companies were small, working closely with research institutes, universities, public procurement agencies, governmental organizations, and other companies. The emerging *R&D-intensive network-based* industrialization was not directly related to exploitation of natural resources, but many of the early companies in the emerging path developed and produced technologies related to Norway's resource endowment (telecommunications, satellites) or produced inputs to resource-based industries (automation systems, detection and communication for fishing and shipping, etc.).

The oil and gas sector provided a profitable domestic market for high-tech companies that could assist in solving challenges posed by the natural environment and political regulation of offshore petroleum production. The emergence of this third path of industrial development thus became an important element of the innovation structures for small-scale decentralized as well as large-scale centralized form of production, and contributed to the transformation of large parts of the economy during the last decades of the

Box 2.3. R&D intensive network-based path

Production-unit as a laboratory: R&D incorporated in the production process
R&D-intensive rather than capital-intensive
Often organized around projects and problem-solving activities (flexible)
Collaborate with other companies or research organizations in innovation and production *networks* that include public and private organizations
Institutions established as outcome of interactive processes among different types of organizations and the environment; no dominant actor shapes the institutional set-up

twentieth century. Firms in this third path supported industries that enabled other sectors of the economy to remain competitive in the face of intensified global economic competition.

HISTORICAL DEVELOPMENT OF THE SMALL-SCALE DECENTRALIZED PATH

The influential position of the small-scale, decentralized form of industrialization in the contemporary Norwegian economy is the outcome of a historical process dating back to the early industrial development of the nation. As late as the middle of the twentieth century, most of Norway's population lived in rural areas and worked in industries closely connected to developing and extracting natural resources. 30 percent of employment in 1946 was in primary production, including agriculture, fishing, forestry and mining (Historisk Statistikk 1994: 237). Most production units were small, and controlled or owned by families or groups of families in local communities. During the nineteenth century the majority of the population was farmers, many of whom combined farming with employment in fisheries along the coast and forestry in inland regions. Norway may be described as a society where a majority of the population were landowners (although a rural proletariat of renters grew during the mid nineteenth century) or had sufficient access to natural resources (especially through fishing) to provide a basis for family income. With few exceptions, ownership was widely distributed and locally controlled. Norway was a society dominated by an independent small-scale "petit bourgeoisie" that was linked to primary industries extracting natural resources (Sejersted 1993).

Norwegian productivity and economic growth during the first half of the nineteenth century was linked to improved efficiency in agriculture, particularly mechanization and the introduction of new crops such as the potato. Economic change in Norway during this period was linked to innovative activities that relied on knowledge from local and external sources. Local blacksmiths established small workshops to supply farms with equipment and machinery, and small-scale capital goods firms supplied the traditional export sectors.

Similarly to the Industrial Revolution in Great Britain, Norwegian industrialization included technological change in large parts of the economy, including old industries like fisheries, agriculture, forestry, and mining, as well as the emerging manufacturing sectors producing textiles and capital goods. To a large extent the productivity improvements and economic growth resulted from small-scale or marginal improvements that did not demand radical organizational or social change.

INFORMAL KNOWLEDGE AND SOCIAL LEARNING

Innovation within the localized and small-scale economy was based on learning processes that involved local as well as international interactions among people, as well as *collective or cooperative forms* of resource allocation within local communities to initiate new production activities. This reliance on the local community for industrial development in Norway has been labelled *localism* (Kjeldstadli, Myklebust, and Thue 1994). The importance of local civil society, local organizations, and institutions was reflected in the many formal and informal institutions for sharing knowledge within the local community. Learning a skilled trade was common among young people from the earlier phases of Norwegian industrial development, and this artisan-based tradition supported the establishment of small independent workshops.

There are also many historical examples of "learning by doing" or "learning by using" by workers that resulted in incremental technical improvements, some of which were patented. This type of incremental innovation can improve productivity significantly over a long period of time, and a number of Norwegian companies established specific institutions to encourage employees to participate in this type of innovative activity (Wicken 1984).

The key sources of innovations in this path were the workshops of capital-goods producers. The mechanical workshops employed people with broadly applicable technical skills, *mekanikus*. Although they had no formal education, most had wide knowledge of various aspects of mechanical engineering. These workshops became small but important organizations for improving companies' production technology and for solving other costumers' technical problems.

The learning processes linked to this path were largely specific to local traditions, problems, and knowledge bases. But international sources of knowledge also were important in cases where local knowledge did not suffice for solving problems of great importance for the local economy. Norway is located nearby the early-industrializing regions of England and Scotland, as well as commercial centres in the Netherlands, Denmark, and Sweden. Systematic search for knowledge—including espionage—was supported both by national political authorities and local municipalities (Bruland 1991). The openness of Norway's innovation structures thus created opportunities for more radical transformation processes.

There was little if any direct relationship between scientific knowledge and innovation in Norway or elsewhere before *c.*1870. But most of the Norwegian population was literate by the beginning of the eighteenth century, and technical and scientific popular literature became available in the mid nineteenth century. Priests in some regions observed that people were engaged not only in reading about new mechanical and scientific inventions, but also discussed how the new knowledge could be used locally.

INSTITUTIONS DEFINING CIVIL SOCIETY IN THE SMALL-SCALE PATH OF INDUSTRIALIZATION

The self-employed farmer (more accurately, the peasant) became the hero of the Norwegian nation during the nation-building period of the nineteenth century. Strong institutions were established to secure the equitable distribution of property among citizens, mainly by regulating the control of and access to natural resources. The core institution was the *odelslov*, which regulated how land could be transferred from one owner to another and basically made land a non-market commodity. Farm land was to be handed down from father to son and could not be sold out of the family. Free access to rich fishing resources along the coast created a large social group of independent fisherman-farmers that owned some land and shares in a fishing vessel and fishing gear. Only in mining and partly forestry did accumulation of capital into few hands take place and challenge the social position of the small-scale independent producer.

The dominating aspect of "localism" in the Norwegian economy was the local savings banks. They emerged during the nineteenth century as a response to the demand for credit from the rural petit bourgeoisie and supported farmers, fishermen, or small industrializts with money for small-scale investments. The banks played an important role in directing resources into local industrial projects, often being the main source for funding of small-scale projects. The local savings banks supported the dominance of the individual or family owned company in the organization of economic activity. Only during the 1990s did the system dominated by small local banks directed towards local industrial development activities break down.

The strong role of local institutions and organizations for industrial development meant that the emerging industrial society was deeply embedded in the existing society. The behavior of entrepreneurs and industrialists was shaped by the values and norms of local society, even when new economic activities also challenged established norms and values. The local basis for funding of a company, as well as the company's reliance on the local community for knowledge and learning, meant that no individual actor could become dominant within the existing social structure. The individual company depended on local collective or cooperative organizations and institutions for survival and for success, relying on them for mobilizing resources, interactive learning, and cooperative organizations for procurement of inputs. The firm had to adapt to local "rules of the game."

The local basis for industrial development made the rural petit bourgeoisie a strong political force within the parliamentary system introduced in Norway in 1884. The Liberal Party (Venstre) represented small-scale industry and local communities, and became the dominant Norwegian political force during the first half of the twentieth century. A Norwegian historian defines the period *c.* 1884–1940 as the Era of the Liberal Party, emphasizing the ideological hegemony of a specific type of policy (Slagstad 2001). The industrialization path characterized by small-scale independent entrepreneurs and business people deeply embedded in local communities and institutions retained a strong position in Norwegian society and politics during first half of the twentieth century.

TRANSFORMING AND REPRODUCING THE OLD PATH

New forms of production emerged internationally from the late nineteenth century and challenged the old path of Norwegian industrial development. The old export sectors were challenged by new technologies, particularly by the development of steam vessels. Modern steam-powered ships challenged the Norwegian sailing fleet, new types of vessels challenged the old wood-and-sail technology of fishing, and modern transport technology contributed to deep crises of agriculture all over Northern Europe. Steam engines also challenged the old structure of the forestry industry that was based on small local sawmills using water power.

These challenges were met in different ways within this small-scale decentralized industrial development path. In many cases the old industries adapted to the new economic environment through restructuring that relied on new technologies and knowledge bases. Agriculture, fishing, manufacturing industry and shipping underwent a transformation between the 1880s and the 1930s, as some of these industries established new institutional set-ups that aided their survival.

Agriculture went through what has been termed "Det store hamskiftet" (The Great Transformation), as a result of which cattle became the main production technology and milk the main product of Norwegian farms. Two key government policies supported the survival of the country's most important industry during the early twentieth century. The Liberal party supported an innovation policy or *productivity policy* to increase efficiency and improve the ability of farmers by to compete. These policies expanded public investments in agricultural science and education in local educational institutions (Landbruksskoler), national research centres and higher education institutions (Ås Agricultural University). Programmes for diffusion of production technology and increased mechanization were introduced. The research centres developed a new breed of cattle for the Norwegian environment that produced milk and meat of good quality (Norsk Rødt Fe), and developed other crops to increase productivity (Nielsen *et al.* 2000). Norsk Hydro produced artificial fertilizers that increased food production significantly. In this way structures and systems for diffusing scientific knowledge and science-based technology were introduced into the primary sector.

These innovation-oriented policies nonetheless failed to solve the problems of Norwegian farmers during the interwar period. The price of milk dropped dramatically during the period and by the early 1930s the crisis had become urgent. In response, the Farmers' Party promoted policies (lønnsomhetslinjen) that sought to control the market (Bjørgum 1968), forcing all milk producers to sell the milk to one monopolistic (cooperative) organization, and prohibiting price competition (Furre 1971). The new distribution system increased income for a majority of the farmers while preserving the small-scale and decentralized structure of the agricultural sector, but the distribution system became more concentrated. Agriculture's political success kept the old path alive without major transformation of production during the second half of the twentieth century, and the system for milk production and distribution established during the 1930s still dominates Norwegian agriculture.

Box 2.4. Conflicts between paths—the Battle of Trollfjorden, 1890

Challenges to old forms of production have some times taken dramatic forms, and the dramatic events have shaped long-term development. The battle of Trollfjorden is one of these events. Trollfjorden is located in one of the world's richest fishing ground for cod in Lofoten. In 1890, a British steamship closed off parts of the fiord with a huge net, and invited local fishermen to catch fish for the vessel. The fishermen revolted and attacked the ship physically in a dramatic episode that has been retold to younger generations through the book "The Last Viking" by Johan Bojer. The local fishermen reacted because they saw the incident as a challenge to their social status as self employed. They were invited to become salary earners—a proletariat—for foreign capitalists. After the "Battle of Trollfjorden," the Storting adopted laws that stopped foreign steel-steam-trawl fishing vessels from taking part in the yearly cod fishery at Lofoten. This was the beginning of a long tradition of protecting small-scale fishing vessels in coastal fishing areas, particularly in Northern Norway, that became the main institution for protection of the fisher-farmer lifestyle during the twentieth century.

The transformation of the fishing industry illustrates the mobilization of political, social, technical, and organizational resources to preserve the old structure of production within this sector. Beginning in the 1890s, traditional Norwegian wooden sail vessels were challenged by foreign (British) steel vessels using steam engines. The government introduced laws to protect small-scale coastal fishing, particularly in the cod fisheries of northern Norway. These laws, however, did not prevent economic decline, as Norwegian fishermen, like farmers, experienced a prolonged period of low prices and poverty. A number of public initiatives were taken to improve productivity and production, relying in part on R&D. The Marine Research Institute (MRI) became a world leader in mapping the movements of fish and wiring the information to fishing communities. (Schwach 2002) The development of modern meteorology by Wilhelm Bjerknes of the University of Bergen made it possible for fishermen to improve productivity and reduce the dangers and risks of fishing. (Friedman 1989) MRI worked closely with fishermen and communicated extensively with fishing communities at sea during the fishing seasons.

Modern technology created the basis for a gradual transformation of Norwegian fishing, beginning with the introduction in the early twentieth century of lightweight inexpensive internal combustion engines that could be used on small wooden vessels. Norwegian fishermen and fishing communities had long sought an engine design that was adapted to the small boats in the North Sea region. In the 1890s, light internal combustion engines that were well suited to the Norwegian fishing fleet were developed in Denmark. Thousands of Norwegian fishing vessels quickly introduced small and inexpensive engines, and engines for fishing boats became a major Norwegian growth industry until roughly 1920, relying on the output of a large number of small workshops producing engine designs that were copied from other producers. A rapid learning process improved the quality of the engines, and more efficient engines opened new opportunities for development of the fishing industry (Wicken 1995).

Mechanization of the fleet made it possible to use heavier fishing gear, and larger nets were introduced. The "wood-and-sail" technology was transformed into a "wood-and-engine" technology, but the fishing industry retained its small-scale structure, characterized by low capital costs and vessel ownership dominated by fishermen in rural villages.

The "wood-and-engine" vessels were not able to increase fishing volumes as rapidly as the more costly steel-and steam technology. Britain passed Norway in total fishing volume during the 1930s, but resistance to the "steel-and-steam" large-scale trawling technology remained strong in the Norwegian fishing industry. The increase in fish volumes on international markets reduced prices and, especially during the 1920s and 1930s, many Norwegian fishing communities experienced severe economic crises. Norway's government intervened to support the fisheries with policies that were similar to those developed at the same time for agriculture. The Storting passed a law in 1936 that forced all fishermen to sell the fish to a monopolistic organization (Råfisklaget). At the same time the industrial capacity to expand processing of fish (fish meal, fish oil) increased, providing an alternative market for fish during periods with very low international prices.

Although the size of the vessels has grown and a fleet of modern, larger fishing ships for ocean fishing gradually emerged, Norwegian fishing vessels continue to be owned by the fishermen, constructed by yards along the coast, and reliant on local workshops for engines, fishing gear, processing of their catch, etc. The fishing industry (including important parts of Norwegian aquaculture—see Chapter 8) remains an important part of the Norwegian coastal economy and continues to improve its efficiency through local learning and incremental innovative improvements.

In Norway's manufacturing sector, the small-scale path was strengthened during the interwar period through a dramatic structural transformation. The recessions during parts of the 1920s (currency crises) and during the 1930s (international recessions) forced a number of established factories to close down or reduce production, creating large-scale unemployment (as high as 30 percent for union members) within the manufacturing workforce. In parallel with this reduction in urban manufacturing production, a large number of small, rural workshops were established in a number of sectors (wooden industry, clothing, furniture, metal products, etc.) that took advantage of economic as well as technological opportunities. Social crises and weak unions in rural areas kept wages low, and the emergence of cheap, small, flexible, and efficient machinery using electric motors reduced capital costs (Sejersted 1982).

This rural industrialization process was supported by public institutions. The government established a state institute for technology (Statens Teknologiske Institutt) in 1916 that diffused knowledge and technology to small companies, and local offices to support small start-up companies with market and product competence (småindustrikontor) were established in many regions. The development of rural industries during the interwar period meant that a large part of Norwegian manufacturing industry acquired the characteristics of the small-scale decentralized development path (Refsdal 1973).

The small-scale decentralized path retained an important position in the Norwegian economy. Norway is far from an urban society, and the economy remained characterized by a large number of small companies. The small-scale decentralized form of industry was modernized and had increased its productivity during the twentieth century, relying on new technologies and new knowledge bases. Technologies regarded as core elements of the Second Industrial Revolution, electricity and the combustion engine, modernized old industries. These developments enabled the small-scale development path to survive, supporting a large group of independent producers in resource-intensive sectors and enabling Norway's rural petit bourgeoisie to retain considerable political power.

LOCK-IN AND THE CREATION OF NEW INDUSTRIES

For some of the old industries the transformation during the early twentieth century resulted in more radical change and a move away from the old path. This was the case for much of the Norwegian shipping industry, which relied mainly on sailing ships until the First World War, by which time they were no longer competitive. The decline of sail-based ocean shipping had a great impact on Norway's coastal rural economy, as demise of the old shipping industry meant the disappearance of much of the small-scale, coastal shipbuilding industry that had specialized in wooden sailing ships.

The technological transformation in shipping proved to be much more radical than in fisheries. Modern ships used steel ships with steam or diesel engines, and the new Norwegian shipping industry based on these technologies differed from its predecessor. The new shipping companies were established in urban areas, mostly around Oslo and in Bergen. Many of these companies were not based in rural or coastal communities, but emerged from interactions between individual Norwegian entrepreneurs and large international corporations. This transformation is most evident in the dynamic new sector of oil transport. Many of the international oil producing companies had traditionally transported their oil with a fleet of wholly owned oil tankers. During the interwar period, however, the leading international oil firms began to contract with independent companies for oil transport, and Norwegian entrepreneurs assumed a prominent role in this segment of the shipping industry.

Another new, large-scale and capital-intensive industry with links to ocean shipping that emerged during the interwar period in Norway was whaling. Entrepreneurs in the Oslo fjord region raised capital to construct large steel ships with steam engines and modern capital-intensive technology to hunt for whales in the Antarctic region. The industry was extremely profitable, and in a short period of time a large oceangoing whaling fleet was established. There were connections between whaling and shipping, establishing a group of capitalists that operated with modern technologies on a global level. Both industries were based on Norwegian knowledge and traditions of sea transportation and natural-resource exploitation.

PATH CREATION: THE LARGE-SCALE CENTRALIZED PATH

The period between 1880 and 1920 has been labelled by some historians the Second Industrial revolution, characterized by the industrial application of new sciences and technologies, including electricity, chemistry, the internal combustion engine, and mass production (Landes 1969). Other historians have emphasized the emergence of large-scale production units and organizations (Chandler 1990) or large technological systems (Hughes 1983). Norwegian historians have argued that the transformation found in many other Western countries, including Sweden, Germany, and the USA, was not reflected in the Norwegian experience (Sejersted 1993).

Sogner has challenged this interpretation of Norwegian economic history, arguing that the new technologies, industries, and social groups characteristic of the Second Industrial Revolution in other nations also became prominent in Norway. According to Sogner, the development of these industries enabled a wealthy urban bourgeoisie to gain influence in society and politics, and exercise considerable power over Norwegian industrial and economic development throughout the twentieth century (Sogner 2001, 2002, 2004). The emergence of this new developmental path was linked to the exploitation of Norway's natural resources. During the two first decades of the twentieth century, the development of electricity production was the impetus for large-scale industrialization, and during the last two decades of the twentieth century, oil and gas played the central role.

Even before the twentieth century, Norway was home to a few large companies linked to extraction of natural resources, particularly in mining. Mining relied on capital-intensive production technologies and was the basis for small-scale urbanization in isolated regions. Silver mining made Kongsberg one of the largest of these small urban communities in Norway during the eighteenth century (Berg 1988), and other mines exploited deposits of silver, copper, sulphur, iron, and nickel. The growth of Norway's mining industry also influenced higher education in Norway, through the establishment by the Danish state of the Mining Academy (Bergverksakademiet) in Kongsberg in 1757.

The emergence of the large-scale form of industrialization in Norway during the early twentieth century was the outcome of technological innovations that turned Norway's abundant waterfalls into a source of electrical power. Beginning in the 1870s, innovations in what became known as electrochemistry and related technologies made possible the use of electricity for the production of metals (steel, aluminum, others) and chemical processes (including wood processing). Many investors in Norway and elsewhere in Europe saw the economic potential of Norway's waterfalls and established large-scale electric-power generation facilities as the basis for large-scale industry projects.

The emerging new path within Norwegian economic development was connected with two linked but differentiated economic processes. Some of Europe's largest electricity-generation works were constructed in Norway (Thue 1994). Although it relied on large-scale enterprise, as was the case in other nations, the development of electric power took a different form in Norway. The main users for Norway's large-scale electricity-generation works were not cities, since consumer demand was insufficient to utilize the vast supply and since most large waterfalls were far away from urban centres. Instead, investment in electricity-generation systems was concentrated in isolated rural sites that could support mining and metals-processing industries. In this respect, the pattern of development of Norwegian hydroelectric power resembled the late-twentieth-century pattern of investment and development in Norwegian oil and gas (Chapter 7).

The expansion of electricity production in Norway created a search for users of vast amounts of energy, and thus provided a second path of development for large-scale centralized industrialization in Norway. Norwegian hydroelectric power became the basis for electrochemical industry (fertilizers), wood processing (paper) industry, and electrometallurgical metals processing (aluminum, ferro), as well as large petrochemical plants during the last part of the twentieth century. The exploitation of electricity in the nineteenth and early twentieth centuries, as well as oil and gas in the late twentieth century, outside of major Norwegian urban centres promoted a geographically decentralized pattern of development of large-scale industries in Norway that was also supported by government policy during the second half of the century.

As was noted earlier, cheap electricity transformed Norway's mining industry. New electricity-based refining technologies made the exploitation of low-grade mineral sources profitable, and the new methods for mining these deposits were introduced on a broad scale in Norway in the early twentieth century. Since mines also were located outside of Norway's main urban areas, this wave of innovation in mining strengthened the geographically decentralized pattern of development of large-scale production units within the country. A number of small industrial towns were founded and older small mining towns expanded as a result of the construction of one or a few new factories, mines or electricity works. Rjukan, Notodden, Odda, Sauda, Høyanger, Arendal, Kristiansand, Sarpsborg, Skien-Porsgrunn, Varanger, Ny Ålesund, Sulitjelma, Mo i Rana, Mosjøen, Folldal, Årdal, Sunndal, and others are examples of the industrial towns established in various parts of the country (Wicken 2005).

NEW SOCIAL GROUPS BENEFITING FROM THE LARGE-SCALE INDUSTRIAL PATH

Innovation processes within the emerging large-scale path relied on links among higher education, formal science, and industry. In Norway, *engineers, scientists, and investors* became important actors in the development of new processes and companies during the first decades of the twentieth century. Beginning in the 1890s, Norwegian individuals and companies who were well informed about European technological developments attempted to achieve control over Norwegian hydropower resources, in the expectation that the potential industrial applications of the electrical power would make these sites valuable. One such entrepreneur was the engineer Sam Eyde, who managed to get control over the hydropower rights associated with a number of large waterfalls. Eyde owned a consulting engineering firm and hoped to profit from the construction of hydroelectric generating stations at these sites. He accordingly also promoted the potential of these sites among potential users of their hydroelectric power. Sam Eyde was representative of one of the new social groups linked to large-scale industrialization. He belonged to a family of shipowners and had received his engineering education in Berlin. This combination of an upper-class family background and foreign technical education was common within a large group of young men involved in Norway's large-scale industrialization path of the late nineteenth and early twentieth centuries. Norway had no national civil engineering university until 1910, which meant that many of the nation's engineer-entrepreneurs of the early twentieth century were trained abroad, were well aware of foreign technological developments, and maintained technical and economic contacts in Europe.

In addition to engineers, scientists many of whom were educated at Norway's only university in Oslo, became involved in R&D and, occasionally, management of the emergent large-scale Norwegian industrial enterprises of the early twentieth century. Their involvement in these activities was consistent with a tradition within the Oslo faculty of direct involvement with corporate and industrial technology development. During the period *c*.1890–1920, Oslo

University had strengthened its faculty and research in the natural sciences. Much of the new research linked theoretical development and practical applications, and often was based on close collaboration with scientists in other countries (see Chapter 3). A classic example of this research approach was Birkeland's series of experiments on the Northern Lights, which resulted in his development of an electric gun. In order to finance his research, he established a company to market the gun internationally for military purposes. The gun was not profitable, but Birkeland's commercial efforts resulted in the development of a relationship with Sam Eyde, who exploited the electric gun in an industrial process to attract nitrogen from the air. As this famous case suggests, the involvement of university faculty with industry during this period less often involved the exploitation of basic scientific theories than the application of their engineering skills.

Eyde's entrepreneurial efforts attracted another new social group into Norwegian industry: large international banks and investment groups. The Swedish Wallenberg family and their bank (Enskilda) were important investors in the development of waterfalls and large-scale industry. They were involved in the establishment of companies for exploitation of the Birkeland-Eyde technology and succeeded in attracting the French Parisbas bank to invest in early production plants of Norsk Hydro, which was established in 1905 (Andersen 2005).

The establishment of Norsk Hydro was crucial for long-term industrialization in Norway. The case illustrates the dependence of this type of industrial development on relationships that spanned national borders, and highlights Norway's developing technological competences in a number of areas. Small companies like Eyde's consultancy firm, Birkeland's marketing company, law firms, financial organizations, etc., collaborated with one another (and with foreign firms or investors) in establishing new industrial companies and in developing new production processes. They also collaborated with the university in Oslo (and NTH after 1910) as well as with political authorities. Sam Eyde's establishment of Elkem in 1904 created an organization that served as an entrepreneurial as well as a technological experiment for the new Norwegian industries. Elkem "spun out" a number of new electricity based companies[3] and developed new processes for the electrochemical industry. The firm's most important innovation was the Söderberg electrode, introduced during the First World War, which was widely adopted in electricity-based smelting processes, including aluminum, throughout the world (see Chapter 6, Sogner 2004). A number of other Norwegian companies became involved in construction and technological development within electricity production and the electricity-based industries.

Box 2.5. The politics of Norwegian electricity—competition between paths

A major conflict in the Norwegian society was how the emerging electrical energy resource should be used to promote social development. Two major ideologies competed. Most politicians and a majority of the people preferred the development of small-scale and local electricity works, financed by local communities. This pattern of development would support the small-scale decentralized form of industrialization. Supporters of this approach argued that electricity could provide light for households and become an energy source for small-scale electric farms and industry.

The opposing group saw large-scale electricity works as the basis for large-scale industrialization, relying if necessary on foreign capital and more centralized planning and coordination. This approach was supported by representatives from large-scale industry, as well as engineers and technocrats in all political parties. Gradually this ideology gained became influence in the Labour Party, and Labour governments supported this form of industrialization after the end of the Second World War (Thue 1994).

The establishment of the new industries was closely linked to knowledge and resources from other countries. As was previously noted, the open character of the Norwegian economy and innovation system were crucial to the establishment of the new path. In cases such as aluminum, most of the technology, investments and skills exploited within Norway were transferred from abroad, and firms active in these sectors had relatively weak links with the overall national Norwegian economy (see Chapter 6). But these processes of inward transfer of technology in particular relied on the Norwegian innovation system's capacity to absorb and incorporate knowledge and resources from abroad into local institutions and organizations.

ESTABLISHMENT OF A NEW INSTITUTIONAL SET-UP FOR THE LARGE-SCALE INDUSTRIAL PATH

Partly because of the strong position of groups associated with the small-scale decentralized path of industrialization in Norwegian policy and ideology, public opposition to large-scale industrialization was particularly significant during the early 1900s. Strong support for large-scale industrial development by other social and political elites, however, meant that Norway's government gradually introduced a number of new institutions to promote the large-scale path of industrialization.

The development of the electrical industry sector was supported by a series of public and political initiatives. The 1899 law regulating public procurement gave priority to national companies in public (municipalities and state) construction of electricity works. The law was updated in 1921 in order to strengthen the use of procurement policies in "infant industry" policies toward the capital goods and construction industries.

The most important policy affecting the large-scale industrial path was the Concessions Laws, which was a central tool in national industrial policy between 1906 and 1917. The laws regulated the ownership of natural resources like waterfalls, and required that non-public owners return ownership of their sites to the state after sixty years (hjemfallsrett), based on the view that natural resources belonged to the society (represented by the state or municipality). The laws established a governmental preference for control of hydropower sites to reside with Norwegian citizens during a crucial period for the establishment of large-scale industrialization based on electricity. Although they favoured national control of a critical natural resource, the Concession Laws did not discourage the import of capital and technology for the new sectors (Lange 1977).

As Chapter 7 discusses at length, the Concessions law was used extensively more than three-quarters of a century later, in the development of Norway's offshore oil and gas industry. In oil and gas, the concessions system enabled the Norwegian government to "encourage" foreign-owned oil companies to use Norwegian suppliers, contribute to training Norwegian companies and individuals, and involved Norwegian firms in technology development.

The government also regulated the development of energy-related large-scale industrialization by controlling access to key inputs, such as electricity and capital. From the Second World War until 1990, the state controlled the construction of new electricity works and oversaw a system of long-term contracts with energy-intensive companies that ensured electricity supplies at stable, relatively low, prices. The state also supported investment through agreements with major international corporations in core industries like aluminum, direct public funding, and ownership of leading Norwegian companies such as Norsk Hydro, ÅSV, Statoil, Jernverket, and mining firms. An example of this policy was the collaboration between ÅSV and Raufoss Ammunisjonsfabrikk to produce aluminum products for the car and construction industries.

The public sector also supported this path of development through funding for institutions devoted to training, education and research in related technological fields. The establishment in 1910 of the National Technical University in Trondheim (NTH) is one example of this policy (see Chapter 3), as is the founding by the Norwegian state during the First World War of a new institution (Råvarelaboratoriet)[4] to promote research on natural-resource development. Not until the post-Second World War period, however, did Norway establish research organizations and institutions that promoted collaboration between large companies and public (or semi-public) research organizations. The reconstruction of NTH after 1945 also included a reorganization to strengthen the ability of the university to support Norwegian industrialization in areas like chemistry, metallurgy, and electricity (Hanisch and Lange 1985). The close relationship between Norsk Hydro and the public research institutes SI and IFA and the later interaction between national oil companies (Statoil, Hydro), and Norwegian research institutes (IFE, RF, and Sintef) further illustrate the public sector's promotion of technological development in this sector (see Chapter 3).

Large parts of the Norwegian capital goods industry, ICT, knowledge of intensive business services, and the public research institutes focused their marketing efforts on large-scale companies within Norway's resource-based industries. (Hauknes 1998) as particularly pronounced with the advent of the offshore petroleum sector. In a short period of time all of the larger Norwegian shipyards moved into the profitable offshore production sector. Only the smallest yards in some regions on the Western coast continued to focus on shipbuilding and after the 1980s, many of these shipyards became involved in the development and construction of special ships for the offshore sector, i.e. supply ships and LNG ships (see Chapter 7). By the 1990s, most of this industry had become part of the offshore sector. A significant share of the revenues for Norway's engineering and technological industries, as well as the Norwegian knowledge-intensive services sector, now are derived from the offshore sector and the same is true of most of Norway's high-tech industry.

THE ROLE OF SCIENCE IN INNOVATION AND THE INDUSTRIAL LABORATORY

Learning processes in the large-scale centralized form of industrialization differed from those associated with the small-scale development path discussed earlier. The large-scale companies depended on formal and science based knowledge, and on more formal types of interorganizational collaboration, but the role of state institutions and organizations remained crucial. Higher education and science entered industry in new forms after the late 1800s through formal training for would-be inventors; through organizations outside the company (such as universities) that provided formal training and research; and by the creation of bodies of empirically grounded, codified scientific and technological knowledge internal to the firm (Bruland and Mowery 2004).

Some Norwegian companies in the large-scale process industries established in-house R&D organizations during the pre-1940 period, and by the outbreak of the Second World War, approximately 400 people worked in Norwegian industrial laboratories. Norsk Hydro remained the most important R&D performer in Norwegian industry throughout the first half of the twentieth century, focusing on two main innovation strategies: (i) developing new processes and new natural resources, and (ii) improving existing processes. Each of these strategies involved different types of external interactions. The development of new processes or products dealt mainly with electrochemical process technologies. An example of this strategy is the lengthy effort by Norsk Hydro to produce magnesium from sea water, a project that began in the interwar period and continued for many decades. This technology development strategy included collaboration between the company's laboratory and scientists in external organizations (Andersen and Yttri 1997).

The second strategy, improving efficiency in processes already in use within the company, demanded close collaboration between company researchers and production personnel. These projects often involved the company's core competences and therefore less frequently relied on collaboration with external organizations (Andersen and Yttri 1997). The success of these projects depended to a large extent on good internal relationships among managers, engineers, and workers.

Norsk Hydro, Statoil and other large resource based companies are the largest R&D-performing companies in the Norwegian economy and along with other firms in the large-scale industrial sector are important actors in industrial research policy. Beginning in the 1960s, however, a new type of company emerged that was smaller but far more R&D-intensive, signaling the appearance of the "R&D-intensive network-based" developmental path. These companies flourished during the 1970s and 1980s in Norway, and after a difficult period during the 1990s, remain important industrial R&D performers.

PATH CREATION IN "ENABLING" INDUSTRIES: THE R&D-INTENSIVE NETWORK-BASED PATH

Beginning in the early 1960s, various groups linked to Norwegian industrial policy argued that Norway's future growth could not be based solely on the resource-based and energy-extensive industries, i.e. on the large-scale centralized industrial path. According to this argument, which was presented in a document from the industrial research council (NTNF 1964), new industries were needed to compensate for the reduction in growth in the old, mature industries.

The NTNF document argued that Norway had to invest more heavily in R&D overall, but further proposed that investment in selected technologies could support the creation of new, faster-growing industries. The policy introduced in the mid 1960s promoted emerging technologies like electronics, computers, automation technology, and telecommunications, and the idea that these technologies would become new industries gained support during the 1980s. Based on this belief, public policies supported re-industrialization strategies that attempted to exploit emerging technologies and knowledge areas (Targeted Technology Areas, Chapter 4).

The policy initiatives of the mid 1960s sought to support the growth within Norway of a new form of industrial production that was already well established in the USA. In the USA, the new form of production represented a fundamental change from an economy closely linked to its natural resource endowment (as US economic development through most of the pre-1940 period had been) to an economy that more "intensively exploited a burgeoning US 'endowment' of scientists and engineers" (Bruland and Mowery 2004). This change was illustrated by the emergence of new industrial sectors like information technology that were not directly linked to natural resources and the natural environment. The dynamics of these industries involved different types of learning and knowledge than earlier types of industrial production. This industrial development strategy for Norway was the goal of an influential group called *the modernizers.* (Chapter 4 discusses the role of the "modernizers" in the evolution of post-1945 Norwegian technology policy.)[5]

INNOVATION: THE PRODUCTION UNIT AS A LABORATORY

The new form of industrial production involved formal knowledge and research in innovation processes in a more radical way than was true of the older science-based industries. Innovation in the large-scale, science-based industries involved industrial laboratories that were separated from production operations. In the new form of industrialization, however, the production unit itself became the laboratory. In parts of the new emerging sectors of ICT and biotechnology, for example, most employees have higher formal education and a large part of the activity within the firm is characterized as "R&D."[6] Science was no longer confined to the laboratory, but technical thinking and knowledge pervaded all aspects of the company. This shift was reflected as well in a production organization that relied on 'projects' (consultancy companies, lawyers, architects, etc.) that demanded flexibility and continuous change in firms' internal organizational structures, along with sustained relationships with other firms and partners.

Three early Norwegian electronics and computer companies illustrate the extent to which the new "high-tech" companies were an extension of laboratory activities. Norcontrol, established in 1965, produced control systems for ships, AME (also founded in 1965) produced semiconductors, and Norsk Data, which was established in 1967, produced computers. All three companies originated in research projects in public research institutes and universities (respectively, SINTEF/NTH in Trondheim; Sentralinstituttet in Oslo; and the Defence Research Establishment at Kjeller). The R&D intensity of these companies remained extremely high for a long period of time (Basberg 1986), reflecting their status as laboratories organized as commercial companies.

The creation of these firms was based on a broader public policy framework to promote spinoffs from universities and research institutes. Funding from the research council (NTNF), the Development Fund (Utviklingsfondet), public R&D contracts, and public procurement systems, enabled spinoff companies to support their technology development activities and to find initial markets in the public sector (see Chapter 4). The companies retained close contacts with the public research institutes and universities, and developed links with other laboratories, with public procurement agencies and with other public sector organizations (NTNF, Utviklingsfondet). Many of these firms also collaborated with one another, as in Norsk Data's development of computers for the Norcontrol[7] (see Chapter 10). The Norwegian companies that emerged within this developmental path thus relied on interaction among various "higher education/high tech" organizations.

R&D INTENSIVE INDUSTRY AND THE TRANSFORMATION OF OLD PATHS

Few if any of the Norwegian companies in the "R&D intensive" developmental path successfully moved into mass production, instead remaining small-scale enterprises that continuously changed their products and often developed tailor-made solutions for other companies or public users. The role of the companies was similar to that of the mechanical engineering companies from the nineteenth century, which played an important role in the innovation system for small-scale industrialization (Rosenberg 1976). Large parts of the ICT industry in Norway emerged as producers of technologies for solving problems in other sectors, particularly those related to Norway's geography and resource endowment (e.g. mountainous terrain, extensive reliance on fisheries and shipping, etc.). The early electronics research projects and production became important for the modernization of fisheries, telecommunication, and eventually, to the modernization of offshore energy production. An excellent example of this type of firm is Simrad, which has developed technologies deployed in fisheries, shipping and the offshore oil sector. (Sogner 1997) (see Chapter 10 on the Norwegian ICT industry for more detail).

Introduction of new process technologies and extensive use of ICT, combined with new forms of organization and management, created new challenges and opportunities for firms and sectors linked to Norway's older industrialization paths. Although some parts of large-scale industry were challenged by these new types of organizations and technologies, some older small-scale industries were revitalized by them. For example, the use of R&D and scientific knowledge as well as ICT, close collaboration between producers and between users and producers, and new forms of organization of production, revitalized the shipbuilding and ship equipment industries, enabling them to develop new forms of "flexible production" (Andersen 1997).

The emerging R&D-intensive form of production also supported performance improvements in sectors that were not linked to resource-based industries. Norway's computer industry provided the public sector with equipment and software for rationalization of administrative processes, and devoted considerable resources towards the development of defense applications. In retrospect, these emerging technologies appear to have been effective instruments for solving challenges in other sectors, although they did not meet the more expansive expectations of the "modernizers" for the transformation of Norway's economy.

The strategy for industrial expansion and reindustrialization based on new technologies areas like information technology, biotechnology, new materials, and technologies for the oil and gas sector, dominated Norwegian industrial policy for a relatively brief period during the 1980s. Indeed, by this time, the emerging ICT and biotechnology industries were already developing links with Norway's two largest export sectors; oil and gas and fish. As Chapter 10 points out, many Norwegian IT companies turned away from export markets to focus on the growing domestic oil and gas industry.

CONCLUSIONS: THE NORWEGIAN INNOVATION SYSTEM

This historical description of the Norwegian national innovation argues that it should not be regarded as a homogenous one, but should be seen as the outcome of multiple path-dependent processes of historical evolution and interaction. A specific "innovation structure" has been established for each of these three paths of development, and each of these structures may be regarded as a *layer* of the overall national innovation system. Each layer comprises specific types of organizations and institutions, relying on separate knowledge bases and often involving different social groups.

Three paths with corresponding layers have been identified and described:

Small-scale decentralized path
Large-scale centralized path
R&D intensive network based path

The three main paths have distinctly different relationships to R&D:

Small-scale decentralized industries do not perform R&D
Large-scale centralized industries perform in-house R&D in separate laboratories
R&D intensive network based industries perform R&D in the ordinary production

These characteristics create different interactions among companies and other organizations in innovation processes.

The small-scale decentralized industries rely on Norway's *public knowledge infrastructure* (Smith 2002) for technology and related knowledge. In addition, firms in these industries depend on scientific and formal knowledge embedded in physical equipment and inputs into the industry.

The large-scale centralized industries have a strong formal knowledge base, including scientific personnel in laboratories, and these experts often constitute a core group in the development of new products and processes. The laboratory scientists in these firms regularly collaborate with colleagues in universities and research institutes, and collaboration frequently develops into closer relationships that blur the line between the industrial lab and the research institute/university. In Norway, a large part of the technical-industrial research institutes (including SINTEF) focus their R&D on this industry (see Chapters 11 and 12). Firms in this sector also collaborate with various parts of the engineering industries in designing new process technologies.

Firms in the R&D-intensive developmental path rely on the "production unit as a laboratory," where R&D and production processes overlap and business activity often is organized on a project basis. A large share of the employees of this type of industry has advanced technical education, and these firms interact closely with other firms or organizations that use and develop science-based knowledge. Norway's public sector and policy institutions have been important part of the innovation infrastructure for this sector, and are often involved in development and innovation processes.

Such heterogeneity in the structure and evolution of different "layers" of national innovation systems is hardly unique to Norway, although the importance of each path and layer may vary considerably among nations, and the relationships among layers of national innovation systems differs because of contrasting historical processes and contemporary economic structures. In Norway, the small-scale decentralized form of industrialization is still dynamic and fishery (including fish farming) is one of the country's main export sectors (see Chapter 8). But the dominant form of economic organization in modern Norwegian industry is the large-scale centralized structure, reflecting the important role of the nation's natural resource-based industries.

The R&D-intensive network-based developmental path has enjoyed mixed success in Norway, but has become an *enabling sector* for both the small-scale and large-scale paths of Norwegian industrialization (Pol *et al.* 2002). The offshore sector has become a major market for firms in many high tech industries and knowledge-intensive business services, large parts of Norway's engineering and capital goods industries, and the research institute sector. The demand from Norway's resource-based industrial sector for knowledge and other inputs, as well as sustained political support for their activities, have resulted in a specific structure for interactive learning in the Norwegian innovation system: the offshore oil and gas industry, along with other large natural resource-based industries attract resources and attention from Norway's domestic "enabling" industries. Much of the innovative activity in Norway's economy today is linked to learning processes within this part of the economy.

NOTES

1. Since the 1980s a large part of the literature discussing the Industrial Revolution in Britain has focused on the importance of small-scale—workshop type—production in addition to the traditional emphasize on the factory as the core

institution. This was certainly also the case for large parts of Scandinavian industrialization (Bruland 1991).

2. The concept of path creation is rather recently introduced into the discussion of innovation, and will here be used as the beginning of a process which over time creates a new form of production drawing on new forms of knowledge, new forms of organizations, as well as involving new social groups. The concept has also been used in analysis of how companies develop new products which give the company new opportunities for development within new sectors (Garud and Karnoe 2002) and also how micro innovations may influence wider national development and create a new long term development (Schienstock 2004). Mowery and Rosenberg (1998) show how a technological-industrial innovation (production of exchangeable components) created the basis for a new direction in American industrial history. The idea that new technologies and innovations have created basis for new forms of industrialization is well known in economic history, i.e. inherent in the concept of 'industrial revolutions' (Bruland and Mowery 2004).
3. Arendal Fossekompani, Bjølvefossen, Titan, Det Norske Nitridaktieselskap, Arendal Smelteverk, grong Gruver.
4. Literally: Laboratory for Raw Materials.
5. A group of Norwegian scientists and engineers became involved in wartime research institutions in the UK and the USA during the Second World War, and after the war tried to introduce the knowledge and technologies that they had developed during wartime into the Norwegian context. The modernizers argued that these emerging technologies would contribute to the modernization of Norway's economy and society, improving performance in old industries like fisheries and process industries, as well as in such key components of public infrastructure as telecommunication, defence, and public administration. These new technologies also were expected to become the basis for future growth sectors and export industries. As Chapter 4 discusses in greater detail, this ideology became influential in Norwegian politics from the mid 1960s onwards. Its influence peaked during the 1980s, when it was widely accepted that the emerging technologies would become the basis for Norwegian reindustrialization. Since the early 1990s, however, this ideology had become much weaker in Norwegian politics.
6. We should note that the first Frascati manual for 1963 explicitly expresses that "R&D" is a different concept from "science". However, in most innovation literature the two concepts are used more or less synonyms.
7. The state has played a crucial role in the development of this type of industry, not only in funding R&D and subsidizing companies, but in particular as "initial market" or "first customer" for new products. This is also part of the reason that this path involves different type of actors and organizations compared to earlier paths, as state agencies have been involved directly in the formation of the path.

REFERENCES

Andersen, H. W. (1997) "Producing Producers: Shippers, Shipyards and the Cooperative Infrastructure in the Norwegian Maritime Complex since 1850," in C. F. Sabel and J. Zeitlin, *Worlds of Possibilities. Flexibility and Mass Production in Western Industrialization*, Cambridge University Press.

Andersen, K. G. (2005) *Flaggskip i fremmed eie. Norsk Hydro 1905–1945*, Oslo.

Andersen, K. G. and G. Yttri (1997) *Et forsøk verdt. Forskning og utvikling i Norsk Hydro gjennom 90 år*, Universitetsforlaget Oslo.

Basberg B. L. (1986) "R&D Performance in Norwegian Electronics Companies 1960–1975," Working Paper no 26 *Norsk elektronikkindustri* 1945–1970, Oslo.

Berg, B. I. (1988) "Gruveteknikk ved Kongsberg Sølvverk 1623–1914," *STS Rapport* no. 37, Trondheim.

Bjørgum, J. (1968) *Venstre og kriseforliket*, Oslo.

Bruland, K. (ed.) (1991) *Technology transfer and Scandinavian industrialization*, New York.

Bruland, K. and D. C. Mowery (2004) "Innovation Through Time," in J. Fagerberg, D. Mowery, and B. Nelson (eds.), *The Oxford Handbook of Innovation*, Oxford.

Chandler, A. D. (1990) *Scale and Scope: The Dynamics of Industrial Capitalism*, Cambridge Mass.

Friedman, R. M. (1989) *Appropriating the Weather: Vilhelm Bjerknes and the Construction of a Modern Meteorology*. Ithaca and London: Cornell University Press, 1989.

Furre, B. (1971) *Mjølk, bønder og tingmenn*, Oslo.

Garud, R. and P. Karnoe (2002) *Path Dependency and Creation*, Lawrence Erlbaum Associates.

Hanisch, T. J. and E. Lange (1985) *Vitenskap for industrien: NTH—en høyskole i utvikling gjennom 75 år*, Oslo.

Hauknes, J. (1998) "Norwegian Input-Output Clusters ad Innovation Patterns" *STEP Report 15/1998*, Oslo.

Historisk Statistikk (1994) (Historical Statistics (1994)), Statistics Norway, Oslo-Kongsvinger 1995.

Hughes (1983) *Networks of Power*, Baltimore.

Kjeldstadli, K., S. Myklebust, and L. Thue (eds.) (1994) *Forminga av industrisamfunnet i Norden fram til 1920*, TMV report no. 5.

Landes, D. S. (1969) *The Unbound Prometheus: Technological Change and Industrial Development in Western Europe from 1750 to the present*. Cambridge University Press, Cambridge.

Lange, E. (1977) "The Concession Laws of 1906–1909 and Norwegian Industrial Development," *Scandinavian Journal of History*, 2/1977.

Mowery, D. C. and Rosenberg, N. (1998) *Paths of Innovation: Technological Change in Twentieth-Century America*, Cambridge.

Nielsen, T. H., A. Monsen, and T. Tennøe (2000) *Livets tre og kodenes kode : fra genetikk til bioteknologi : Norge 1900–2000*, Gyldendal, Oslo.

NTNF (1964) *NTNF's forskningsutredning 1964*, Oslo.

Pol, E., P. Carroll, and P. Robertson (2002) A new Typology for Economic Sectors with a View to Policy Implications, *Economic Innovation and New Technologies*, vol.11(1), pp. 61–76.

Refsdal, A.-O. (1973) "Nyetablering og krise—en undersøkelse av bedriftsdannelse innen industri, håndverk og handel under krisen i 1930-åra," Master Thesis in History, University of Oslo.

Rosenberg, N. (1976) *Perspectives on Technology*, Cambridge University Press, Cambridge.

Schienstock, G. (2004) "From Path Dependency to Path Creation: A New Challenge to the Systems of Innovation Approach," in G. Schienstock (ed.) *Embracing the Knowledge Economy: The Dynamic Transformation of the Finnish Innovation System*, Cheltenham, UK: Edward Elgar Publishing Ltd.

Schwach, V. (2002) *Havet, fisken og vitenskapen: fra fiskeriundersøkelser til havforskningsinstitutt 1860–2000*, Oslo.

Sejersted, F. (1982) (ed.) *Vekst gjennom krise. Studier i norsk teknologihistorie*, Universitetsforlaget Oslo.

—— (1993) *Demokratisk kapitalisme*, Universitetsforlaget Oslo.

Slagstad, R. (2001) *De nasjonale strateger*, Oslo.

Smith, K. (2002) "Innovation Infrastructure," Working Paper at UNU/Intech July 30, Maastricht.

Sogner, K. (1997) *God på bunnen: Simrad-virksomheten 1947–1997*, Novus Forlag, Oslo.

—— (2001) *Plankeadel: Kiær- og Solberg-familien under den 2. industrielle revolusjon*, Oslo.

—— (2002) "Det norske næringsborgerskapet under den andre industrielle revolusjon," *Særtrykk Handelshøyskolen BI*, no. 35.

—— (2004) *Creative power: Elkem 100 years: 1904–2004*, Oslo.

Thue, L. (1994) *Statens Kraft 1890–1947. Kraftutbygging og samfunnsutvikling*, Cappelen, Oslo.

Wicken, O. (1984) "Learning, Inventions and Innovations. Productivity increase and new technology in an industrial firm," *The Scandinavian Economic History Review*.

—— (1995) "Norsk fiskerihistorie: politikk i møte med regionale kulturer, STEP report 17/1994, Oslo.

—— (2005) "Diverse Regional industrialization: Norway during the first half of the twentieth century," in K. Bruland and J.-M. Olivier (eds.), *Essays in industrialization in France, Norway and Spain*, Oslo 2005.

3

Public Sector Research and Industrial Innovation in Norway: A Historical Perspective

Magnus Gulbrandsen and Lars Nerdrum

This chapter deals with the *historical role* of *public research organizations* or *public sector research* (PSR) for industrial growth and innovation in Norway—and the changes in this role over time. Public research organizations include research institutes and higher education institutions, and date back to the nineteenth century. Our main concern is the rationale and contributions to industrial innovation of PSR, but we also touch on research and innovation policies (discussed in more detail in Chapters 2 and 4).

Norway is a small and open country that has followed international trends in research and innovation policy and institution-building. Templates and inspiration for new institutions and organizations have most often come from abroad, e.g. university and sector-specific institute models from Germany, Sweden, and the US. Nonetheless, the Norwegian system has developed an internationally unique structure of strong sector-oriented research institutes, in stark contrast with Sweden, where such institutes have not developed into a large separate sector.

Some authors have claimed that the recent emphasis in many industrial economies on university-industry relations is not new but rather represents a return to an earlier form of knowledge production (Martin and Etzkowitz 2000; Martin 2003). The Norwegian case, however, reveals a more complex story. Tensions between different models of knowledge production have always been present within Norway in particular, as Chapter 2 points out, because of the Norwegian innovation system's three different *layers* that involve different institutions and approaches to innovation. Different public research organizations have been established to support each layer, and some organizations have become centres for interaction among all three layers.

Table 3.1. Central public R&D organizations in Norway referred to in the chapter

Acronym	Name (English translation)	Started	Comments
CMI	Christian Michelsen Research Institute	1930	Located in Bergen. Science and technology activities were later separated as Christian Michelsen Research (CMR) in 1992 (owned jointly by CMI and UiB).
FFI	Norwegian Defence Research Establishment	1946	Located at Kjeller close to Oslo. Departments in other cities like Bergen and Horten at various points in time.
IFA/IFE	Institute for (Nuclear) Energy Research	1951	This was a spin-out from FFI. After it was decided not to build commercial nuclear plants in the 1970s, the name changed to Institute for Energy Research (IFE). It maintains two nuclear reactors at Kjeller close to Oslo and in Halden.
KV[1]	Kongsberg Weapons Factory	1814	Perhaps the most important "National Champion" of the industrial policy of the first three post-Second World War decades. It was dissolved in 1987 but the military activities were continued and now form the "Kongsberg Group" with 4,200 employees. Other civilian activities are found e.g. in Kongsberg Automotive (2,100 employees). These firms are private and listed on the Oslo stock exchange.
NLH	Norwegian College of Agriculture	1897	Located at Ås close to Oslo. University status from 2005, changed name to University of the Life Sciences.
NTH	Norwegian Institute of Technology	1910	This technological university was merged with the rest of the University in Trondheim, forming NTNU in 1996.
NR	Norwegian Computing Centre	1952	Located in Oslo. Famous for its role in the development of the Simula object-oriented computer language.
NTNF[2]	Norwegian Research Council for Scientific and Industrial Research	1946	Together with four other research councils, NTNF was merged into the Research Council of Norway in 1993.
NTNU	Norwegian University of Science and Technology	1996	The full merger between NTH and the other parts of the University in Trondheim.
RCN[2]	Research Council of Norway	1993	The result of a merger of the five research councils.
RF	Rogaland Research	1973	Located in Stavanger close to the college which became the University of Stavanger in 2005. The city is a major location for oil and gas industry and R&D.
SI	Central Institute for Industrial Research	1950	Located in Oslo next to the university, merged with SINTEF in 1993 under the latter name.
SINTEF	Foundation for Industrial and Technological Research	1950	Located in Trondheim next to NTH, several departments in Oslo. Many different parts, e.g. marine technology MARINTEK and the electronics lab Elab.
TF	Telephone Agency's Research Institute	1967	Located in Oslo, earlier at Kjeller close to Oslo. Established as a public research institute but later the R&D unit of privatized telephone company Telenor.
UiB	University of Bergen	1946	
UiO	University of Oslo	1813	

[1] KV is not strictly a public research organization (PRO) but was state-owned for a long time and constituted an important partner for many of the PROs.

[2] NTNF and RCN are not strictly PROs either as we use the term in this chapter, but they are important public organizations for support of industry-oriented R&D.

This chapter is organized chronologically, with sections dealing with the dominant policies and institutions in each of several historical periods. The first section describes the development of Norway's higher education system and the emerging complementary belief in the transformative power of science around 1900, a discussion that is followed by an account of the first research institutes. The next sections analyze the experience of World War II and the renewed belief among policymakers that science itself was an important source of practical ideas ("science push"), and conclude with an examination of Norway's contemporary system of industrial R&D support that includes mission-specific and sector-specific institutes. A table of acronyms and organizations is found in Table 3.1.

THE ORIGINS OF THE HIGHER EDUCATION SYSTEM IN NORWAY

Compared to other countries, Norway was a higher education latecomer. Although the idea of a university emerged as early as 1629, before UiO was established in 1813 Norwegian students had to go to Copenhagen to become priests or judges (Collett 1999: 13). As with every other college, university or research institute that followed, there were expectations that the university would help reach important societal and economic goals in Norway. This is perhaps obvious but nevertheless an important point: the government and the public have expected *practical* benefits from all the public research organizations that have been established. Such a utilitarian component has been a key part of many other countries' higher education and research policies as well, not least the US (cf. Mowery and Rosenberg 1993).

The two most important public research organizations in Norway before Second World War were the University of Oslo (UiO) and the Norwegian Institute of Technology (NTH) in Trondheim, which we discuss in more detail below.

THE UNIVERSITY OF OSLO—NORWAY'S HUMBOLDT UNIVERSITY

The foundation of the University of Oslo benefited from a large-scale private fundraising effort in the first decade of the nineteenth century. More than 3,600 individuals contributed, and many of the university's buildings during its first 120 years were financed entirely by private donations. Similar processes led to the establishment of other Norwegian universities and colleges. In Bergen, industry, local banks and individuals worked hard to get a university to the city. Even the University of Tromsø, which is probably the least industry-oriented university in Norway, was first promoted by the industrialist and politician Hans Meyer at the beginning of the twentieth century. Local savings banks played important lobbying roles in all of these cities (Sejersted 1993: 170).

But with the exception of a mining-oriented degree, UiO did little teaching that was deemed relevant for industry (Collett 1983: 8). Some practical tasks were allocated to the university, e.g. UiO assumed responsibility for the scientific part of the national geographical and geological surveys in the 1830s (Schwach 2000: 28) which helped to fund natural science (Collett 1983: 32). During its first 120 years the University of Oslo nevertheless was more and more influenced by German "Humboldtian" ideals like "academic freedom" and training as a source of personal development and enlightenment. Close collaboration with industry found little support, and the university's training in mining declined (Hanisch and Lange 1985: 21). Indeed, the university as a whole scarcely grew during the late nineteenth century, as only two new professorships were established between 1877 and 1890. By the turn of the century, most of the laboratory equipment at UiO was more than forty years old (Collett 1983: 20).

Examples of collaboration between UiO and private firms emerged again in the early twentieth century. A professorship of "insurance mathematics" was established in 1913, followed by other professorships and a degree organized in partnership with insurance companies (Collett 1999: 120). Even more important was the increasing role of the university in developing the welfare state, e.g. supporting research in macroeconomics for economic policy, hygiene, and other medical disciplines for improving the population's health, and education for improving schools and social equality. Although the university displayed little interest in industrial collaboration, several well-known professors with interests in both basic and applied research (Sejersted 1993: 149) combined groundbreaking basic science with remarkable practical contributions. This served as a counterweight to the Humboldtian ideas.

According to Sejersted (1993: 141), the academic project that produced the most practically important applications was Vilhelm Bjerknes' creation of modern meteorology, a scientific achievement in its own right that had large and positive consequences for the productivity and safety of agriculture, fishing and shipping (see Friedman 1989). Johan Hjort's fundamental work in oceanography also had a broad practical impact, not least in the mapping of stocks of fish. His charting of sea life also led to the development of a shrimp trawl for use in deep waters (Schwach 2007). Hjort was a key figure on the Fishery Board (1900)—the predecessor of the Institute of Marine Research in Bergen—and a cofounder of the now closed-down Whale Research Institute. Positions outside the poorly funded university provided Hjort and others access to much more substantial resources than an academic position alone would have (Schwach 2000; Collett 1999: 141).

The best-known example of industrial impact is related to UiO physics professor Kristian Birkeland. He had achieved fame at the end of the nineteenth century for his theory that the Northern Lights are due to electromagnetic radiation from the sun, and a few years later he invented a method for fixating nitrogen from air that could be used for the production of fertilizers (*ibid.* p. 109; cf. also Andersen and Yttri 1997). Birkeland teamed up with engineer and entrepreneur Sam Eyde, and supported by foreign capital, they set up the company Norsk Hydro (Hydro in short) in 1905. Throughout his career Birkeland worked on both fundamental and more practical problems. His fifty-eight patents and role in establishing three spin-off firms signify his importance as an early academic entrepreneur.

Hydro was a major success and soon became the country's largest company, a position that it held for more than 100 years. It is a prime example of a science-based firm and became a meeting place for science and technology, theorists and practitioners (Andersen and Yttri 1997: 27; Sejersted 1993: 139–40). Arguably, it showed that the "science push" model worked, at least in the view of some leading figures. UiO professor/rector Waldemar Brøgger—the most active participant in Norwegian research policy debates for decades (Collett 1983)—argued that the autonomous advancement of basic science would inevitably lead to industrial applications as well as more general social welfare and "happiness" (Kvaal 1997: 66). Even the more practically inclined professors at the technical university NTH embraced the science push model (Kvaal 1997: 92). The belief in the transformative power of science, however, mainly led to the creation of *applied* organizations such as the technical and agricultural universities (see below), rather than expanded financial support for basic research.

Despite the very low levels of basic research funding, this period arguably constitutes the first "golden age" for Norwegian academic research, during which several individuals made long-lasting marks on international science. Brøgger emphasized the practical benefits of science, not least the Hydro case, in arguing for more public and private funding for fundamental research, but he was unsuccessful. Even the famous Birkeland received minimal research funding. Professor Brøgger referred to the National Research Council in the United States and the Carlsberg foundation in Denmark as models for such research support (overlooking the fact that the National Research Council supported little if any academic research) and argued for the "duty" of the state and wealthy citizens to support science (Collett 1983, e.g. pp. 29–30). Indeed, many of the leading professors who collaborated with industry and/or worked to ensure the practical benefits of their research, may have done so because of a *lack* of funding for basic research (Friedman 1989 makes this argument about Bjerknes).

THE SLOW BIRTH OF THE TECHNICAL UNIVERSITY

Most universities and colleges have been built on existing organizations: museums, technical schools and/or societies for the advancement of science. The first higher education provider and industrially relevant school in Norway was the small Mining School at Kongsberg, founded in 1757 (Collett 1999: 24–5). The subsequent decline of mining education in the mid nineteenth century at UiO intensified the debate about creating a national institute of technology for industrially relevant research and higher education. The idea was not new; inspired by the Ecole Polytechnique in Paris, all of the Nordic countries desired this kind of organization at the beginning of the nineteenth century (Hanisch and Lange 1985: 12). The first Nordic institute of technology was established in Stockholm in 1826, and the second in Copenhagen in 1829. Norway's did not appear until almost a century later.

A group of scientists from the University of Oslo initiated the first campaign in the 1820s to establish a "Higher technical school" in Norway, arguing that R&D would create new industries and that a national supply of engineers would support Norway's economic development (*ibid.* pp. 14–15). But the proponents found little support for their faith in industrialization in a country whose economy was based on agriculture and fisheries. The Parliament, dominated by representatives of these rural trades, turned down the proposal, arguing that the Norway's tiny manufacturing and resource-processing industries by no means could justify the high costs of creating a technological university.

During the second half of the nineteenth century, new industries nevertheless grew, particularly around Oslo, and lobbying for a technological university was renewed as the demand for technologically skilled candidates outstripped the supply from the lower-level technical schools. But the supporters of a technical university once again were limited to members of the politically weak and small urban bourgeoisie. Although the Government made several concrete proposals to the Parliament about establishing an institute of technology, the

Parliament rejected it five times before the 1890s. At that point, the demand for skilled technical labour had grown so large that the issue could no longer be ignored. A geographical struggle emerged among the three cities Bergen, Oslo, and Trondheim, all which hosted technical schools that sought upgrading. The industrial interests wanted the new organization in Oslo, where most of the industry was located, but they had to give up the location battle to be able to get a technical university at all (*ibid.* p. 32).

Norway's Parliament voted to establish the Norwegian Institute of Technology (NTH) in Trondheim in 1900, and the Institute was opened by King Haakon in 1910, later than most other European and many South American countries. It is noteworthy that Norway established academic institutions for public R&D and training in agriculture and fisheries (the Norwegian University of Agriculture, NLH, 1897, and the Fishery Board, 1900), supporting the small-scale decentralized industry layer of the innovation system, before establishing a technical university supporting large-scale centralized industry (Chapter 2; also Sejersted 1993: 140).

Expectations were high with respect to the training and to the R&D activities at NTH. Initially, only the training activities were successfully implemented; industrially oriented R&D was difficult to establish (Hanisch and Lange 1985: 69). NTH professors saw no conflicts between scientific work and practical contributions, but national industry was technologically weak and focused mainly on quick answers to small technical problems. Some professors had substantial consultancy activities in the 1910s—professor Olav Heggstad employed 30 people in his water power consultancy company. Organic chemistry Professor Claus Riiber, with a background from work in breweries in Denmark and Norway, tried to establish a spin-off firm that failed, and he later went into basic research. Attempts at starting a joint lab with chemical company Elkem also failed, and several professors left to work for industry.

NTH imposed greater restrictions on faculty consultancy activities in 1921, leading most of the remaining professors into basic research with little or no industrial collaboration. This ensuing period of academic isolation was often viewed as problematic by the faculty, many of whom were motivated by research at the intersection of basic science and practically oriented technology. But the NTH faculty increased their scientific competence, enhancing their capability to meet new industrial needs two decades later (*ibid.*).

During the 1930s, NTH tried to approach companies with proposals for collaboration on R&D, relying on a new category of professors with excellent academic credentials who were eager to help out with industrial development. This strategy of "technology push" received little support in the firms, as the so-called "engine case" highlights (cf. Hanisch and Lange 1985: 121–4).

Many Norwegian ships used a primitive type of electrical engine produced by more than 120 Norwegian firms. NTH's public experiments demonstrated that diesel engines provided superior performance. Since no Norwegian producers of diesel engines existed, however, a switch to production of diesel engines would inevitably lead to a difficult restructuring that few established engine firms might survive. Although NTH Professor R. Lutz argued that firms could purchase licenses to make engines, he was accused of "helping the foreigners" and "not being able to understand user needs and the special national context." Only Aker licensed diesel-engine technology, and it was the sole survivor within the small marine engine industry. Norwegian industry arguably lacked sufficient trust in the validity and value of professorial advice.

Despite these problems, the University of Oslo, along with NTH and the more specialized colleges, benefited from a tendency in the years before the Second World War to view science as a source of rationalization, systematization, and other improvements in industrial production and society in general (Sejersted 1993: 160). Chocolate factory owner Johan Throne Holst provided generous support for research in social sciences and in other disciplines such as nutrition physiology (Collett 1999: 140). Shipping magnate Wilhelm Wilhelmsen funded a department of bacteriology at UiO in 1937, and in 1953 shipowner Anders Jahre established a fund for the advancement of science that to this day sponsors the most prestigious academic prize of the University of Oslo (*ibid.*). Many of these private gifts came in response to hard work by particularly entrepreneurial professors (*ibid.* p. 141).

Industrial pressure was less significant in the establishment of the universities in Bergen, Trondheim (in the mid-1990s merging with NTH to form the Norwegian University of Science and Technology NTNU) and Tromsø. Some of Norway's engineering colleges, a few of which trace their history back to the technical schools of the nineteenth century, have served local/regional needs for engineers with bachelor degrees, but these organizations have rarely carried out substantial R&D. The main exception to this rule is the engineering college in Stavanger, which grew in importance after the town became the "oil capital" of Norway in the 1970s and now is an integral part of the recently established University of Stavanger (2005).

THE ORIGINS OF NORWAY'S RESEARCH INSTITUTES

The first universities in Norway, despite some extremely talented and motivated staff, were relatively weak institutions with little funding and political support. For decades, leading scientists survived on funding for purposes like geological surveys, coastal mapping and housing for poor people. In addition to largely futile attempts at increasing public funding, professors like Waldermar Brøgger lobbied private capitalists in the beginning of the twentieth century for the establishment of private basic research foundations, inspired e.g. by the Carnegie and Rockefeller foundations in the U.S. and the Nobel foundations in Sweden (Collett 1983: 44–9). For Brøgger and other UiO professors, the ideal was Andrew Carnegie's claim about the "obligations of wealth" rather than Wilhelm von Siemens' statements about the necessity of publicly supported research institutes oriented to industry (*ibid.* p. 48). But neither new basic research organizations nor large private foundations were established in Norway during this period.

Instead, the belief that technological development should be under the control of users and the demands of production had many followers in Norway (Sejersted 1993: 146). Organizations that resembled independent research institutes in fields such as geographical and geological surveys and oceanography/marine sciences already existed in the late nineteenth century. Independent engineer Axel Krefting had suggested the establishment of a public chemical industrial laboratory to help spur a chemical industry as early as 1893, but with little response (Collett 1983: 80). Inspired by countries like the UK, Germany, and the US, the question of establishing industrially directed public research institutes was formally put on the policy agenda in 1919 (Kvaal 1997: 86; also Collett 1983: 91–3). Six institutes, each of which was designed to serve a particular industry, were proposed by the Ministry of Industrial Supply right after WW1. The proposal argued that these organizations should respond to user needs yet be knowledgeable about the academic state-of-the-art in relevant fields. A desire for increased self-sufficiency in the aftermath of a world war that caused shortages of many raw materials was a central factor behind the Ministry proposal.

NTH, however, wanted a multidisciplinary "central institute" to be established that could support all industries, based on the Mellon Institute in the United States (Kvaal 1997). As expected, the NTH professors wanted this "transfer mechanism" located in Trondheim, next to the technical university. This proposal set off yet another geographical battle, replete with arguments about whether applied research should be close to industry or close to academic science that echoed those heard when NTH was established (cf. Gulowsen 2000). But plans for most institutes collapsed in the face of the economic downturn in the 1920s.

A few early research institutes were formed as industry cooperatives rather than public organizations. Many industries considered sharing the costs of a research institute, but only two managed to implement the plans. The paper industry started a research institute in Oslo in 1925 and the canned food industry founded another one in Stavanger in 1931. Economic problems put other initiatives on hold until after the Second World War, when organizations were established in Oslo for user-controlled R&D related to breweries (1946), herring oil and flour (1948), wood technology (1949), textiles (1949), leather tanning (1950), potato flour (1950) and brick works (1952). A shipbuilding institute and a cement and concrete institute also were founded in Trondheim during this period. Many of the cooperative industry research institutes encountered financial problems during the 1950s and were absorbed by the two large contract organizations SI and SINTEF (Gulowsen 2000).

In addition to the institutes discussed above, the "State Raw Materials Laboratory" was started in one of the University of Oslo's buildings in the early 1920s as the first public research institute with an explicit mission to help create new industry. The laboratory was a scientific success but failed to meet the expectations of industry, moved to Trondheim in 1952, and became part of the Norwegian Geological Survey. Another contract research institute, the Christian Michelsen Institute (CMI), was established in Bergen in 1930 by a foundation dedicated to the advancement of science. Although plans for upgrading the Bergen Museum to a university had been around for a long time, Norway's second largest city got a cross-disciplinary contract research institute sixteen years before it got a university. This sequence once again highlights the strength of the belief in Norway in the importance of supporting applied research in established industries or resource-based sectors.

INDUSTRIAL RESTRUCTURING AND A LACK OF R&D SUPPORT INSTITUTIONS

A few large Norwegian firms did establish close collaborative relationships with university faculty and research. Inspired by the Birkeland/Eyde success and supported by government, Norsk Hydro and Elkem established very ambitious R&D projects for several decades from around 1910 to exploit Norwegian natural resources. University professors such as mineralogy/geochemistry professor Victor Goldschmidt, "one of the greatest scientific talents ever," were involved as consultants or part-time staff by Elkem (Sogner 2004: 57–8). Hydro also had many contacts at universities in other countries (Andersen and Yttri 1997: 40–2).

With few exceptions, these attempts to create new product lines based on Norwegian natural resources failed completely. Hydro's efforts in coal, titanium, kali, plastic, uranium and artificial fabrics, among other technologies, failed to reach commercialization. Sejersted (1993) argues that Hydro's strategic agreement during the 1930s with the German chemical giant IG Farben did not allow Hydro to concentrate its efforts sufficiently. Andersen and Yttri (1997: 302–4) also indicate that these ambitious R&D projects did not get enough funding. Hydro's foreign competitors used hundreds and occasionally thousands of people in the development of a single technology, in contrast to Hydro's deployment of tens of people on comparable projects. The company spent around one percent of its turnover on R&D, far less than, e.g. its German competitors (Gulowsen 2000).

The Hydro example illustrates another important characteristic of the Norwegian innovation system during the prewar period: even the most technically advanced Norwegian companies spent modest amounts on R&D by comparison with their foreign competitors. Moreover, the majority of Norway's industrial companies were small and technologically weak (Sejersted 1993: 155), perhaps because of the long delays in creating Norwegian institutions of higher technological education (Hanisch and Lange 1985: 20). The general public was sceptical towards science and theory, and prime minister and Shipping Federation President Aanon Knudsen expressed in 1919 that most tasks in society did not require higher level training at all (*ibid.* p. 100).

Although Norwegian engineers received recognition for their work on hydro-electric power stations and the development of big industry, many were nevertheless forced to go abroad to find employment. Few companies hired scientists and engineers. But some of those who did made important innovations in the first decades of the twentieth century, e.g. related to paper production and vegetable oil processing (Sejersted 1993: 147). Public research organizations were not directly involved here. The arguably the most important radical innovation in the period between the world wars, the Söderberg electrode, was developed in Elkem's own R&D labs. This innovation is still unimaginable without the Birkeland-Eyde process, which contributed to creating private R&D laboratories that made it possible for people like NTH professor Peder Farup to find interesting full-time work in Elkem (Sogner 2004: 61).

Following a painful restructuring period, Norwegian industrial activity increased significantly during the 1930s. Many larger companies went bankrupt, but a vibrant small firm movement supported growth in industrial employment of almost 7 percent annually between 1933 and 1939 (Sejersted 1993: 181). The mean size of Norwegian companies decreased: in 1930, 27 percent of industrial employment was found in firms with less than 50 employees; but by 1948 this share had increased to 42 percent (*ibid.* pp. 181–2).

This change in Norwegian industrial structure proved to be an important precondition for the later establishment of research institutes. The restructuring of industry also resulted in greater adoption by new firms of modern production methods (Hanisch and Lange 1985: 138). During the second half of the 1930s, NTH enjoyed a sharp increase in private funding, often for a certain department or laboratory with "no strings attached," and new contacts expanded with industries that previously had little interest in university partnerships. The most important case of the latter is probably the pulp and paper firm Borregaard, which provided a professorship in 1936. Even the Confederation of Norwegian Industry, which had sided with the engine producers in their critique of NTH during the early 1930s, raised a large sum of money for the university's 25th anniversary in 1935.

During the late 1930s, Norwegian companies enhanced their abilities to utilize external knowledge and to define and outsource technical problems, developments that were essential to the growth of the post-war research institutes that sought to respond to industry demands. A major element that nevertheless was lacking in the Norwegian innovation system during this period was the intra-firm industrial laboratory, although Hydro, Elkem and a few other large firms had established such organizations. Application-oriented and industry-friendly public researchers found few partners in industry and few domestic users of new technologies. Most of these individuals continued to do basic research, even as they tried to maintain their research in areas they believed might be relevant for the nation's future. One example of this "applications-oriented basic research" is the experiments in the 1930s with feeding of fish larvae by professor Hjort and colleagues, which represented some of the earliest work in the technology of fish farming (see Chapter 8).

The pre-1940 Norwegian innovation system also lacked the public institutions to support R&D in industry, the research institutes, or academia. The State Raw Materials Committee, established in 1917 and led by professor and Elkem consultant Goldschmidt, convinced the Ministry of Industrial Supply to fund some research projects. It was followed by the Central Committee for Scientific Collaboration in Support of Industry in 1918, renamed the Council for Applied Science in 1921. These committees functioned as a meeting place between industrialists, professors and government representatives, potentially able to mobilize resources from many sectors of society.

Norway did not differ greatly during this period from many other countries in the relations between public research and industry. Scientific research was viewed as crucial to economic and political development, just like in other countries. But in Norway this did not translate into long-term and significant basic research funding from public sources as in Germany or private foundations as in Sweden and the United States. Norway was a latecomer in building institutions like universities, research institutes and research councils, and there were few industrial laboratories. Many professors survived on funding for geological surveys and mapping of fish stocks, the latter activity an area in which Norway was one of the first countries to undertake science-based investigations. Important advances, including the development of modern meteorology and oceanography, emerged from this period of minimal support for basic research. Indeed, Norwegian researchers in research institutes or universities may never have faced such strong incentives for getting involved in user-oriented work as they did in the first decades of the twentieth century. Nonetheless, by 1940, Norway had developed scientific competencies in a number of disciplines central to the country's geography and natural resources. Although some meeting places between public researchers and industry existed, the capability to absorb and utilize the knowledge was nonetheless still lacking.

THE SECOND WORLD WAR AND THE EMERGENCE OF THE IDEOLOGY OF RESEARCH-DRIVEN GROWTH

Norway was occupied by German forces during 1940 to 1945. The inability of Norway's military to resist German forces successfully in April 1940 created a strong political impetus for modernizing and improving national defense after the war ended. The wartime experience also contributed to the development of a postwar ideology of industrial modernization (see Chapter 4) with corporatist characteristics that emphasized collaboration among sectors and groups in society (Hanisch and Lange 1985: 182).

In 1946, the government-appointed Vogt committee presented a Green Paper on research in technology and natural science. The paper quoted the 1945 report (*Science: The Endless Frontier*) by Vannevar Bush, director of the wartime Office of Scientific Research and Development in the United States, arguing that a country should seek to be independent from others in developing new scientific knowledge. It also embraced the rhetoric of "science/technology push"—that scientific and technological breakthroughs are the major sources of innovation. The University of Bergen was created in April 1946 and the NTH budget was doubled the same year, indicating that the government for the first time was willing to allocate substantial funds to academic R&D (Kvaal 1997: 397).

MILITARY R&D AS AN ENGINE FOR INNOVATION

As in many other countries in the late 1940s (Larédo and Mustar 2001a), Norway's public R&D efforts initially emphasized defence and nuclear research. These activities were concentrated in one important research institute established just after Second World War: the Defence Research Institute (FFI 1946). The nuclear activities were separated from FFI in 1951 to form the Institute for Nuclear Research (IFA, later Institute for Energy Research IFE; cf. Njølstad and Wicken 1997; Njølstad 1999). This year Norway became the first country after the United States, the United Kingdom, France and the Soviet Union to build a nuclear reactor (Njølstad 1999), and despite its small size the country by 1975 was the seventh largest weapons exporter in the world (Wicken 1983).

Outside the defence sector, a new public–private partnership, the Norwegian Research Council for Technology and Natural Science (NTNF), was founded in 1946 with crucial political support from Norsk Hydro (Andersen and Yttri 1997). NTNF was given considerable freedom and was not subject to normal bureaucratic control mechanisms and directives. The Federation of Norwegian Industry agreed to contribute half of the council's funding (Hanisch and Lange 1985: 182). Key employees in the new research council and institutes were a highly select assembly of people, "consisting of well-educated young men closely coupled with illegal intelligence work and often forced to leave the country because of their link with the resistance movement" (Wicken 1994: 16), several of whom had worked in British wartime R&D projects like radar development. The Council's political influence benefited from the appointment of resistance leader Jens Christian Hauge as Minister of Defence.

FFI developed technologies and relied on private companies to produce them for military applications and, occasionally, for civilian markets as well (Ørstavik 1994). Its first success was in sonar technology (Njølstad and Wicken 1997: 69–70), and the firm chosen to produce the technology, SIMRAD, used it in products for Norway's domestic fishing fleet and subsequently in other markets (see Chapter 2 and Sogner 1994). FFI also played a central role in the early development of Norway's computer industry in the late 1950s. Nonetheless, the institute struggled to find capable private partners in its technological projects and, after some trial and error, became a key strategic partner for the two state-owned weapon companies, Kongsberg Weapons Factory (KV) and Raufoss Ammunition Factory (RA) (Njølstad and Wicken 1997: 443). Wicken (1983: 27) argues that the postwar development of these two companies, KV in particular, would not have been possible without the "technological level at FFI and the organization's ability to transfer technological knowledge to industrial production."

Other public R&D organizations, such as the (eventually successful) cybernetics/automation department at NTH (Kvaal 1994), which initially faced great difficulties in persuading industry about the potential of automation, used KV as a vehicle for demonstrating the practical and commercial merits of its technologies (Wicken 1989). Hence, defence-related R&D organizations came to play an important role in the Norwegian postwar innovation system as a testing ground for ideas and technologies from different research groups and disciplines (Wicken 1992). As was noted above, the role of defence-related R&D organizations was enhanced by their strong links with such politically influential individuals as defence minister Hauge, KV director Bjarne Hurlen (whom Hauge hand-picked) and FFI director Finn Lied. They were central nodes in a network whose influence extended far beyond the defence sector (see Chapter 4).

INDUSTRIAL RESEARCH INSTITUTES AND KNOWLEDGE TRANSFER

Shortly after its creation, NTNF took up the earlier idea of creating a large multidisciplinary industry-oriented research institute in Oslo. The Central Institute for Industrial Research (SI) was founded in 1950, and was intended to have close relations to the Faculty of Mathematics and Natural Science at UiO. NTH faculty and administrators objected to the creation of SI, arguing that technological industry-oriented research should be located near their institute of technology in Trondheim. Their campaign against SI failed, but a few months later NTH started its own "central institute," SINTEF, which the Ministry reluctantly accepted in order to avoid another devastating interregional battle. The Christian Michelsen Institute in Bergen also included technological R&D in its natural science department from the early 1950s.

NTH now entered a new era as an entrepreneurial and proactive university: "The pressure from competing groups released a will to act and ability for renewal that has later been one of the technical university's main characteristics" (Hanisch and Lange 1985: 213). SINTEF proved to be adept at handling large numbers of industrial contacts, and influenced research and teaching at NTH (*ibid.* 246). The income from industry contracts made the university less dependent upon ministerial routines and budgeting processes. Positions could be created for talented researchers without need for the Ministry to create new professorships. Companies partnering with SINTEF gained access to a wide range of engineering graduates. SINTEF also built new labs, often co-located with NTH departments. The institute expanded to around 2,000 employees in 1993, when it merged with its less entrepreneurial and successful competitor SI, which at that time had around 300 employees.

Norsk Hydro supported the new institutes in various ways but did not rely on them for its own R&D (Andersen and Yttri 1997: 154). The company supported the establishment of SI and supplied more than 50 percent of the institute's private funding during the 1950s (Gulowsen 2000: 30 and 110). Hydro donated deuterium ("heavy water") worth six million NOK to nuclear institute IFA for its reactor start-up in 1951 (uranium came from a partnership with Dutch research organizations), and the SINTEF materials section received half a million NOK from the firm as a start-up gift (Andersen and Yttri 1997: 155). Norsk Hydro's actions were motivated by its interest in strengthening Norway's overall technological capabilities, rather than its desire for assistance in specific projects.

The first Norwegian R&D statistics, published in 1963, confirm the important role played by the institutes and the modest role played by private industry. Close to one-half of total R&D in that year was carried out by the institute sector. Barely 200 firms were involved in R&D, and their spending constituted only about 1 percent of their turnover (Skoie 2005: 208). Industry reported performing slightly more than 1,000 R&D man-years in 1963 (less than one percent directly publicly funded); compared to, for example, around 14,000 in 2005 (4.2 percent of which were publicly funded).

The NTNF research council was concerned about the low levels of industry-performed R&D and developed a strategy document in 1964 that advocated increased investment in education and research (see Skoie 2005: 208). In industrially relevant fields, the council suggested a division of labour in three areas. First, the public should take responsibility for long-term R&D related to "infrastructure" (defence, telecommunication, roads, railways etc.) and "national strategic areas" like the exploration of natural resources and energy supply. Second, private firms should take responsibility for R&D close to production and new product development. Finally, a third area should be characterized by "collaboration and sharing of work." This included the build-up of competence in new areas like automation as well as large projects involving many different firms or industrial sectors. Government followed up the recommendations, in a number of ways, including the establishment of an industrial development fund and expanding its use of R&D contracts in public procurement, by allowing NTNF to support industrial research directly, as well as providing financial support for the formerly private industry collective research institutes. Most of these new mechanisms were placed outside of NTNF, reflecting policymakers' opposition to a stronger role for the NTNF council, unlike, e.g. the strengthening of TEKES in Finland (Skoie 2005: 212).

Responsibility for innovation policy in Norway was thus spread out on a number of actors.

Norway's network of industry-oriented institutes is similar to those of other European countries, which also developed industry- and mission-oriented research institutes during this period. Germany's Fraunhofer institutes, which in 2007 employed 12,500 people, are similar to SINTEF in their mixture of old sector-specific institutes and institutes linked to university departments (cf. Meyer-Krahmer 2001). The Dutch TNO institute, which maintains close links to technical universities, also resembles SINTEF (see van der Meulen and Rip 2001). The Finnish VTT, which began its existence as a technological service institution, also has evolved into a multidisciplinary technological institute of the SINTEF type (Sörlin 2006). Particularly strong linkages between institutes and universities may be a unique feature of the Norwegian system.

UNIVERSITIES, INSTITUTES, AND INDUSTRY

As was noted above, Norwegian universities received a significant and steadily increasing basic funding for the first time after 1945. Technological research however, remained concentrated at NTH and in the institutes. The other universities' incentives and opportunities for industrial collaboration therefore were minimal during the first three decades after Second World War and, as in many other countries (Martin 2003), there was little interaction between Norway's universities and industry.

Nevertheless, individual professors interested in practical applications had ample opportunity through mission-oriented institutes (FFI, IFA), institutes related to geology and marine research and institutes more directly tied to industrial needs such as SINTEF and SI. The activities initiated and controlled by the researchers arguably were as important for industrial development as short-term contract work (Moe 1999; Gulowsen 2000).

In fact, many projects in the institutes in the postwar decades required high scientific and technological competence. Most of the new fundamental results in electronics and reactor technology came from Norway's research institutes, not the universities (Ørstavik 1996). Boundaries were blurred; e.g. SINTEF became involved in teaching and the operation of "supercomputers" at NTH (Gulowsen 2000). Ideals for the institutes were technological universities like Stanford University and the Massachusetts Institute of Technology. The research institutes also helped bring different actors in the innovation system together to work on state-of-the-art technological initiatives with industrial potential, a good example of which may be the Penguin Missile project, probably the largest concentrated R&D effort ever in Norway.

Norway's public research organizations were able to work with all of the different layers of the nation's innovation system (see Chapter 2 for a discussion of the three "layers" within the Norwegian innovation system). The small-scale decentralized industries found public partners within SINTEF, the collective industry institutes, and in other organizations such as the Institute for Marine Research. For example, the postwar revitalization of Norwegian shipbuilding relied on strong networks among ship-owners, yards, and suppliers. Two key innovations from the early 1950s, pitched propellers and passive stabilization of vessels, were developed in collaboration with SINTEF and the NTH Ship Model Tank (Andersen 1997: 469). As Chapters 2, 4, and 7 argue in greater detail, Norway's offshore oil industry became a very important market for the shipyards during the 1970s, and the SINTEF/NTH cluster made important contributions, since most shipbuilding firms preferred to use external sources rather than to build in-house R&D units (*ibid.*).

Many more firms in the large-scale centralized industries of e.g. chemicals, metals, pulp and paper had their own R&D laboratories and co-operated with NTH, the agricultural college and other universities directly. Finally, the R&D-intensive network industries, which received significant public R&D funding from defence and electronics programs, relied in particular on the defence research institute, FFI.

Over time, Norway's public research infrastructure became very heterogeneous and differentiated, in part as a response to the very different needs of various industries and firms. In an evolutionary framework, institutes created new variety but also constituted a selection mechanism for innovation. Through their contact with the universities and with the international research frontier, they were able to suggest new opportunities and new solutions to companies. Since they depended on client and public funding with many strings attached (regional, industrial, disciplinary and technological), and they had a strong technology push perspective, the institutes often transferred a single technological solution to an entire industrial sector or group of firms.

THE CREATION OF A NATIONAL INNOVATION SYSTEM

Motivated in part by the ideology of "technology-based industrial growth," (Chapter 4), Norway created a strong innovation support structure in the 1960s and the 1970s. This innovation system included publicly funded research, government loans for technology-based industries, tax deductions for R&D, technology development-oriented public procurement policies, public planning agencies with responsibility for technological development, public R&D contracts and structural control of high-tech industries. Although these mechanisms were not the result of a single policy, the end result was a system at the national level with many linkages between firms and public research organizations. Research policy and large parts of industrial policy were decentralized and sector-based—each ministry had the initiative and responsibility for R&D within its field of operation. Most ministries were responsible for one or several research institutes that served an array of goals that included more than technology development alone.

Beginning in the 1960s, most large industrial companies in Norway established formal and regular collaboration with NTH, other universities and relevant research institutes. Hydro's fertilizer division in 1968 formalized its links to the Norwegian Agricultural University (NLH), with which it had enjoyed informal links for more than sixty years (Andersen and Yttri 1997). The new Hydro aluminum division had large-scale collaboration with IFA, SI, SINTEF and NTH, and regarded the two latter as world-class within metallurgy (*ibid.*). Hydro's magnesium division, however, was based completely on technology developed in-house. Companies like Elkem used NTH and SINTEF for process innovations (Sogner 2004). Pharmaceuticals firm Nycomed, now a part of General Electric, developed strong ties to the University of Oslo, while Hydro's oil and gas division entered a long-term relationship with the university and research institutes in Bergen. The divisionalization of large companies and decentralization of industrial R&D had positive effects for the public research system, as the companies became less self-sufficient in technology and expanded their reliance on external sources of knowledge (Gulowsen 2000: 110).

Many smaller high technology companies also deepened their linkages with public research organizations during this period. For example, SIMRAD, founded on FFI-developed technology, broadened its collaborative relationships beyond FFI and the Institute of Marine Research to establish new links with maritime electronics R&D programs at NTH and SINTEF (Sogner 1997: 100, 173). With increasing national ties between industry and particularly the research institutes, the organizations co-evolved into a strongly integrated system. Below we describe the development in two important sectors: electronics/computers and oil/gas (for more details, see Chapters 10 and 7).

ELECTRONICS: THE MODERN AND TOP-DOWN PLANNED INDUSTRY

As Chapters 2 and 4 note, the political power of Norway's "industrial modernizers" meant that the electronics industry received massive public support for R&D and other activities from the 1960s through the 1970s. NTNF's influential 1964 report on electronics switched the focus of public policy from supporting research institutes to funding a few companies more directly (*ibid.*), following a "national champion" strategy that was similar to those of France and other European countries (Larédo and Mustar 2001b). The electronics industry became a major institute partner, performing more than 50 percent of its R&D in institutes like SINTEF, SI, FFI, and CMI (Basberg 1986). However, as Chapter 4 documents, Norway's national champions failed to become successful international corporations.

On the civilian side, Norway's national telephone company gradually became a central actor in the electronics research system, establishing a central laboratory in 1967—*Televerkets Forskningsinstitutt* (*TF*), located next to FFI just outside of Oslo (see Collett and Lossius 1993). TF was established with a radical R&D strategy that sought state-of-the-art technological performance rather than incremental improvements. FFI and particularly SINTEF Elab were partners and driving forces in the development of TF. Elab was established in 1961 to transfer knowledge from the Electrotechnical department at NTH, but the SINTEF division was allowed to initiate its own projects and to take on contracts without the direct consent of the university department (Gulowsen 2000: 160–217). Aasmund Gjeitnes, director of Elab for 23 years, sought large projects of a "breakthrough" type. His governing principle was that research should be equally oriented at industry, at the public sector and at the NTNF research council. Director Nic. Knudtzon at TF had a similar view, initiating a close collaboration that endured for three decades, assisted behind the scenes by the same individuals that figure in many of Norway's other attempts to spark "technology-driven industrialization": FFI director Finn Lied, former minister of industry, who sat on the board of the telephone company; former minister of defence and justice Jens-Christian Hauge and former minister of industry Kjell Holler, who later became the telephone company director.

Perhaps the greatest technological success of the TF–SINTEF Elab relationship was the GSM mobile telephone system. Norway's geography and topography had long created a need for efficient communication among many small fishing vessels close to the coast, with the large merchant fleet worldwide, and now also with the offshore oil platforms. The Nordic telephone companies had in 1981 gained international recognition for developing the NMT mobile phone system, the result of a public–private collaboration that spanned national borders. In Norway, TF and Elab worked intensely to develop a solution that was 100 percent digital, which would greatly increase the system's capacity. The Norwegian solution was deemed a clear winner of the technical competition by the trans-European group overseeing the development of a new mobile telephone architecture. No Norwegian telecommunications equipment firms were in a position to exploit the standard through production of compatible equipment, partly because of the foreign ownership of the leading Norwegian firms. As a result, other countries' firms could develop, produce and market the equipment, a feature of the Norwegian GSM architecture that made it more popular among European telecommunications equipment firms. The main benefit for Norway was that the selected system was well suited to the country's particular challenges.

Public research organizations also played a major role in the development of a national industry within the computer field, as Chapter 10 discusses in greater detail. NTH was central, not least in establishing the companies Computer Techniques and Norcontrol at the end of the 1960s. In addition to NTH, FFI, the astrophysics department at the University of Oslo, the Central Institute for Industrial Research (SI), Norsk Regnesentral (NR) in Oslo, and CMI in Bergen also played a part of the computer industry's development (Wicken 1989; Kvaal 1994). Key company Norsk Data (ND), founded in 1967 as a spin-off by three engineers from FFI, became a large mini-computer company, but was out of business by the end of the 1980s. The same fate befell the other major Norwegian computer-involved firm, the defence company KV (see Chapter 10).

THE OIL AND GAS INDUSTRY: NOT STARTING FROM SCRATCH

Public research organizations in different cities played major roles in the development of technologies for Norway's oil and gas industry. Particularly strong linkages developed between Statoil and the institute Rogalandsforskning (RF) in Stavanger and between Hydro and the University of Bergen and the Bergen-located research institutes. NTH, SINTEF, the Geological Survey of Norway, SI, IFA/IFE and others contributed important innovations. The "Technology Agreements" policy adopted by the Norwegian government (see Chapter 7 for further discussion) increased funding from the international oil companies for R&D in Norway's institutes and universities. Approximately half of this funding from international oil firms went to Norwegian research institutes, with SINTEF accounting for the largest single share.

Even before the large-scale development of offshore oil and gas, Norway's public research organizations in fields such as geology, marine technology, materials and electronics laid an important intellectual foundation for the industry, creating a small pool of knowledge that could be expanded when the need arose in the early 1970s. Some institutes moved into new areas of R&D. IFE (nuclear research) is a good example, using expertise accumulated through nuclear R&D to enter electronics, computing and petroleum technology, areas of R&D that soon became as important as its traditional activities. One assessment of IFE's oil and gas R&D estimates that its work in such areas as reservoir modelling, process automation and corrosion problems saved the oil companies of "hundreds of millions" in reduced material and maintenance costs (Njølstad 1999: 398). Njølstad (1999: 522–3) argues that IFE's build-up of scientific competencies in the 1950s and 1960s was necessary for these advances, and suggests that the decision to spend 5 million NOK in 1947 on Randers' reactor project (the initial core project at IFE) may have been the best public R&D investment ever in Norway, even though commercial nuclear energy was abolished.

Technology and "goodwill" agreements transformed the R&D demand side for Norwegian public sector research organizations. Seven of the nine largest purchasers of R&D services from SINTEF in 1987 were oil companies, spending 410 million NOK for SINTEF R&D while all other large customers supported R&D in the amount of 110 million NOK (Gulowsen 2000 p. 76). The 1970s were an important turning point for Norway's research institutes, which became increasingly dependent upon a few large customers and accordingly more vulnerable to changes in the organization and extent of these flows of private R&D funding.

The oil and gas industry's share of total Norwegian private R&D spending has been decreasing since the mid-1980s and dropped below 10 percent in the 2000s. One reason for this apparent drop, however, may reflect a tendency for more of the knowledge and technology development in the industry to be carried out in firms that are classified in industries other than oil and gas, e.g. electronics, software, and materials. Many important support companies for the oil and gas sector are spin-offs from research institutes like Rogalandsforskning in Stavanger.

TENSIONS WITHIN THE PUBLIC R&D SECTOR IN THE NEW MILLENNIUM

For accounts of recent developments regarding university-industry and institute-industry relations, see Chapters 11 and 12. Here, we briefly discuss some recent developments, many of which are not unique to Norway. One such change in higher education is "massification," an enormous increase in student numbers. The funding of university research and the number of faculty members depend on the number of students, and the volume of university research in Norway has more than doubled since the early 1980s. Two colleges have been upgraded to universities, and more will probably come as close to 100 specialized colleges were merged into twenty-six larger state colleges in the 1990s, several of which aspire to become universities.

New regional institutes with a social science profile have been established in the 1980s. Liberalization of policy and increased weight on user control within the institute sector has created financial problems for many institutes, and the sector has become a somewhat less visible actor in the research system. All five research councils, including university-oriented NAVF and technical-industrial NTNF, were merged into one research council in 1993. This has probably contributed to competition and strains between the institutes and the higher education institutions.

The late 1970s and 1980s saw increased tension between universities and institutes, not least between NTH and SINTEF. The latter became dissatisfied with professorial control of its activities; which limited its development to strategies that would not threaten NTH (Gulowsen 2000: 68). Director Stenstadvold never challenged the NTH professors, fearing that they would find other outlets for their technology transfer activities (*ibid.* 62). For their part, some NTH professors expressed displeasure with the way SINTEF dominated the contract research market. In 1980 the research institute was reorganized into a private, independent foundation. Presently, almost all research institutes are either foundations or limited companies with universities and local/regional authorities as main shareholders.

Efforts by some institutes to weaken their links with Norwegian universities was a mixed blessing, as many industrial firms collaborated with research institutes in order to improve their recruitment of university graduates (*ibid.* 194). Some of these firms shifted their R&D strategies to deal more directly with the universities, which for their part lacked experience with this form of interaction. A few years into the new millennium, however, these patterns are being reversed in some cases, as Norwegian universities seek to exercise greater control over the research institutes in their vicinity. The University of Stavanger has merged its R&D activities with the Rogaland research institute (RF), and a new phase of NTNU/SINTEF collaboration has been catalysed by the creation of a parallel system for commercialization of research results. The University of Bergen has gained control of the Christian Michelsen Research Institute, and the University of Tromsø of the Norut institute group. The competition between the sectors for public and private funding continues, and may increase further, since all Norwegian universities have started technology transfer offices in 2004/2005, expanding their activities in a domain formerly under the control of the institutes.

CONCLUSIONS

Within the European context, little is unique about the origins and structure of the Norwegian public research system and its relation to industry, although it must be emphasized that some key institutions, such as stable research funding agencies and industrial laboratories, originated in Norway well after their appearance in other nations. And because of its small size, Norway has not entered the full array of scientific fields as substantially as larger countries. Norwegian policymakers have always stressed the need to set priorities, and even the largest projects within Norway have been fairly small by comparison with large-scale programs in other industrial economies.

Historically, research policy has been a secondary priority in Norway by comparison with other European countries. Norway's resource-based industries have for decades drawn the attention of policy makers and industrialists away from industries that are more dependent upon R&D and innovation. R&D has been seen as an instrument to achieve other goals, including industrial growth, regional development, increasing quality and efficiency in schools, housing, and the public sector, etc., rather than an activity that inherently merits support.

Interestingly, Norway was one of the first nations to utilize science for practical purposes such as mapping of natural resources and experiments related to fish and agriculture. But fundamental research in the natural sciences and technology grew very slowly in Norway, although individual professors sometimes succeeded in combining groundbreaking basic research with practical projects of great significance to the nation's economic and social development. The establishment of technical university NTH in 1910 marked the start of a public research effort directed at industrial ends. The first post-Second World War decade spawned a host of new government organizations, including a research council where industry had a clear voice (NTNF). After the Second World War the universities received increased basic funding for the first time and new research institutes came to fill the role the university professors had played before 1945. A heterogeneous infrastructure of research organizations designed to produce sector-specific knowledge production has grown continuously over the last century. Although some of these organizations were influenced by institutes in Sweden, the United States, the United Kingdom, and elsewhere, the resulting Norwegian system of publicly funded

R&D perhaps most closely resembles that in Germany, which also includes an array of comprehensive universities, technical universities and specialized colleges, along with many different types of research institutes.

Different types of Norwegian public research organizations have been established to cater for the needs of the three different "layers" of the innovation whose development was discussed in Chapter 2, e.g. the college of agriculture and industry-specific institutes for Layer 1, university departments as partners of the industrial labs in Layer 2, and commercialization-oriented institutes for Layer 3. But the research institutes and at least some of the higher education institutions now provide "meeting places" for institutions and actors from all of these layers to get together and collaborate. The growth of such "meeting places" is partly due to the long-time dominance of NTH/SINTEF in R&D and training and partly due to the emergence of the oil industry, which created strong incentives for the research groups that had focused on small, R&D-intensive firms and industries within Norway to turn their attention to the needs of the nation's large resource-based firms.

As has been pointed out in this chapter, as well as in Chapters 2 and 4, an important mission of the research institutes established after 1945 in Norway was developing technological solutions for firms with low levels of R&D investment. A string of policy failures, along with changes in the public funding structure and changes in the structure of ownership of Norwegian industrial firms combined to shift the R&D focus of the institutes to incremental innovation. In effect, this shift in mission involved a reduction in the gap between user needs and scientific dreams that reflected adaptation in the research agenda of the institutes to accommodate the needs of existing firms while convincing these firms to use sophisticated technologies. This process contributed to reinforcing the structure and the incremental innovation approach of Norwegian industry. The survival to the present day of many of Norway's traditional industrial companies, including many with low levels of self-financed R&D investment, may be taken as evidence of the success of Norwegian policy. Still, it has been argued that the limited development of new high-tech industries within Norway (alternatively, the collapse of some such as IT) reflects a failure to break out of a systemic "lock-in" that affects the overall economy. Nonetheless, even the surviving firms in Norway's "small-scale" industry sector do not fit the stereotype of "low tech" enterprises—fish farming, oil/gas extraction and transportation, and metals production all use highly sophisticated methods and technologies. Many of these new applications reflect efficient adoption and adaptation of existing technologies that have been developed in partnership with public sector research or with companies that owe their existence to such research.

Of course, the most important contribution of public sector research (PSR) in Norway, as in other industrial economies, is providing firms with a flow of human resources. Universities and institutes contribute to a "public reservoir of competences" (Larédo and Mustar 2001a: 504) that includes general knowledge, specific solutions, interpretations of problems and mobility of students and experienced researchers. But Norwegian PSR also has played an active role throughout the twentieth century in creating new technologies, commercial activities, and firms.

REFERENCES

Andersen, H. W. (1997) "Producing Producers: shipping, shipyards and the cooperative infrastructure of the Norwegian maritime complex since 1850," in Sabel, C. and Zeitlin, J. (eds.), *World of possibilities. Flexibility and mass production in western industrialization*, Cambridge: Cambridge University Press, 1997, pp. 461–500.

Andersen, K. G. and G. Yttri (1997) *Et forsøk verdt. Forskning og utvikling i Norsk Hydro gjennom 90 år*, Oslo: Scandinavian University Press.

Basberg, B. L. (1986) "R&D Performance in Norwegian Electronics Companies, 1960–1975," Notat nr. 22 serien Norsk elektronikkhistorie 1945–1970, Teknologihistorieprosjektet, NAVF-NTNF.

Bush, V. (1945) *Science—The Endless Frontier. A Report to the President on a Program for Postwar Scientific Research*, National Science Foundation, Washington, DC.

Collett, J. P. (1983) *Videnskap og politikk. Samarbeide og konflikt om forskning for industriformål 1917–1930.* Oslo: Universitetet i Oslo, Hovedoppgave i historie.

—— (1999) *Historien om Universitetet i Oslo*, Oslo: Scandinavian University Press.

Collett, J. P. and B. O. H. Lossius (1993) *Visjon—Forskning—Virkelighet. TF 25 år*, Televerkets Forskningsinstitutt, Kjeller.

Friedman, R. M. (1989) *Appropriating the weather: Vilhelm Bjerknes and the construction of a modern meteorology*, Ithaca, NY: Cornell University Press.

Gulowsen, J. (2000) *Bro mellom vitenskap og teknologi. SINTEF 1950–2000*, Tapir Forlag, Trondheim.

Hanisch, T. J. and L. Even (1985) *Vitenskap for industrien. NTH—En høyskole i utvikling gjennom 75 år*, Oslo: Scandinavian University Press.

Kvaal, S. (1994) "Servoentusiastene og visjonen om et moderne Norge," in Wicken (ed.), *Elektronikkentreprenørene. Studier av norsk elektronikkforskning og -industri etter 1945.* Oslo: Ad Notam Gyldendal.

—— (1997) *Janus med tre ansikter. Om organizeringen av den industrielt rettede forskningen i spennet mellom stat, vitenskap og industri i Norge, 1916–1956*, Doctoral dissertation, Trondheim.

Larédo, P. and P. Mustar (eds.) (2001a) *Research and innovation policies in the new global economy*, Cheltenham: Edward Elgar.

——— (2001b) "General conclusion: three major trends in research and innovation policies", in P. Larédo and P. Mustar, *Research and innovation policies in the new global economy*, pp. 497–509.

Martin, B. R. (2003) "The changing social contract for science and the evolution of the university," in A. Geuna, A. J. Salter, and W. E. Steinmueller (eds.), *Science and Innovation. Rethinking the Rationales for Funding and Governance* (Cheltenham: Edward Elgar).

—— and Henry Etzkowitz (2000) "The origin and evolution of the university species," *VEST Journal for Science and Technology Studies*, 13, 9–34.

Meyer-Krahmer, Frieder (2001) "The German innovation system," in P. Larédo and P. Mustar (eds.), *Research and Innovation Policies in the New Global Economy. An International Comparative Analysis*, Cheltenham: Edward Elgar, pp. 205–52.

Moe, Johannes (1999) *På tidens skanser*, Trondheim: Tapir Forlag.

Mowery, David C. and Nathan Rosenberg (1993) "The U.S. National Innovation System," in R. R. Nelson, (ed.), *National Innovation Systems. A Comparative Analysis*. New York/Oxford: Oxford University Press, pp. 29–75.

Njølstad, O. (1999) *Strålende forskning. Institutt for energiteknikk 1948–98*, Tano Aschehoug, Oslo.

—— and O. Wicken (1997) *Kunnskap som våpen. Forsvarets forskningsinstitutt 1946–1975*, Tano Aschehoug, Oslo.

Schwach, V. (2000) *Havet, fisken og vitenskapen. Fra fiskeriundersøkelser til havforskningsinstitutt 1860–2000*. Bergen: Havforskingsinstituttet.

—— (2007) "Et rekefiske på vitenskapens grunn," *Historisk tidsskrift*, 3/2007.

Sejersted, F. (1993) *Demokratisk kapitalisme*, Oslo: Scandinavian University Press.

Skoie, H. (2005) *Norsk forskningspolitikk i etterkrigstiden*, Oslo: Cappelen.

Sogner, K. (1997) *God på bunnen. SIMRAD-virksomheten 1947–1997*, Oslo: Novus forlag.

—— (2004) *Skaperkraft. Elkem gjennom 100 år 1904–2004*, Oslo: Messel Forlag.

Sörlin, Sverker (2006) *En ny institutssektor—en analys av industriforskningsinstitutens villkor och framtid ur ett närings-och innovationspolitiskt perspektiv*, Stockholm: Ireco.

van der Meulen, Barend and Arie Rip (2001) "The Netherlands: Science policy by mediation." In Larédo, Philippe and Philippe Mustar (eds.), *Research and Innovation Policies in the New Global Economy. An International Comparative Analysis*, Cheltenham: Edward Elgar, pages 297–324.

Wicken, Olav (1983) "Vekst og våpen. Norsk militærproduksjon som industripolitisk virkemiddel i 1960-årene," FHFS notat nr. 8/1983, Forsvarets høgskole, Forsvarshistorisk forskningssenter.

—— (1989) "Norsk datahistorie," *Teknisk ukeblad*, Oslo: Ingeniørforlaget.

Wicken, O. (1992) "Kald krig i norsk forskning," Institutt for forsvarstudier, IFS Info nr. 6, 23 sider.

—— (1994) "Elektronikkrevolusjonen," in O. Wicken (ed.) (1994) *Elektronikkentreprenørene. Studier av norsk elektronikkforskning og—industri etter 1945*, Ad Notam Gyldendal, Oslo, pp. 10–23.

Ørstavik, F. (1994) "Forskningsingeniører i blandingsøkonomien. FFI som kraftsentrum i norsk teknisk forskning," in O. Wicken (ed.) (1984) *Elektronikkentreprenørene. Studier av norsk elektronikkforskning og—industri etter 1945*, Ad Notam Gyldendal, Oslo, pp. 28–46.

—— (1996) "The hierarchical Systems Paradigm in Technological Innovation. A Sociological Analysis of Technology Creation and Technology Policy in the Field of Mini-Computers in Norway." Oslo: University of Oslo, PhD thesis.

4

Policies for Path Creation: The Rise and Fall of Norway's Research-Driven Strategy for Industrialization

Olav Wicken

R&D (Research and Development) has played a central role in Norwegian public industrial policy for a relatively short period. Before 1963, there was little interest in linking technological innovation with national industrial strategy. During the mid 1960s, attempts were made to connect public research more closely to industrial development, and the state became more heavily engaged in funding industrial R&D. During the 1980s, governments increased public funding for industrial R&D substantially, and for a short period of time R&D became a core element in national industrial policy. However, in the early 1990s the situation again changed, and public funding for industrial R&D lost its significance in national industrial strategies.

This chapter argues that the rise of support for publicly funded R&D as a core element of industrial policy was closely linked to the emergence of new technologies, and to a new ideological basis for industrial policy that sought to encourage new industries based on these technologies as the foundation for Norway's economic growth. Beginning in the 1960s, electronics, computers, telecommunication, industrial automation technologies, etc., were seen as the key ingredients in the modernization of older industries and the development of new industries. By the 1980s, Norway could boast many examples of rapidly growing "R&D driven" companies, and there was wide acceptance that *research-driven industrialization* was crucial for future economic growth. This policy was the core element in a strategy of path creation and a R&D intensive network based path (see Chapter 2) of the economy.

The ideology of research driven industrialization was first articulated during the 1960s by a group of people with links to the Labour party, Norwegian research institutes and universities, and the defense sector. This small group of "modernizers" became influential in political decision-making in a number of areas, and supported the creation of public funding and other forms of support for research-driven industrial growth. (Wicken 1994) These policies were strengthened during the 1980s as the ideology of R&D-based growth dominated Norwegian industrial policy debates. The modernizers sought to support the *R&D-intensive network-based* path of development that was discussed in Chapter 2 in order to support economic growth that was not dependent on Norway's natural resources.

R&D ENTERS INDUSTRIAL POLICY

The new interest in R&D for industrial development that emerged during the 1960s was linked to the wider politics of technology during the Cold War, specifically the growth of public funding for R&D and education by the US and Western European governments in response to the Soviet Union's launch of the Sputnik satellite in 1957. The US in particular increased defence R&D funding and soon entered into a race with the Soviet Union that focused on space technologies.

The effects of Sputnik on Norwegian policy were complemented by the revision of national science and technology priorities in Norway's close ally, Great Britain, during the early 1960s. In 1963, British Prime Minister Harold Wilson gave his famous speech on the 'White Heat' of technology-driven industrial revitalization,[1] which became linked with a wider discussion in British media on "Linking science with industrial growth".[2] The discussion in Britain was closely linked to developments in the OECD, which began to argue that all member countries should adopt science policies that were directed towards industrial development.[3] The OECD issued a series of reports on member states' industrial science policies and on the 'technology gap' between America and Europe throughout the 1960s. Equally influential may have been the development and definition of the concept "Research and Development"—R&D—by an OECD committee in 1963, a concept that was soon used to promote "R&D policies" in Western economies (OECD 1963).

The debate in Britain during 1963, along with a 1963 OECD report on science policy in Norway,[4] sparked a discussion within the Norwegian Ministry of Industry on how to organize and regulate public R&D for industry.[5] R&D had, of course, been supported by Norwegian industry and by Norwegian industrial policy for many years before the 1960s. Much of the public R&D policy focused on firm's characteristic of the *small-scale decentralized* path of resource-based industrial development in Norway that was discussed in Chapter 2. The state established a public research infrastructure for both fishing and agriculture in the nineteenth century, and many of the research institutes founded during the twentieth century also contributed technical knowledge and support for this type of industrialization. Some publicly funded R&D also supported innovation in the *large-scale centralized* form of industrialization described in Chapter 2. Both the Technical University (NTH) in Trondheim and research institutes like SI, IFA, CMI and Sintef collaborated with laboratories and management in many of the large scale process industry companies.

Nonetheless, R&D was not a central element of policy for these industries during the 1950s. This was particularly true for the politically important large-scale process industries (e.g. aluminum, electrochemicals) during the 1950s, when R&D support focused on incremental improvements in existing processes, rather than a transformation of the industrial and economic structure of Norway.

The 1960s witnessed renewed interest in public funding for R&D support systems for both the small-scale and large-scale Norwegian industrial sectors. To support the small-scale decentralized form of industry, a new financial fund for collective industrial research (Bransjeforskningsfondet) was established in 1967, and during the 1980s a number of institutions for promoting diffusion of new technologies among small companies were introduced. The support system for the large-scale centralized form of industry became more complex, including changes in education and university research in response to demands from the emerging oil sector during the 1970s.

Policymakers' views on the contributions of R&D to established industries in Norway did not change radically during the 1960s. Some new institutions were established to support research in large-scale companies, and funding of the public research infrastructure for small-scale industries and mechanisms for the diffusion of new technologies, also increased. A more radical change in public industrial R&D policy was associated with the emergence of the idea of 'research-driven industrialization' based on the new technologies of the 1960s. This belief in economic transformation through strategic R&D investments focused on the creation of *R&D-intensive network-based* industries, the third development path discussed in Chapter 2. A number of new institutions were established to promote these new technologies and the new developmental path.

Between 1963 and 1967, the Norwegian government established a number of institutions to support R&D in companies and increased public R&D funding. Within this short period of time the technical-industrial research council (NTNF) was reorganized and new institutions for the distribution of R&D funding to companies were introduced. The Development Fund was established in 1965, the Fund for Collective Industrial Research was established in 1967, and a subsidy system for public R&D contracts was introduced in 1967.

MARGINAL INNOVATION R&D IN LARGE-SCALE CENTRALIZED INDUSTRIES

R&D played a marginal role in the Norwegian industrial policy for large-scale centralized industrialization that was introduced after the end of the Second World War. This strategy sought to transform the traditional industrial structure dominated by SMEs (small-scale decentralized industrialization) into a more modern one dominated by large-scale and capital-intensive corporations (large-scale centralized industrialization, discussed in Chapter 2). The Ministry of Industry was established in 1947, and the first Minister of Industry, Erik Brofoss, was the architect of the large-scale industrialization policy.[6] The policy sought to build up large-scale production units that could exploit Norway's cheap electricity and other natural resources. The Ministry of Industry focused its policy on building state-owned companies or attracting international investment in areas like metal production (aluminum and steel), energy (electricity and coal), minerals (mining), and fishing (the filet industry). The government worked closely with the management of the major state-owned companies, like Norsk Hydro and ÅSV (for a discussion of public policy in aluminum, see Chapter 6). The emergence of the oil sector in 1970 reinforced this policy, which focused on *capital* as the strategic input factor for industrialization, treating R&D investment as only marginally important (Jensen 1989).

A number of proposals to establish new resource-based large-scale industries linked to the steel industry, the metals industry, oil refineries, copper mining, and carbide production were discussed by governments during and after the 1950s.[7] Some of the industrial projects required public R&D investments, as was the case with proposals to exploit Spitsbergen's coal deposits for the production of coke, as well as the proposal to establish the steel works in Narvik.[8] In this industrial policy, R&D was an instrument to solve problems in production processes, but was not a driving force in industrialization. Within the Ministry of Industry, R&D policy was marginalized, and there was little if any interaction between the Ministry and the research council responsible for industrial research. NTNF was not part of the government's industrial policy.

Still, NTNF established close relationships with individuals from various industrial companies who served on NTNF's board and committees. The main proponent for this strategy was the director of NTNF during 1946–80, *Robert Major*. The Ministry of Industry, however, regarded the representatives of industry that sought to collaborate with NTNF as an unusual group of industrial managers, interested in science and technological research, but not representative of their companies or industries.[9]

Beginning in late 1963, the Ministry of Industry reduced NTNF's 'splendid isolation,' trying to coordinate the research council's strategy with Ministry policy and linking NTNF more closely to industry. The Ministry opposed the use of NTNF funds to build up a large research institute sector, instead preferring that public resources support R&D in company laboratories. The Ministry also wanted to give companies greater influence over how the formulation of NTNF research programmes and over the distribution of funds among projects and companies.

To implement this policy, NTNF introduced a system for direct funding of R&D within companies in 1965 (PIR, Prosjekter i Industriell Regi). This became an influential part of NTNF's policy until the mid-1970s, and after 1990 was the major public programme for supporting industrial R&D. *Johan B. Holte,* the research director and later the administrative director of the country's largest industrial R&D performer, Norsk Hydro, was influential in defining NTNF strategy for public R&D policy towards large-scale industries. He argued that a small country like Norway lacked the financial resources for developing new large-scale industrial processes, and that the necessary technology had to be procured from abroad. In this view, Norwegian public R&D funding should be used to improve production marginally through small-scale projects. Not surprisingly, Holte's recommendations echoed the strategy of Norsk Hydro at the time, and his views shaped NTNF's R&D policy in the process industries.

In parallel with its reorganization of NTNF, the Ministry of Industry established an alternative organization in 1965 for funding industrial development projects, the Development Fund (Utviklingsfondet). The Development Fund was a response to NTNF's recommendation that a new organization be created within the council (or that an independent commercialization company be established) to ensure that new technologies developed by the research institutes were commercialized. NTNF found that existing firms were not interested in producing research-based products and technologies, and argued that Norway needed this type of organization.[10] The NTNF proposal triggered opposition from the Ministry of Industry, which feared that the establishment of a commercialization organization would further isolate researchers and scientists from commercial concerns. The Ministry developed an alternative scheme, in which politicians and representatives from "core" companies made decisions on public industrial R&D strategies.

Erik Brofoss became the first chair of the Development Fund, and the director of the Norwegian Defence Research Establishment (FFI), Finn Lied, was named to the Fund's board. Other board members represented politically important industries in the resource-based sector and shipbuilding. The main strategy of the Development Fund was to use its support for industrial R&D to create "a rational industrial structure" for each industry, based on the argument that Norway's industrial development was hampered by a lack of companies with sufficient financial and managerial strength. Just as other European governments were doing at the time, the Development Fund sought to use its financial and political leverage to develop "national champions" (Sogner 1994)—in all sectors of the economy, including those dominated by small-scale industry. One example of this strategy was the attempt to transform the politically important shipbuilding industry. The Aker mek.Verksted firm received financial support that was designed to make it the dominant firm in Norway's shipbuilding industry. This strategy included Aker's establishment of corporate R&D laboratories as well as collaboration with public research institutes on projects such as automation of ship navigation and operations (Andersen 1986).

The Development fund distributed most of its funding to selected national champions in shipbuilding,[11] process industries, and electronics. During the 1965–77 period, the Development Fund supported projects in 287 companies

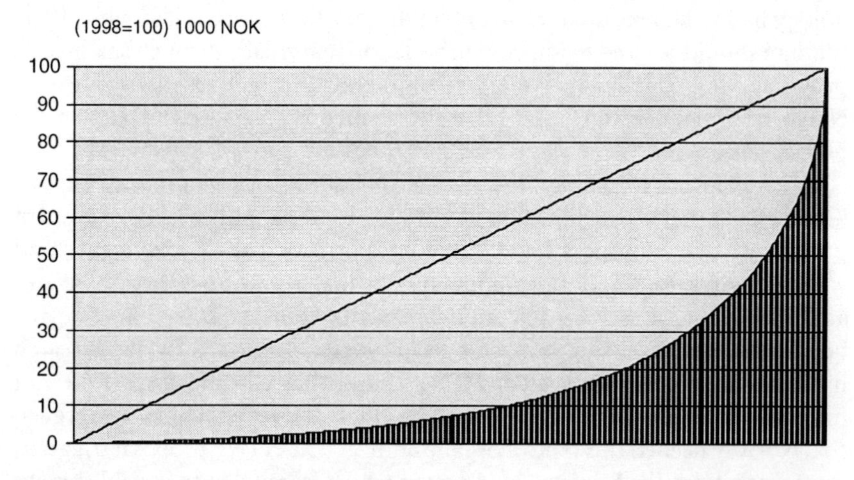

Figure 4.1. Cumulative distribution of funding from Utviklingsfondet (Development Fund), 1965–1977.

Note: Figure 4.1 shows cumulative distribution between companies receiving support from Uviklingsfondet for the period 1965–77. The figure shows that 60 percent of total number of companies received appx. 10 percent of total funding, while 24 companies received 50 percent of total funding.

(see Figure 4.1). Twenty-four of these 287 firms, 8 percent of the total number, received half of the total funding awards during the period.

DIFFUSION OF KNOWLEDGE: R&D IN THE SMALL-SCALE, DECENTRALIZED INDUSTRIAL SECTOR

Labour's industrial policy of the 1950s and 1960s provided minimal support for the small scale decentralized path of industrialization, a posture that was reflected in its R&D initiatives during the period. The nonsocialist political parties had opposed this policy during the 1950s, although the Conservative party became more supportive of Labour's philosophy on industrial development during the 1960s.

Despite this indifference from the central government, a substantial infrastructure of research institutes designed to support individual industries was established after 1950. At the technical university (NTH), the research council (NTNF), and within industry, initiatives were taken to establish laboratories for R&D in industries where company-funded R&D was nonexistent or marginal. These sectors included construction,[12] shipping and shipbuilding,[13] and electricity production.[14] NTNF also supported research projects organized by industrial associations for industries with low levels of self-financed R&D investment and conducted analyses of R&D strategy in individual industries, such as textiles and clothing (NOU 1973).

In 1965, a nonsocialist government consisting of four parties gained power and promoted a more supportive policy toward small-scale industry in Norway that included a new policy for R&D support in these sectors. Since the interwar period, a number of small-scale industries had established collective laboratories to solve industry-wide problems. Among the first labs of this type were those in the canning and paper industries, and during the post-Second World War period additional collective labs were established at the initiative of companies or industrial associations. The new government introduced a specific funding system (Bransjeforskningsrådet, Collective Industrial Research Council) to support this type of industrial research in industries with limited in-house R&D. The classic collective research laboratories focused on problems and challenges that were common to all or many of the companies in the industry, and normally avoided projects that favoured one company at the expense of a competitor in the same industry.[15]

A longstanding part of government R&D strategy for small-scale industry was public funding for the diffusion of science-based knowledge or technology. Norway's resource-based industries in the small-scale sector (fishing, seafood, agriculture, forestry) had long-established systems for collective research and for diffusion through education, consultancy organizations, seminars and conferences, training programmes, etc. The Marine Research Institute and the higher education institutions and agricultural research institutes played a core role in linking industry to science and research. Nevertheless, public support for R&D in the small-scale decentralized sector remained marginal until the 1980s.

PATH CREATION: R&D-INTENSIVE, NETWORK-BASED INDUSTRIES

As was previously noted, the new industrial policy instruments and ideology of the 1960s became closely linked to the R&D-intensive, network-based industries described in Chapter 2. An increasing share of public R&D funding that flowed through the research council, the Development Fund, and other public R&D contracts were allocated to ICT firms during and after the 1960s. Many of Norway's newly anointed national champions were ICT firms. For example, Simrad, Norway's major producer of maritime electronics, received significant funding through all of the programmes established to support R&D in industrial companies. Norsk Data, Kongsberg Våpenfabrikk, and Noratom-Norcontrol also received funding from these sources. As Chapter 10 describes in detail, Norway's ICT industry during the 1960s and 1970s became a central focus of the Norwegian industrial R&D policy that sought to accelerate economic growth through the creation of new industries, and significant resources were used to support the development of high-tech companies in this area. The state also invested substantial resources in a public research infrastructure, developing research institutes and university research directed towards this industry, and expanding education in the area.

The role of defence in industrial research strategy

The emergence of a defence-related industrial R&D strategy in Norway was closely linked to the international political and military challenges of the late 1950s. The Norwegian Defence Research Establishment (FFI) had since its establishment in 1946 collaborated closely with the British and American militaries. The new post-Sputnik emphasis on science and technology in Western defence gave FFI a central role in the inward transfer to Norwegian firms and government agencies of knowledge from the USA and led to expanded financial support from the USA and NATO. National defence also provided a strong argument for building up Norwegian national research capabilities. During the 1960s FFI and the affiliated Institute for Nuclear Energy Research became the largest research institutes in Norway. The main representative of the new approach of the FFI was *Finn Lied,* who became director of FFI in 1957 and remained in this position until 1982 (Njølstad and Wicken 1997).

NATO and US defense policy in the late 1950s sought among other things to build up an independent defense industry in Western Europe. Norway had a very limited defence industry during the 1950s that consisted mainly of three state-owned companies (one for shipbuilding, one for ammunition, and one for general weapons production). The defence minister of the first post-Second World War governments, *Jens Chr. Hauge,* supported the use of NATO funds to modernize Norway's defence companies. A powerful member of the board of Kongsberg Våpenfabrikk (KV), Hauge supported the transformation of KV into a modern high-tech defence company, and in 1957, appointed *Bjarne Hurlen* as director of the company (Wicken 1987a). Hurlen and Hauge remained the key managers of KV until 1986, and Finn Lied, the leading Labour industrial policy strategist from the 1960s to the 1980s, was on the KV board for most of this period. These three representatives of the Norwegian industrial-technological system became strong proponents of science-based industrial policy (Wicken 1994).

A technology-driven system for developing Norway's defence industry was established during the 1960s. FFI and KV obtained support for expanded R&D from the Ministry of Defence (FFI's owner) and the Ministry of Industry (KV's owner). In the late 1950s, FFI also began to receive substantial financial support from the USA for the development of a ship-to-submarine missile system (Terne III) and long-distance subsea detection systems for submarines (Bridge). The missile system was produced by KV and the Ministry of Defence became the sales organization for the system within NATO.

Increases in Norway's defense R&D and procurement budgets created new opportunities for national production of defence equipment. The largest investment was the construction of a modern navy (Flåteplanen 1963), which for the first time involved Norwegian companies in large-scale sales to Norwegian defence forces. Hauge, Lied and Hurlen turned KV into an instrument for industrial policy, arguing that the company should improve its product and production technologies, and diffuse the resulting ideas and practices to other parts of the economy (Wicken 1984, 1987b). For a quarter of a century KV was responsible for production of technologies that had been developed in Norwegian research institutes, most frequently in FFI. The technologies included computers, numerical control machinery, fire control systems, gas turbines, and weapons systems (the Penguin), as well as sub sea equipment and other products (Njølstad and Wicken 1997; Wicken 1988).

FFI and KV became leaders within Norway in a new type of production where science and research played a core role for success. To a large extent the defence industry during the early 1960s created a path for policymaking and direction of this type of industry that remained influential in the following decades. Hauge, Hurlen, and Lied represented a tradition in social democracy in supporting the importance of state-owned production units and public control of core knowledge bases. They tried to monopolize Norway's emerging high-tech industrialization by making KV the core player in technology-driven or research-driven industrialization.[16]

The alliance between FFI and KV influenced the wider policy system supporting a science- and technology-driven policy strategy. In 1960, FFI introduced systems evaluation techniques for assessing new technologies, an analytic technique pioneered by the RAND organization in the USA. Its reliance on systems analysis helped FFI to become an important instrument for defence planning.

Lied and his colleagues also influenced strategic planning in the technical-industrial research council (NTNF). Lied was at this time a member of NTNF's executive board and therefore a member of the group that during 1963–4 undertook an evaluation of NTNF. The ministry of industry intervened and demanded that three more persons should be part of the evaluation committee. Two of them were Jens Chr. Hauge and Bjarne Hurlen. The military-industrial system (Wicken 1990) thus had considerable influence within the emerging R&D industrial policy.

A RESEARCH DRIVEN STRATEGY

The 1964 NTNF evaluation report that was discussed above had argued that Norway's future economic growth could not depend solely on resource-based industries; the nation had to develop new industries based on new technologies, and R&D would play a crucial role in such a strategy. This argument was consistent with the thinking of the representatives of the military-technological system that emerged in the late 1950s, and it soon became widely accepted by scientists and research engineers in Norwegian universities and research institutes.

Norway lacked research-based industries and had very few R&D-intensive companies in 1963. The main exceptions to this characterization were firms with strong historical links with defence activities, including Simrad and Nera, both of which are discussed in greater detail in the chapter on Norway's ICT industry (Chapter 10). There was also a small company that specialized in nuclear-energy technology (Noratom), but other than these firms, research-based industry in Norway was almost nonexistent.

Beginning in 1964/5, Norwegian industrial R&D policy promoted R&D-based new firms and industries, and the FFI-KV alliance was a core part of the strategy. The Norwegian military nevertheless resisted pressure from the FFI-KV interests to purchase the weapons systems and technologies that it produced. To overcome the military's reluctance to purchase domestically developed products, the Ministry of Defence (headed by the industrializt Grieg Tiedemand) developed procurement regulations that forced the Norwegian military to consider the industrial implications for Norway of their weapons purchases (Wicken 1987b).

The nonsocialist government also supported the introduction of a subsidy system for public R&D contracts that encouraged Norwegian government purchasers of new products to pursue development projects with Norwegian industry. The system of public R&D contracts and related public procurement policies became an instrument for the promotion of high-tech or research-based industrialization, especially for the ICT industry (see Chapter 10).

The industrial research council, NTNF, became a strong promoter of industrialization based on emerging technologies such as electronics, computers and telecommunication (ICT). NTNF allocated a growing share of its budget towards these technologies, and the research council established large projects that involved the newly established companies of the research-driven path as well as older companies that were national champions. The automation-computing committee within NTNF became a particularly strong supporter of research-based industry (Kvaal 1994), and companies like the minicomputer producer Norsk Data, Kongsberg, Norcontrol, and Simrad were the primary recipients of NTNF funding in this area. The public support system also supported the entry of national champions like Aker (numerically controlled machinery and semiconductors) and Hydro (ship automation) into publicly funded automation research projects (Wicken 2000; see also Chapter 10).

Not all supporters of a research or technology driven strategy for industrialization supported the strong role of defence R&D and procurement policy within this strategy, especially the central role of Hauge, Hurlen and Lied, and two controversies erupted in 1967. A group of researchers wanted to leave FFI to set up a private company for commercial production of a minicomputer design that they had developed, but Lied objected and transferred the prototype to KV for production. The young researchers challenged Lied, enlisting support from within FFI and from NTNF, and their departure from FFI led to the establishment of Norsk Data, a symbol of research-based industrialization and an alternative to the state-driven strategy of KV-FFI.

A similar development in telecommunications research took place in 1967. FFI had initiated the establishment of a national research laboratory for telecommunications (TF), supported by the Minister of Transport and Telecommunications Håkon Kyllingmark. But the director of TF, Nic Knudtzon, did not support the state-driven strategy of FFI-KV, and TF chose to collaborate with both national (Nera) and international companies (Elektrisk Bureau, controlled by Ericsson and STK, controlled by ITT)[17] (Collett and Lossius 1986).

The policies that emerged after 1964 in Norway, including the new strategy following by the NTNF, the buildup of KV as a commercialization company, the development of a telecommunications R&D system, and other instruments supported a research-driven industrial strategy that was similar to those of other European economies. Norway's strategy differed somewhat from those of at least some other European governments, however, in its heavy reliance on promotion of military technologies. KV-FFI, under the supervision of Hauge, Lied, and Hurlen, shaped the operation of the emergent policy system that emphasized state ownership and control. Over time, however, parts of the public support system broke with this ideology and promoted a more liberal strategy based on private ownership of companies.

By the late 1970s, a substantial policy structure had been established to promote R&D–intensive, network-based companies within Norway. A significant number of new companies had been established and some of the old companies had entered high-tech industries. The main companies were Kongsberg Våpenfabrikk, Norsk Data (computers), Simrad (marine electronics), Norcontrol (automation systems), and AME (semiconductors) in addition to the larger telecommunication companies (EB and STK). By this time Norway had a large number of small high-tech companies with growth potentials.

These emerging industries were supported by a substantial publicly funded knowledge infrastructure. In addition to the national industrial research council (NTNF), the public research organizations within both the defense sector (FFI) and the telecommunication sector (TF) purchased research services from Norwegian companies, research institutes, and universities. A significant research base within ICT developed as many research institutes expanded their capacity and competence within this area. FFI, TF, IFA, CMI, Sintef (for example Elab), SI, Norsk Regnesentral, and others developed wide competences in regulation technology, communication, computing, semiconductors, etc. Public procurement also promoted R&D in companies through R&D

contracts between companies and public agencies, especially in the defense and the telecommunication sectors.

RESEARCH-DRIVEN RE-INDUSTRIALIZATION STRATEGY (*c*.1978–90)

During the 1980s the research-driven industrial policy reached its peak. A 1979 report by a committee headed by Finn Lied that presented ideas for a new industrial policy (NOU 1979) reiterated some of the ideas from the 1960s, emphasizing the role of emerging technologies for future industrial development. These ideas had by this time been widely circulated in international analyses of the economic recession of the 1970s, as some scholars claimed that a new Industrial Revolution had begun that would be characterized by the emergence of new growth sectors based on radical technological innovation. (Mensch 1979; Freeman and Soete 1982). This interpretation suggested the possibility of *re-industrialization*, based on emerging technologies that could become the basis for the next long-term growth period. Future growth of the overall economy depended on the growth of new sectors based on new technologies.

In many ways this re-industrialization strategy based on new technologies and investment in R&D resembled the ideas of the Norwegian modernizers of the 1960s. There was, however, a major change in the political context that influenced the conditions under which new industries could be established in the 1980s and contrasted with the 1960s. Although Hauge, Lied, and Hurlen had proposed a state-driven policy during the 1960s, the Lied committee in 1978 argued that the international context had changed significantly, making a more liberal policy necessary for the research-driven industrial strategy. The more liberal views were supported in a 1981 policy paper on national R&D policy (NOU 1981: 30) that advocated a liberalization of the research system in order to improve interaction between the large industrial research institute sector and companies, and to increase industrial R&D. With the support of nonsocialist governments during 1981–86 (Willoch I and II), Norway adopted a more liberal industrial ideology.

Industrial R&D in Norway benefited during the 1980s from increased public spending. During 1983–93, public R&D spending increased from 5,500 to 9,000 mill. NOK in fixed (2000) prices, i.e. more than 80 percent. The growth in public R&D during the second half of the 1980s was linked to the decision of the Brundtland government to increase public R&D by five per cent as part of the Technology Targeted Area (TTA) policy (St.prp. 1, 1986–87). The

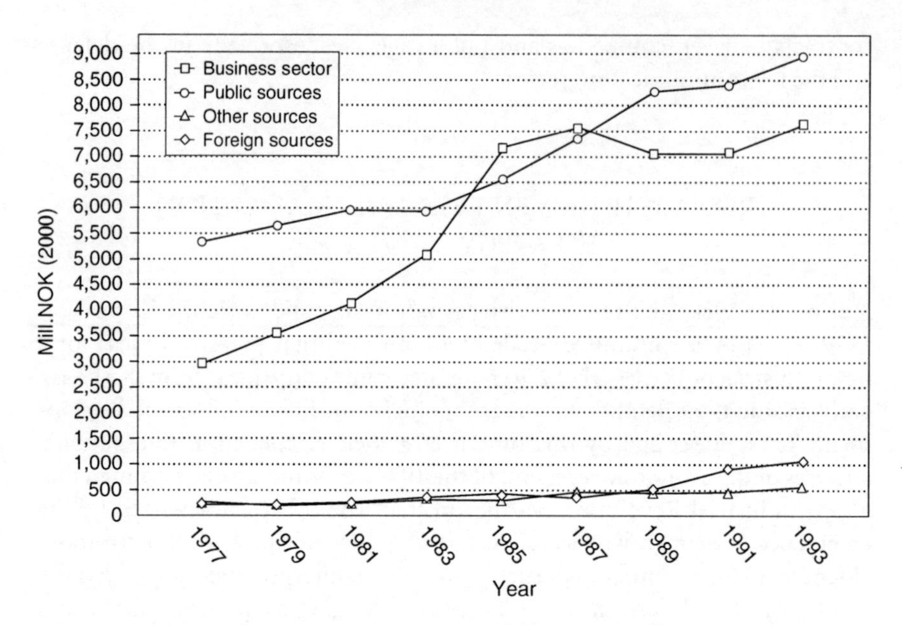

Figure 4.2. R&D by funding agency, million NOK (fixed prices, 2000).

TTA strategy was a technology- and R&D-driven policy that relied on public investments in R&D, and a substantial part of the funding increase was linked to industrial R&D (see Figure 4.2).[18]

Implementing a research-driven re-industrialization strategy

Like its predecessors during the postwar period, the new Norwegian policy was part of a wider international trend in industrial policymaking. During the mid-1980s, many OECD countries introduced programmes that were similar to the Targeted Technology Area (TTA) policy, based on the idea that governments should select potential growth sectors and move resources to these industries. The main areas selected by European countries were linked to emerging technologies like IT, biotechnology and new materials. The background for the re-industrialization strategy in Europe was the rapid de-industrialization (reduction in manufacturing-industry employment) that characterized the 1970s. "De-industrialization" in Norway differed somewhat from that of much of Western Europe, however, as declines in Norwegian manufacturing employment occurred only in the late 1970s.

The research-driven industrial strategy of the 1980s was supported by both nonsocialist governments (Willoch I and II 1981–6, Syse 1989) and the Labour government (Brundtland I, 1986–9). New technologies and R&D had achieved a more dominant ideological position in industrial policymaking than at any time before or since. The Ministry of Industry started to plan the strategy in 1982 and presented a plan for expanding industrial public R&D funding in 1984. The plan gave priority to the three new technology areas that internationally were accepted as important; information technology (IT), biotechnology, new materials. However, the government also decided that Norway should support the development of technology for two industries that were especially important to its economy and society: fish farming and offshore oil and gas production.[19]

This policy framework reflected the existing economic and political realities in Norway, including the political influence of various industries. Expanded support for emerging technologies was well received in large parts of the research communities and in Norway's small but rapidly expanding high tech industry. But the older sectors of the economy remained influential, and developments during the 1980s proved that technological opportunities within industries based on natural resources were significant. The highly technical offshore sector represented opportunities for industrial growth within the *large-scale centralized path*, while the emerging fish farming sector represented the long tradition of the *small-scale decentralized* form of industrialization (see Chapter 2 for more detailed discussion of each of these developmental paths). All of the main paths of Norway's economic development were represented in the new policy; and no type of industrialization was excluded.

Re-industrialization strategy for small-scale industry: fish farming

As Chapter 8 discusses in detail, systematic experimentation in farming of salmon and trout had started in Norway during the 1970s and by the 1980s, the technology appeared to have considerable promise for exploitation in the coastal economic regions. The development of the new industry was closely linked to the traditional fishing sector, which was the source of most capital, technology, and knowledge. Most early Norwegian fish farmers had backgrounds in the coastal or ocean fisheries, and most innovations were the outcome of close interaction between local fish farmers, mechanical workshops, banks, and other knowledge sources (Berge 2006).

Formal R&D played a marginal role in the early developments of the new industry. Early discussions during 1972–1977 of government policy for the emerging industry conceptualized R&D as a servant, support function or problem solver for the industry. (Mariussen 1992) During the early 1980s, however, the role of R&D in public policy to support fish farming policy changed, largely at the behest of the industrial research council; NTNF. Fish farming now was redefined as a technological foundation for establishing industrial food production in the ocean, rather than small-scale production of salmon and trout, and the industry was renamed mariculture (*havbruk*).[20] The fish farming industry was redefined as a future potential growth sector and a targeted technology area.

As the political saliency of fish farming grew, three ministries and three research councils competed to control the development of the new industry. In addition to the industrial research council, NTNF, there was also a research council for agriculture (NLFR) and one for fishing (NFFR), which created a diverse system for funding of fish-farming research. Researchers could apply for resources from multiple sources, and competition among the three research councils for control of the emerging industry produced rapid growth in public R&D funding. In 1973, funding of R&D relevant for fish farming from all Norwegian government sources amounted to *c*.1.5 million NOK. By 1984, the three research councils alone spent 17 million NOK on R&D for fish farming, and this amount increased to 31 million in the following year and to 75 million NOK by 1988. During the 1988–1992 period, annual research-council spending on R&D for fish farming averaged *c*.125 million NOK (Mariussen 1992).

A number of large Norwegian companies also became involved in research-driven fish farming during the 1980s, further expanding R&D funding and technological innovation within the sector (see Chapter 8). Among the most important of these large-firm entrants was Norsk Hydro which selected biotechnology, fish farming, and new materials as core areas in a research-driven diversification strategy that relied on R&D (Andersen and Yttri 1997). Hydro established R&D strategies and organizations for fish farming, and became a driving force for research in the area.

By the late 1980s, political documents argued that success in Norway' fish farming sector depended on R&D and public investment in R&D, a view that was also reflected in some of the major industrial companies. But the belief in research-driven innovation in fish farming did not last long. By the early 1990s, Norsk Hydro had abandoned its research-based diversification strategy and had reduced its involvement in fish farming. When the industry experienced a downturn during the early 1990s, scepticism grew concerning the role of R&D in the industry's future development. One evaluation of R&D policy in fish farming argues that cutbacks in public R&D funding reflected a loss of confidence by Norwegian policymakers in the research-driven strategy for industrial development in this sector (Mariussen 1992:12). At least in fish farming, R&D once more was reduced to being the servant of industry.

Re-industrialization strategy for R&D intensive industries: IT

Information technology (IT) was a core technology in research-driven industrialization. A large share of Norway's public R&D funding was directed towards this group of technologies, and IT received by far the largest share among technology fields of NTNF funding during the latter part of the 1980s, accounting for more than 50 percent of total funding allocated to targeted areas. The sector's share was especially high during 1986–88 (NHD 1990).

Policymakers' optimism over the prospects for this emerging technology was linked to the IT sector's rapid growth and the economic success of a number of Norwegian companies (see Chapter 10 for a more detailed discussion). Kongsberg Våpenfabrikk (KV) became one of the country's leading high-tech companies and a symbol of Labour's strategy of developing new industries with state ownership and control. Norsk Data (ND) also grew rapidly during the 1970s and 1980s, and was by the mid 1980s a leading European producer and exporter of minicomputers. As a private company registered on the stock exchange in both Oslo and New York it symbolized the entrepreneurial opportunities of the more liberal economy of the 1980s. A large number of other Norwegian ICT companies also enjoyed substantial growth until 1987 (see Chapter 10).

Just as had been the case for Norway's Development Fund (see above), public R&D funding for companies during the 1980s continued to favor a select group of national champions in the ICT and other sectors. Between 1978 and 1987 NTNF distributed 50 percent of its total funding for corporate R&D to just thirteen companies (Figure 4.3). The five largest projects in NTNF's corporate programme during the 1980s were all in the IT sector (see Table 4.1).

Norsk Data (ND) enjoyed strong political support during this period and until the late 1980s had a near-monopoly on computer sales to public sector organizations, for which it received high prices. The other main national champion, KV, received annual operating subsidies from the government to cover its financial deficits, a signal of the continuing strength of the military-technology system within Norway in the 1980s.

NTNF's strong support for Autodisplay typifies the application of the dominant "research-driven reindustrialization" ideology of the 1980s within the ICT sector. The project was based on a plan by an entrepreneur who founded a company in 1985 to develop a new intelligent automobile instrument panel. The plan called for the establishment of a large plant in Sandefjord in 1991 in collaboration with an Italian producer of car instrumentation that could make Norway a more important player in the European and international car industry. It was the largest project ever supported by the research council, and received additional funding from the pan-European Eureka

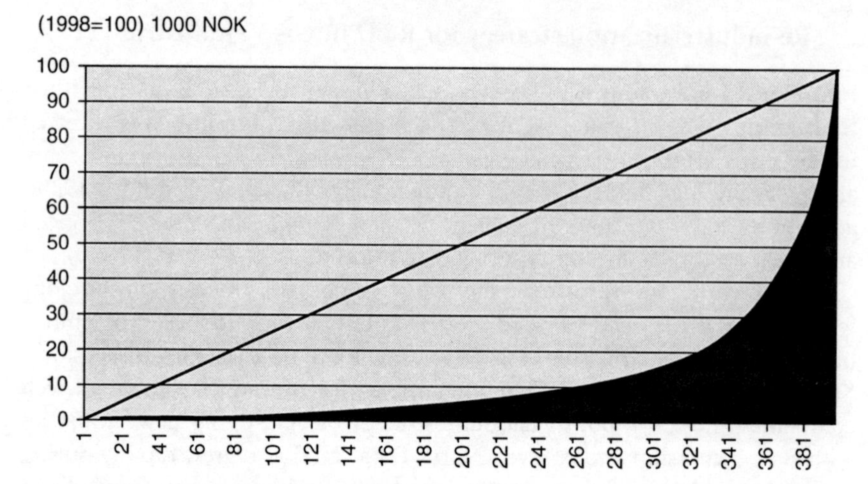

Figure 4.3. Distribution of funds from NTNF (Nyskaping i næringslivet/Innovation in industry), 1978–1987.

Note: The figure suggests that 13 companies received 50 percent of total funding by NTNF 1978–1987.

Table 4.1. The five largest projects in NTNF's corporate program during the 1980s

Company	Project	Details
Autodisplay	Instruments for car industry	25 M NOK
Norsk Data	Software	22 M NOK
Dolphin Server	Orion superserver	14 M NOK (the follow-up of ND)
Sensonor	Integrated sensors (including car industry)	9 M NOK
Norsk Data	Archival programmes	4 M NOK
Norcontrol Simulation		2 M NOK

Source: NHD 1990: 106.

project Prometheus. It failed and production never started. Other NTNF-supported projects, including SensoNor, Norcontrol and Dolphin, sought to develop new technologies that would serve as the foundations for new industries.

The belief that IT research projects could develop new technologies that would form the basis for the growth of new or old companies underpinned the R&D fundings and related strategies within NTNF, in defense-related R&D (FFI) as well as telecommunications R&D (TF). As a result, public R&D contracts were directly primarily towards IT projects.[21] This ideology and support policy for R&D notwithstanding, the most important influence on high-tech industrial development in Norway during the 1980s was the growth of offshore oil and gas production. Beginning in the late 1980s, Norway's "goodwill" policy (see Chapter 7) mandated the creation of links between international oil companies and national R&D organizations. This policy strongly shaped the direction of the IT/ICT industry.

Re-industrialization strategy for large-scale technologies: R&D in the offshore oil and gas sector

The development of a petroleum sector in Norwegian economy was not based on the idea of a research-driven industrial growth. Here, production came first, and strategies for R&D followed only when the industry was well established.[22] The introduction of R&D as part of oil policy emerged in 1978 when the government introduced the Technology Agreements that are discussed in greater detail in Chapter 7.[23]

The Goodwill Agreements (GWA) became very influential instruments in increasing industrial R&D in Norway. Oil companies paid around 80 percent tax on the revenues from the North Sea, and the companies' investment in Norwegian R&D was therefore subsidized by the state. The oil companies' collaboration with Norwegian companies and research institutes had a great impact on the research system. Companies and research institutes turned towards the offshore sector, as is apparent from R&D spending data. The oil and gas sector rapidly increased its procurement of R&D from external sources, and its spending on contract R&D reached 700–800 million NOK during the first half of the 1980s. By comparison, all manufacturing industry procured R&D services from external organizations for *c.*250–450 mill NOK (Statistics Norway). The surge in demand for R&D from the oil and gas sector had a crowding-out effect, as the expansion in petroleum related R&D corresponded with a period of reduction of industrial R&D in other sectors (Smith and Wicken 1992). The GWA strongly influenced industrial R&D in the longer term as well, as oil companies reported that they invested 5.800 million (1990 prices) NOK through approximately 1500 different projects during 1979–91.

The rapid increase in procurement of R&D services in the offshore sector that began in the late 1970s influenced the development of universities, research institutes, and much of Norway's emergent high-technology industrial sector, especially in ICT. Norway's ICT industry found a lucrative market for technology development in the large oil and gas sector, and a substantial part became incorporated as problem-solvers into the large technological system developing in the North Sea (see Chapter 7 for a more detailed discussion).

The growth of foreign oil companies' role within the Norwegian R&D infrastructure occurred just a few years before a more market-oriented policy was introduced for the research institutes. The new policy mandated that institutes should become independent organizations. Although institutes would continue to receive a part of their income from the research council, they were expected to obtain the remainder of their income from projects with private and public sector organizations. The new policy encouraged research institutes to seek new markets for their services, and several expanded their activities in the offshore sector. Three institutes became closely linked to the Oil Directorate (OD) (which became a new de facto "research council"): Rogalandsforskning in Stavanger, Institute for Energy Technology at Kjeller outside Oslo, and SINTEF in Trondheim. These institutes, and SINTEF in particular, conducted R&D to support the strategy of Norway's Oil Ministry and Oil Directorate, which focused on increasing revenues from the oil fields. The most important parts of the programme sought to improve knowledge regarding flows of gas and oil in pipes, and raise the percentage of offshore oil reservoirs that could be extracted.[24]

THE BREAKDOWN OF THE R&D-DRIVEN INDUSTRIAL STRATEGY

The dominance of the R&D-driven strategy in public industrial policy broke down during the early 1990s. The change in political thinking took place inside the Labour government of Gro Harlem Brundtland as she became prime minister for the third time in 1990. The "research-driven strategy" for industrialization was abandoned in favour of a "user-driven strategy" for industrial research and development. Existing industry became far more influential in industrial R&D policy making, and representatives of R&D-intensive, network-based industries became less influential.

The breakdown of the research-driven strategy resulted from changes in different parts of industry during the economic crises of 1987–93, as well as learning from the experience of the policy instruments introduced during the late 1980s. The evaluations of the "Targeted Technology Areas" criticized the implementation of that policy, particularly the lack of good instruments for realizing the objectives, and this criticism made a strategic policy change politically feasible. The main factor behind the breakdown of the research-driven industrialization strategy, however, was a change in the relative political influence of the *large-scale centralized* and *R&D intensive network* industry sectors between 1986 and 1993.

The most important factor in the decline in the political legitimacy and influence of the R&D-intensive network industry sector were its growing problems during 1987–88. The two major national champions in ICT entered into economic crises, and Norsk Data went bankrupt. This was a shock to many as ND had been the model company in the research-driven industrial strategy of the 1980s. In a surprisingly short period of time, the vast network of organizations and individuals around ND disintegrated and very little of the computer industry survived.

Kongsberg Våpenfabrikk (KV) had lost money almost every year since the beginning of the 1970s, and the Labour government (Brundtland 2, 1986–9) finally broke with the core group of modernizers from the 1960s. Hauge-Hurlen-Lied lost control of KV, and the company became involved in an international scandal over the use of KV technology to produce silent propellers for strategic Soviet submarines[25] (Wicken 1988). KV was divided into a number of companies and privatized; only the defence part of KV remained state-owned. The KV privatization symbolized a shift in Labour policies towards a more market-oriented industrial strategy, and a clear signal that the KV model for research-driven industrialization was no longer viable. The weakening of the network around KV and the breakdown of the network around ND weakened the political influence of R&D-intensive industry in Norway. The collapse of Norway's national telecommunications equipment industry in the early 1990s (see Chapter 10), further weakened the belief that future industrial growth would come from public support for R&D in emerging technologies, and the research-based industries accordingly became weaker players in the political system.

The effects on policy of these problems in emerging industries were exacerbated by the availability of an alternative industrial strategy for Norway. The offshore oil and gas sector had been selected as a major growth industry in the 1970s and was subsequently selected as a targeted technology area. The fall in petroleum prices in 1986 created uncertainty over the future profitability of the industry, but by 1990 the industry was again profitable. The state-controlled companies Statoil and Hydro, the offshore engineering companies (Kværner and Aker), supply industries, and other parts of the large scale technological system of the oil industry became the focus of policy.

The strong position of the offshore oil and gas sector effectively displaced the research-driven strategy. Within this industrial policy, R&D was no longer a source of economic transformation, and public R&D spending instead supported problem-solving activities in existing production systems.

Public policy followed a revision in the strategies of the largest resource-based companies. Norsk Hydro abandoned its R&D-driven diversification strategy, focusing instead after 1990 on improving production in its 'core activities' of fertilizers, aluminum, and oil/gas, and reducing R&D in other product areas. Norsk Hydro, Statoil and a few other firms became the new Norwegian "national champions".

The offshore oil and gas sector had by 1990 become a major part of Norway's economy, influencing technology policy debates and the behaviour of R&D-performing organizations and other sectors of the economy. The growth of the oil and gas sector also influenced the last stages of the professional careers of the promoters of the research-based strategy of the 1960s. Beginning in the early 1970s, Jens Chr. Hauge and Finn Lied turned their political interest towards the emerging offshore sector, emphasizing institution building for a state-owned and nationally controlled system for exploiting the new natural resource. Hauge became the first chairman of the core political instrument, Statoil, and Lied soon followed him as the board chair (Hanisch and Nerheim 1992).

The growth of R&D and related activities supporting offshore production also attracted large parts of Norwegian industry seeking to exploit this lucrative market. Norway's industrial policy for the offshore sector (see Chapter 7) and the large scale of investment in the sector influenced the development of the Norwegian R&D infrastructure, the structure of Norwegian manufacturing industry, and some parts of the Norwegian service sector. The sectors of Norway's economy that supplied offshore production activities grew to constitute a significant part of the economy, a reality that influenced the behaviour of companies and policymakers. Norway's industrial R&D strategy shifted in the 1990s to focus on the problems of the resource-based industries, particularly oil and gas. The large investments in the offshore sector during the early 1990s drew research institutes, universities and R&D-intensive companies into deeper involvement in Norway's oil economy. The R&D-driven strategy for industrialization based on emerging industries had come to an end.

The breakdown of the research intensive strategy for industrialization, along with the rapid growth of the offshore oil sector, created a path for Norway's industrial R&D investment and innovation system that contrasted with those of neighbouring economies such as Sweden and Finland, both of which now are among the most R&D-intensive economies in the world. In these countries the R&D-intensive industries remained successful or were able to survive economic downturns, and national policies retained their focus on R&D-driven industrial strategies in industries like ICT, pharmaceuticals, aerospace, and biotechnology. Norway focused on developing resource-based industries in which R&D is not the sole driving force, but instead supports industry performance, acting as a servant in solving problems in large technological systems or as part of wider knowledge flows in the economy. One result of this trajectory of development is the persistently lower level of R&D investment in Norway by comparison with its Nordic counterparts.

NOTES

1. At the Labour Party's 1963 annual conference, Wilson made possibly his best-remembered speech, on the implications of scientific and technological change, in which he argued that "the Britain that is going to be forged in the white heat of this revolution will be no place for restrictive practices or for outdated measures on either side of industry".
2. S. Zuckerman, "Linking science with industrial growth," *Financial Times* 7.11 (1963).
3. OECD published the first report on science policy, the Piganiol report, in 1963 and organized the first ministerial meeting on Science Policy the same year.
4. The director of the research council (NTNF), Robert Major, played an important role in drawing the attention of OECD policymakers to R&D policy. The 1963 OECD analysis was probably ordered by NTNF in order to bring industrial research into the realm of industrial policy. (Interview Robert Major)
5. Cuttings in the archives of the Ministry of Industry "Government drive to modernise Britain," *Financial Times* 7.11. 1963 and S. Zuckerman "Linking science with industrial growth" and "Getting ready to push research", *Financial Times* 15.11. 1963; "Three new bodies should take over from DSIR", NOE-RA, VI, Boks 59, Utvalg i forbindelse med opprettelse av et utviklingsselskap.
6. Brofoss remained the leading industrial strategist in Labour until his death in 1972.
7. The main policy making organization within the ministry was the Negotiation Committee (Forhandlingsutvalget or Gøtheutvalget) headed by the State secretary Odd Gøthe from 1955. This committee discussed new ideas for establishment of new industry and came up with proposals to be discussed by the government. Næringsdepartementet, Utredningsavd, VI, 40; Forhandlingsutvalget for nye industritiltak, *Mappe 1 1955–6,6* Notat MM 29.11.63.
8. NOE-RA, VI, Boks 59: Utvalg i forbindelse med opprettelse av et utviklingsselskap, April 1962: PM om et utviklingsfond.
9. This was the perception of the ministry of industry: "NTNF is today primarily an advisory body for researchers, not for industry. The industrializts that participate in the NTNF do so in their private capacity and this does not

guarantee close contact with industry" NOE, VI, Boks 59: Utvalg i forbindelse med opprettelse av et utviklingsselska 7.3.64.

10. KV became this type of organization between 1963 and 1987.
11. Shipping and shipbuilding also played a central role in the strategy of NTNF. There was a build-up of the research infrastructure for these industries in Trondheim, and NTNF also assigned a high priority to this cluster when handing out R&D support to companies. The ship classification company, Det norske Veritas (DnV), was crucial in the strategy for developing technological competence within this part of the economy (Andersen and Collett 1989), and DnV alone received 17 percent of all money distributed from NTNF to companies during 1967–77. In addition companies like Aker, Moss Rosenberg Verft, Ankerløkken, Bergen Mek Verksted, Wickman motorfabrikk, and Norcontrol all received substantial support.
12. Byggforskningsinstituttet, Norsk Geoteknisk Institutt, Vassdrags- og havnelaboratoriet.
13. Skipsmodelltanken, Skipsteknisk Forskningsinstitutt, from 1972 Norges Skipsforskningsinstitutt.
14. Elektrisitetsforsyningens forskningsinstitutt.
15. By contrast, the industrial institutes controlled by NTNF or linked to NTH were largely consultancy companies that undertook contract work to solve problems within individual companies.
16. With the exception of one year KV never made a surplus. Every year it received (substantial) subsidies from the ministry of industry. It was more a technology-industrial policy instrument than a ordinary industrial company (NOU 1989).
17. The specific direction of Norway's research-based policy strategy became evident in the 1970s when Jens Chr. Hauge made an attempt to reorganize the IT industry that had emerged from the mid 1960s by using policy to encourage development of large firms within Norway's IT sector (Sogner 1994). Hauge's effort failed and the research intensive industry remained small scale with a few exceptions, KV, ND and the two major international telecommunication, STK (ITT) and EB (Ericsson, Norwegian owners from 1974).
18. This is based on NIFUSTEP Statistikkbanken for studies of R&D funding in the yearly state budget, defining the following categories as "industrial R&D": primary industries, manufacturing, energy, transport- and telecommunication, space and defence. State funding in 1983 was 1189 mill NOK, increasing to 3034 mill NOK in 1993. In nominal prices the increase 1983 to 1993 was 155 percent.
19. The targeted technology areas were introduced for the five years period 1986/87 to 1990/91.
20. The term "mariculture" first appeared in US policy documents of the 1970s. See 'Project Mari Culture' by National Oceanic and Atmospheric Administration Sea Grant Office 1970, publication J. Hanson (ed.), *Open Sea Mariculture*, Dowden, Hutchinson and Ross (1972). Based on Mariussen (1992).

21. Pharmaceutical industry was traditionally weak in Norway. However, during the 1980s the small company Nyco made a break through in radiocontrast agent production, becoming a leading international producer of the field. (Amdam and Sogner 1994) The company's growth and profitability supported the idea that R&D could lead to new growth opportunities within research based industries. However, this sector was much weaker in Norwegian politics and science, and was not able to create the same type of support as that of the ICT sector.
22. NTNF established a R&D programme for the emerging offshore sector during the 1970s, mainly directed towards development of production technology for the sector. It was part of the research council's wider strategy for promoting industrial development. However, the programme was rather limited and not incorporated into a wider strategy for the development of the new industry.
23. The TC consisted of three types of regulations called "50 percent agreement", "technology agreements", "goodwill agreements". The latter represents 77 percent of all R&D funding of the system. Data on the Goodwill Agreements based on Wiig (1993).
24. In the 1970s it was expected to harvest app 20 percent of the total oil in the reservoir, while the percentage was appx. 50 at the turn of the century.
25. KV exported numerical control systems for machinery to Toshiba, and Toshiba re-exported the systems to the Soviet Union. This created the deepest crises in Norway's relationship with USA since the Second World War. The government decided to get rid of the Kongsberg name and renamed the company to Norsk Forsvarsteknologi (Norwegian Defence Technologies).

REFERENCES

Amdam, R. P. and Sogner, K. (1994) *Rik på kontraster. Nyegaard & Co—en norsk farmasøytisk industribedrift 1874–1985*, Ad Notam Gyldendal, Oslo.

Andersen, H. W. (1986) Fra det britiske til amerikanske produksjonsidea: forandringer i teknologi og arbeid ved Aker mek. Verksted og inorsk skipgsbyggingsindustri 1935–1970, Trondheim.

Andersen, H. W. and Collett, J. P. (1989) Anchor and Balance. Det norske Veritas 1864–1989, Oslo.

Andersen, K. G. and Yttri, G. (1997) *Et forsøk verdt. Forskning og utvikling i Norsk Hydro gjennom 90 år*, Universitetsforlaget Oslo.

Berge, D. M. (2006) "Havfiske inn i nye næringer", in Bjarnar, O., Berge, D. M. and Helle, O., *Havfiskeflåten i Møre og Romsdal og Trøndelag. Volume 2, Fra fri fisker til regulert spesialist 1960–2000,* Tapir, Trondheim.

Brofoss, K. E. (1993) "Innsatsområdene som forskningspolitisk virkemiddel", *Rapport 4, NAVFs utredningsinsstitutt.*

Collett, J. P. and Lossius, B. (1986) *Visjon, forskning og virkelighet: TF 25 år*, Oslo.

Fagerberg, J., Guerrieri, P. and Verspagen, B. (eds.) (1999) *The Economic Challenge for Europe. Adapting to innovation based growth*, Edward Elgar, Cheltenham.

Freeman, C. and Soete, L. (1982) *Unemployment and Technical Innovation: A Study of Long Waves in Economic Development*, London.

Hanisch, T. J. and Nerheim, G. (1992) "Fra vantro til overmot?" vol. 1 of *Norsk oljehistorie*, Norsk Petroleumsforening, Oslo.

Hanson, J. (ed.) (1972) *Open Sea Mariculture*, Dowden, Hutchinson, and Ross.

Jensen, K. (1989) "Forskning og ny teknologi; fra mulighet til forutsetning. Om moderniseringsmiljøet som pådriver i norsk industriutvikling på 50 og 60-tallet", Master's thesis in history, University of Oslo.

Kvaal, S., (1994) "Forskning og industripolitikk," *Historisk Tidsskrift* 1/1994.

Mariussen, Å. (1992) *Fra vann til hav. Evaluering av havbruksforskningen. Prosessevaluering av havbruk som forskningspolitisk hovedinnsatsområde*, Nordlandsforskning, Bodø.

Mensch, G. (1979) *Stalemate in Technology: Innovation overcome Depression*, New York.

Njølstad, O. and Wicken, O. (1997) *Kunnskap som våpen. Forsvarets Forskningsinstitutt 1946–1996*, Universitetsforlaget, Oslo.

NHD (1990) *Rapport fra IT-utvalget oppnevnt ved kongelig resolusjon 12. januar 1990: Evaluering av nasjonal handlingsplan for informasjonsteknologi*, Oslo.

NOU (1973) 40 *Norsk konfeksjonsindustri og dens konkurranseevne.*

——(1979) 35 *Strukturproblemer og vekstmuligheter i norsk industri.*

——(1981) 30 *Forskning, teknisk utvikling og industriell innovasjon.*

——(1989) 2 *Kongsberg Våpenfabrikk.*

OECD (1963) *The Measurement of Scientific and Technical Activities : Proposed Standard Practice for Surveys of Research and Experimental Development*: Frascati manual, OECD, Paris.

Smith, K. and Wicken, O. (1992) *Olje og Gass som hovedinnsatsområde—Prosessevaluering*, NTNF, Oslo.

Sogner, K. (1994) "Fra plan til marked. Staten og elektronikkindustrien på 1970-tallet", *TMV report series*, no. 9, Oslo.

St.prp. no. 1 (1986–87) Statsbudsjett for 1987, Oslo.

Wicken, O. (1984) Våpenimport eller egenproduksjon. Hvorfor Norge ikke bygde ut militærindustri 1945–1950, FHFS notat 3/1984, Oslo.

——(1987a) "Norske våpen til NATOs forsvar," *Forsvarsstudier 1*, Forsvarshistorisk Forskningssenter Oslo.

——(1987b) "Militære anskaffelser—forsvars-eller industripolitikkö," in Böhme, K.-R. (ed.), *Krigsmaterielanskaffning, Föredrag vid Nordiska historikermötet i Reykjavik 1987*, Meddelanden från Militärhistoriska avdelningen vid Militärhögskolan, 3, Stockholm.

——(1988) "Stille propell i storpolitiske storm," *Forsvarsstudier 1.*

——(1990) "The Norwegian Military-Technology System," in Gleditsch, N. P. and Njølstad, O., *Arms Races: Technological and Political Dynamics*, Sage, London.

——(ed.) (1994) *Elektronikkentreprenørene. Studier av norsk elektronikkforskning og—industri etter 1945*, Ad Notam Gyldendal, Oslo.

——(2000) Forskning, næringsliv og politikk. En historisk fremstilling av norsk næringslivsforskning og politikk, *Working Paper 6, Center for technology, innovation and culture*, Oslo.

Wiig, H. (1993) "Olje mot forskning. En oppgave om goodwillavtalene i norsk forskningspolitikk, og teknologioverføring i FoU-samarbeidene", Master's thesis in political science, University of Oslo.

5

Historical Fingerprints? A Taxonomy of Norwegian Innovation

Fulvio Castellacci, Tommy H. Clausen, Svein Olav Nås, and Bart Verspagen

This chapter uses quantitative indicators to analyze and describe the Norwegian innovation system. This quantitative description of Norway's innovation system enables us to compare its structure with that of other European economies, and we investigate whether the historical evolution of the Norwegian innovation system discussed in Chapter 2 is visible in the system's current structure. Chapter 2 described three historical layers (the small-scale industrialization path, the large-scale centralized path, and the R&D intensive network-based path) in today's Norwegian production and innovation system as representing distinctive approaches to the organization of production and innovation. Our aim is to find out whether these institutional and organizational differences influence contemporary innovation activities in Norway. Our analysis highlights the importance of Norway's economic structure as an influence on its national innovation system. We find that the resource-based nature of its economy has affected innovation within Norway (see also Grønning, Moen, and Olsen, 2008).

The organizational and institutional features of the different developmental paths in the Norwegian economy summarized in Chapter 2 are likely to affect the competitive environment for innovation in Norway. More specifically, the large-scale centralized path and the R&D intensive network based path are each associated with specific competitive environments, with different roles for small and large companies, and different levels of new-firm formation and entry. These different environments are associated with different incentive structures for innovation.

This chapter might be described as a search for the "fingerprints" left on Norwegian innovation by the historical developments described in Chapter 2. A central issue in the chapter thus concerns the role of path dependency in Norway's innovation system. How has its historical development influenced the current state of the Norwegian innovation system and what does this influence imply for the performance of this system or its prospects for change?

This chapter's analysis is based on the premise that innovation is not limited to so-called high-tech sectors, such as ICT-related industries and pharmaceuticals, where formal R&D is very important. In sectors other than high-tech, innovative firms may rely on different sources of knowledge (e.g. external ones) or different forms of learning (e.g. learning-by-doing), neither of which are fully captured in most R&D statistics. The so-called Community Innovation Surveys seek to describe and map this type of innovation, and we draw extensively on them in an effort to map these sources and styles of innovation at the sectoral level. This approach assumes that firms within individual sectors adopt relatively homogenous approaches to innovation (e.g. Pavitt, 1984). We develop a sectoral taxonomy of the Norwegian innovation system and relate the findings of the sectoral discussion to the historical description of Chapter 2.

Section 2 of this chapter presents a quantitative description of the industrial structure of Norway's economy, as complement to the institutional and organizational history in Chapter 2. Section 3 compares Norway's innovation performance with those of other high-income European economies and presents the sectoral taxonomy of the Norwegian innovation system. Section 4 presents our analysis of market structure and entrepreneurship in the major sectors of Norway's innovation system. Section 5 concludes the chapter with a discussion of the implications of our analysis for the Norwegian innovation puzzle.

THE EVOLUTION OF NORWAY'S ECONOMIC STRUCTURE

The three paths of development described in Chapter 2 have had profound implications for the industrial structure of the Norwegian economy. At least two of these three developmental paths involved resource-intensive sectors, most recently the production of oil and gas. This section examines the evolution of Norway's industrial structure and briefly compares it with that of a set of reference countries. As Figure 1.1 in Chapter 1 shows, Norwegian production of oil and gas took off in the 1970s, and we accordingly begin our analysis in 1970. The OECD STAN database is our primary source of data, and both the group of reference countries[1] that we use for comparison with Norway and the sectoral taxonomy in our discussion are dictated mainly by the availability of data from this source.

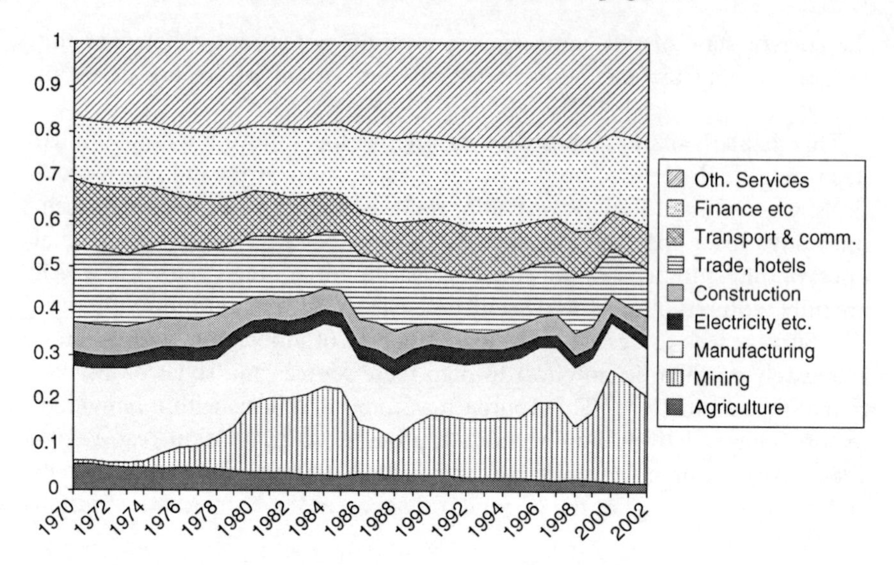

Figure 5.1a. The evolution of sectoral shares of GDP in Norway.

Figures 5.1a and 5.1b present the basic data on Norway's industrial structure, i.e., the evolution since 1970 of the shares of total Norwegian value added (GDP) accounted for by the nation's main economic sectors. We use value added rather than employment in order to highlight the rise of Norway's oil and gas sector, since this highly capital-intensive sector has a much lower share of employment than of value added. By the mid-1980s, the oil and gas sector had grown to account for 20 percent of GDP, and after some fluctuation (largely reflecting the erratic movement of oil prices during the period) has grown again to a new peak of 23 percent in 2003.

As in other high-income industrial economies, the GDP shares of manufacturing and agriculture have declined in Norway since 1970. Although Norway's agricultural sector has long exerted considerable political influence within the nation (not least in Norway's policy toward EU membership), this sector (which includes fish farming) accounted for less than 1½ percent of GDP by 2003. Norway's manufacturing sector accounted for 21 percent of GDP in 1970, but declined to slightly less than 10 percent by 2003.

Two large service-related sectors expanded their shares of Norwegian GDP during the 1970–2003 period. The first of these is business services, within which financial services accounts for the largest share. This sector grew from 14 percent in 1970 to 20 percent by 2003. The other large services sector includes social and personal services (labeled "other services" in the figure),

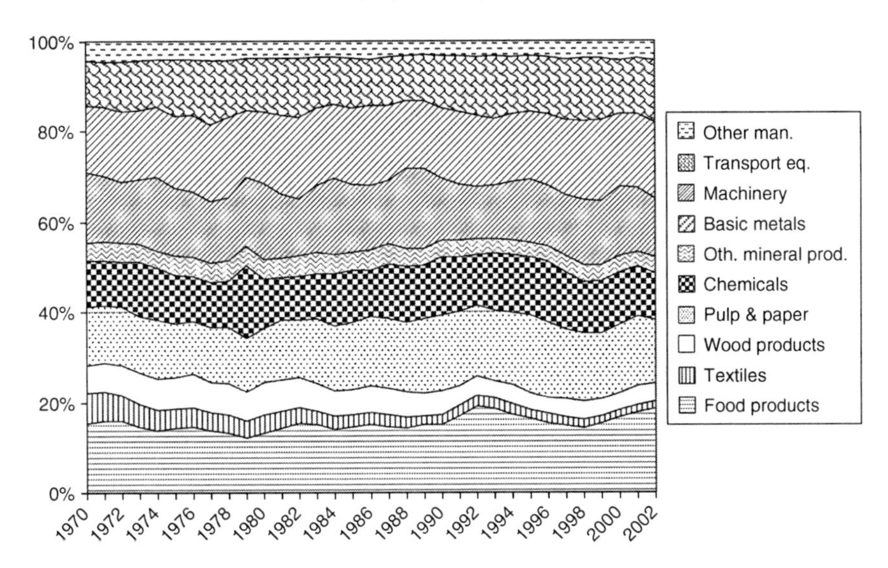

Figure 5.1b. The evolution of sectoral shares of value added in manufacturing in Norway.

including the public sector (e.g. education). This sector grew from 17 percent to 23 percent of GDP during 1970–2003.

Figure 5.1b displays a detailed breakdown of the shares of manufacturing value added accounted for by individual manufacturing industries.[2] By and large, the shares of individual industries within overall Norwegian manufacturing were fairly stable during 1970–2003. None of the sectors experiences a major decline or rise during the period, with the possible exception of textiles, which declines from almost 7 percent to less than 1½ percent of manufacturing by 2003. Figure 5.1b highlights the substantial (especially by comparison with other European nations) share of Norwegian manufacturing accounted for by resource-intensive sectors, such as pulp and paper or basic metals. Taken together with the large share of Norwegian GDP accounted for by the oil and gas sector, these data confirm the portrait of Norway as a very resource-intensive economy.

A similarly resource-intensive economic structure is found in at least two other OECD countries: Canada and Australia. The overall economic structure of these two high-income economies resembles that of Norway, although the specific resource-based activities that loom large in these nations differ from those that are significant in Norway.[3] Surprisingly, the two European economies that are most similar to Norway are Denmark and the Netherlands. The Danish "resource-intensive" sector consists mainly of agriculture and its related processing industries, rather than oil and gas. The large size of the natural-gas producing sector in the Netherlands produces some important similarities between the overall economic structure of the Netherlands and that of Norway. In contrast to Norway, however, Dutch natural gas is much more easily accessible, and as a consequence, the natural-gas sector has developed fewer strong linkages to other sectors in the Netherlands than it has in Norway.

In summary, Norway differs significantly in its economic structure from most other industrial economies, especially those in Europe. The reliance on resource-intensive industries and sectors that has characterized Norway since the earliest years of industrialization has persisted for well over a century, and has contributed to the development of a national innovation system that also contrasts with that of most other industrial economies.

INNOVATION IN NORWAY

Norway compared to other European countries

We use data on innovation outcomes, innovation expenditures, innovation goals, and innovation-related information sources from the European Community Innovation Survey to analyze the innovation activities of Norwegian firms.[4] Most of these data refer to the year 2000 (CIS-3). Figures 5.2a–5.2d compare these innovation indicators in Norway with those from three sets of reference countries: the other Scandinavian countries (Denmark, Sweden, Finland), the other Northwest European countries (the three Scandinavian countries, Netherlands, Germany, United Kingdom, Belgium, and France), and the South European countries (Greece, Italy, Portugal, and Spain). The first two of these three sets of reference countries are most relevant to our discussion, because they contain countries with per capita income levels approaching that of Norway.

Figure 5.2a provides a summary comparison of the innovative behavior of Norwegian firms and enterprises in the three reference groups, based on different measures of the share of firms that innovate. The leftmost bars refer to the share of firms with any innovation activity at all (including failed innovation attempts and ongoing innovation that has not yet paid off). The average Norwegian level is about 80 percent of that in the rest of Scandinavia or the rest of Northwest Europe, and just slightly above the level in South Europe. When the comparison is limited to indicators of successful innovation, Norway once again is lower than the Scandinavian or Northwest

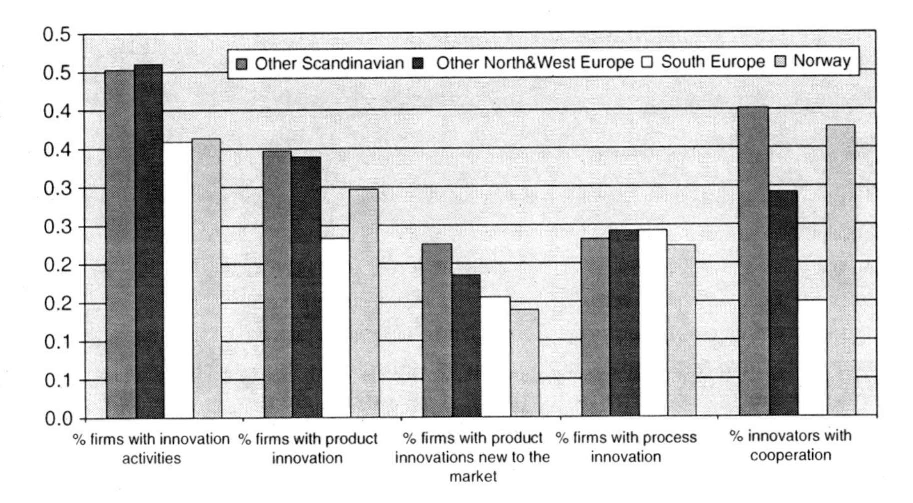

Figure 5.2a. Innovation activities: Norway relative to three reference groups.

European reference groups, and well ahead of the Southern European group. The "product innovation" indicator includes innovations that are new to the firm, but not new to the market (i.e. imitation). When the comparison is limited to product innovations new to the market, Norway falls behind all of the reference groups, including Southern Europe. Norwegian firms also exhibit lower levels of process innovation than firms in any of the three reference groups. Thus, these data highlight the so-called Norwegian innovation puzzle: Norway, a country with a high level of income per capita and high productivity growth, displays a relatively low level of innovation.

Finally, Figure 5.2a provides data on innovation cooperation. The Scandinavian countries exhibit relatively high levels of cooperation (the share of firms that report that they are active in innovation that also report that they cooperate with other firms in innovation), and Norway is no exception to this. The Norwegian level of innovation cooperation is essentially equal to that in the rest of Scandinavia, but higher than in Northwest Europe, and much higher than in South Europe.

Figure 5.2b summarizes and compares data on innovation expenditures, expressed as a share of sales.[5] Innovation expenditures include five different categories, of which only two are related to R&D. The leftmost group of bars in Figure 5.2b includes the turnover of non-innovating firms, while the second set of bars focuses only on innovating firms. This indicator is approximately equal across the three reference groups, and Norway is consistently lower than the average, at roughly 60 percent. Thus, the lower share of Norwegian firms reporting that they are active innovators reflects lower levels of intrafirm

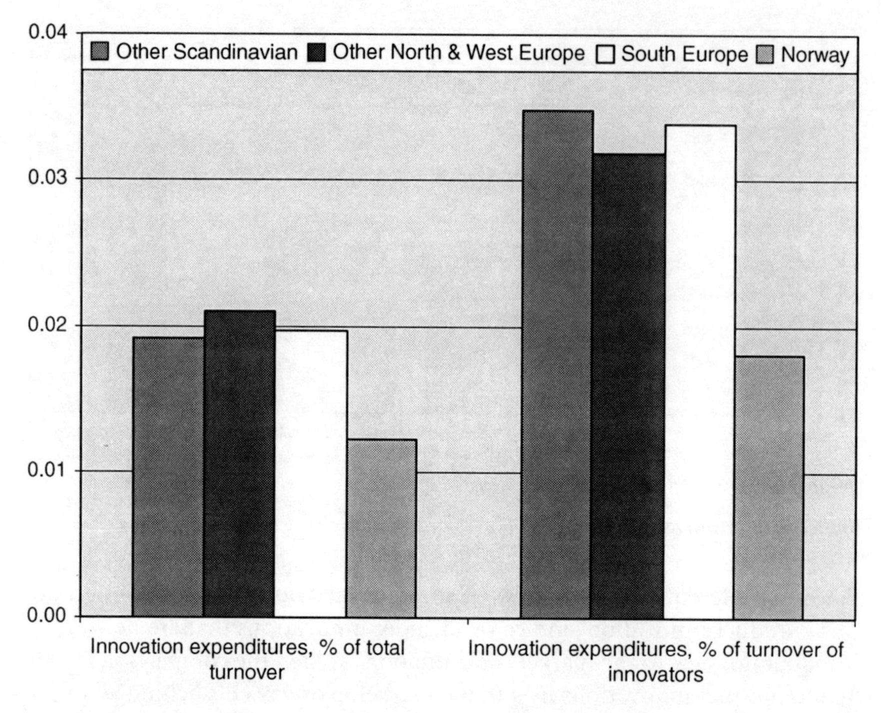

Figure 5.2b. Innovation expenditures: Norway relative to three reference groups.

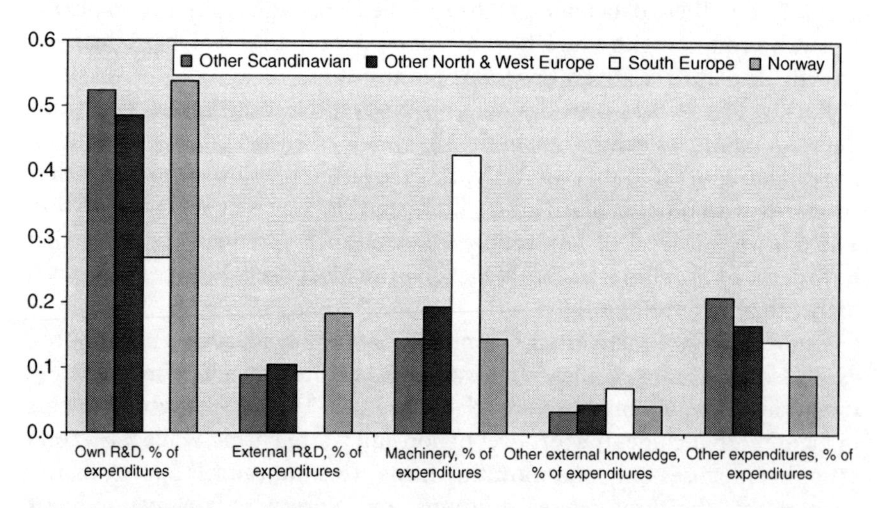

Figure 5.2c. Shares of Innovation expenditures: Norway relative to three reference groups.

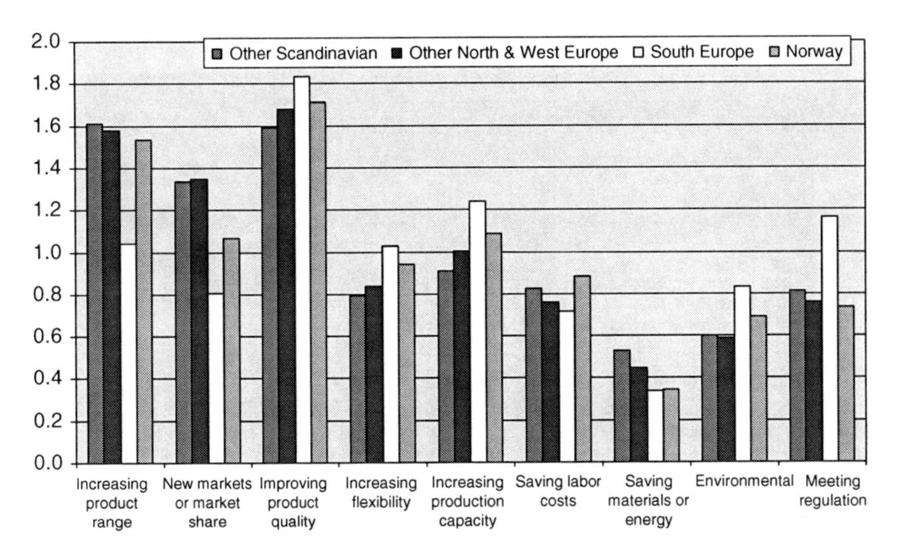

Figure 5.2d. Importance of various motives for innovation: Norway relative to three reference groups.

innovation expenditures (from public and private sources) relative to firm sales in Norway by comparison with other European economies.

Figure 5.2c compares the composition of innovation expenditures, including expenditures on in-house R&D, on external R&D (i.e. funds invested by firms on R&D performed by other entities, including design firms, research institutes, and universities), machinery, external knowledge acquisition (e.g. licenses), and other (training of personnel, design, marketing of new products). Each of these categories is expressed as a share of the total expenditures for innovation-related activities within firms in a country (and hence a country cannot have relatively high values in all categories). Expenditures on external R&D account for a larger share of Norwegian innovation spending than in any of the three reference groups. Expenditures on in-house R&D account for a higher share of Norwegian innovation expenditures than in South Europe, but are roughly equal to the shares in Northwest Europe and Scandinavia. Investment in new machinery accounts for a lower share of Norwegian innovation expenditures than in South Europe or Northwest Europe, and is comparable to the shares observed in the other Scandinavian countries.

Data on the objectives of innovation are summarized in Figure 5.2d, reporting the percentage of innovative firms reporting that a particular motive is highly important. For each country, the variables for the separate effects have been scaled to an average of one, so that the indicators express the relative importance of a motive, rather than absolute differences

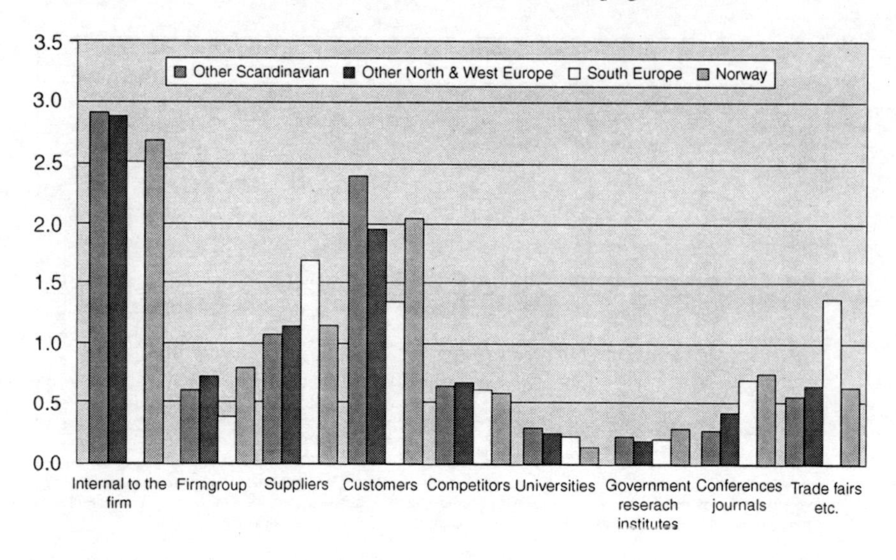

Figure 5.2e. Importance of information sources for innovation: Norway relative to three reference groups.

in its importance among countries. The results for Norway display consistent differences with those for Scandinavia and Northwest Europe. The motive, "Reaching new markets or increasing market share" is consistently reported by Norwegian firms to be less important than is true of innovating firms in Scandinavia or Northwest Europe. Norwegian firms report that two product-related motives for innovation expenditures, "increasing product range" and "improving product quality," assume a level of importance that is similar to the responses from innovative firms in Northwest Europe and Scandinavia. Process-improvement motives, however, are reported by Norwegian firms to be significantly more important in their innovation expenditures relative to the Northwest Europe and Scandinavian reference groups. The Norwegian scores indicate that increasing flexibility, expanding production capacity, saving labour costs, and saving the environment all are more important motives for innovation expenditures than appears to be the case in Northwest Europe or elsewhere in Scandinavia. The comparison of Norwegian and Southern European firms reveals a different set of contrasts in motives for innovation expenditures, as Norway scores high on product-related motives, and somewhat lower on the process-related motives.

The innovation survey also provides data on the information sources that innovative firms report to be important in their activities. This is an area in which Norwegian firms appear to behave quite differently than firms in all three of the reference groups. Figure 5.2e shows that Norwegian firms assign great importance, relative to firms in the other reference groups, to conferences, workshops and journal publications, all of which are knowledge sources that require modest investments of resources, time, or personnel to access. Universities are also a less important source of knowledge in Norway than in other Scandinavian or Northwest European countries, but Norwegian firms assign greater importance than firms elsewhere in Scandinavia or Northwest Europe to government research institutes. These data suggest that government institutes may be partial substitutes for universities in the innovation-related activities of Norwegian firms. Geography could play a role in this, because universities are fewer in number and much more geographically dispersed in Norway than are government institutes. Relative to South Europe, within-enterprise-group information sources are rated as more important for Norwegian firms.

Overall, these data indicate that Norwegian levels of innovation effort, measured both as inputs and as outputs, are considerably lower than the levels of innovation-related inputs and outputs in the European reference countries that are nearest to Norwegian income levels. Indeed, Norwegian firms report levels of innovation-related activity in these data that are comparable to those reported by South European firms, where per capita incomes are much lower than in Norway. With respect to the sources of information used in innovation and the motives for innovation, on the other hand, Norwegian firms more closely resemble their Scandinavia and Western European counterparts, although even here differences exist. The stronger emphasis in Norway on process-improvement motives for innovation and the more important role of the technological institutes stand out as unique to Norway.

Innovation in Norway at the sectoral level

With our conceptual framework of co-evolution of industrial structure, institutions and innovation behaviour of firms in mind, this section examines how these aggregate measures of Norwegian innovative activity are related to the composition of the Norwegian economy. Has the evolution of the Norwegian economy along a resource-intensive path also affected the characteristics of innovation and is this influence visible in today's innovation statistics at the sectoral level? And, if so, does this help us in explaining the low levels of innovation-related activity reported by Norwegian firms? Our sectoral analysis uses data on 60 industrial sectors covering firms in both manufacturing and services at the 3 or 4 digit NACE level (see the chapter Appendix for a list of the sectors included in this analysis). The sectors included in this analysis were selected as those that are economically significant within Norway, based on the nation's pattern of economic specialization. Accordingly, we include as separate entities some sectors that are of little economic significance elsewhere in Europe (e.g. fish-farming). Our choice of sectors for this analysis also is limited to those with sufficient coverage in the Norwegian CIS database.

In addition to the CIS-indicators that were used in the international comparisons discussed in the previous section, we include several additional CIS-indicators in this sectoral analysis that are available only for Norway.[6] These additional indicators include both innovation measures and data on sectoral industrial structure and dynamics. The data on sectoral structure and dynamics are used in the next section's analysis of market concentration. Here, we use the sectoral innovation indicators to examine the different "styles of innovation" among sectors in Norway. This analysis tests the influence on overall innovation-related activities in Norway of the nation's specialization in resource-based activities, and should shed light on the importance of the three layers of the Norwegian innovation system for the broad structure of innovation in Norway.

We use a total of thirty innovation-related variables for each of the sixty sectors to characterize sector-specific "styles" of innovation. Ten of these variables are related to innovation expenditures, and include 4 of the categories used in Figure 5.2b above: own R&D expenditures (which includes intrafirm R&D investment funded by public and private sources), external R&D expenditures, machinery, and externally acquired knowledge. In addition, we separate the "other expenditures" category of Figure 5.2c into three subcategories: training, marketing of new products and design, which makes seven categories/variables. We express each of those seven categories as a fraction of total sales in the sector (including sales of non-innovators). Three additional expenditures variables disaggregate "own R&D expenditures" along two functional dimensions: process R&D as a fraction of total R&D, basic R&D as a fraction of total R&D, and applied R&D as a fraction of total R&D.[7]

The left panel of Table 5.1 provides a summary of these 10 variables. "Own R&D" is the largest expenditures category, followed by machinery and external R&D. But the standard deviations are large, in all cases larger than the mean, which indicates that these 60 sectors vary greatly in terms of innovation activity. In terms of the functional categories of R&D, process R&D accounts for about a third of the total (two-thirds is product-related). Basic R&D is a very small share of total "own R&D expenditures", just over 4 percent, but here too the measure's standard deviation is large. Applied R&D is just below 20 percent, and development R&D accounts for more than 75 percent, of total own R&D expenditures.

Our analysis includes another twenty variables drawn from the Norwegian CIS that are based on innovator firms' reported motives and information

Table 5.1. Summary statistics for innovation expenditures, R&D, innovation effects, and knowledge sources, 60 Norwegian sectors, 2000

	Average	Std. dev
Innovation expenditures categories as a fraction of sales		
Own R&D	1.85	2.92
Machinery expenditures	0.43	0.94
External R&D	0.32	0.50
Design expenditures	0.07	0.11
Marketing expenditures	0.07	0.10
Other external expenditures	0.07	0.11
Training expenditures	0.06	0.06
Innovation effects, fraction of innovators that rank an effect as highly important		
Product quality	0.28	0.14
Increased range of products	0.25	0.20
Production capacity	0.24	0.15
Saving labor costs	0.20	0.14
Flexibility	0.20	0.15
Expanding markets, market share	0.18	0.13
Meeting regulations or standards	0.17	0.12
Saving the environment	0.15	0.16
Saving materials and energy	0.09	0.09
Functional R&D categories as a % of own R&D		
Process related R&D	31.1	20.4
Applied R&D	19.2	17.9
Basic R&D	4.36	11.4
Information sources, fraction of innovators that rank a source as highly important		
Internal sources	0.52	0.18
Customers	0.37	0.19
Suppliers	0.21	0.13
Other parts of enterprise group	0.18	0.14
Meetings & journals	0.12	0.11
Exhibitions and trade fairs	0.11	0.11
Competitors	0.11	0.13
Public research organizations	0.06	0.07
Consultancy enterprises	0.05	0.07
Universities	0.05	0.09
Commercial R&D labs	0.03	0.05

sources. These variables are reported above in Figures 5.2d and 5.2e, although we have disaggregated some of them into more detailed categories. All nine of the reported motives for innovation in Figure 5.2d are included in the analysis, but we supplement the nine information sources reported in Figure 5.2d with innovators' reported rankings (based on the share of respondents ranking the source as "highly important") of the importance of consultancy enterprises and commercial R&D labs as sources of knowledge. The right panel of Table 5.1 provides summary statistics for these variables.

Improvements in product quality, increasing the range of products and expanding production capacity are the most important motives for innovation in Table 5.1, with around 25 percent of the innovators ranking these as highly important. The other motives rank just below this, with 15–20 percent of respondents assessing them as highly important. By contrast, only 9 percent of firms rank saving materials and energy as important. Internal firm sources are the most important knowledge source, ranked as highly important by 52 percent of innovating firms, followed at some distance by customers and suppliers (respectively ranked as highly important by 37 percent and 21 percent of innovating firms). Interestingly, universities and public research institutes, along with consultancy enterprises and commercial R&D labs, rank low as knowledge sources.

We use cluster analysis of the data for these 30 variables to cluster the 60 sectors into a smaller number of groups that share similar characteristics (methodological details on the procedure are available from the authors). Diagram 1 displays the results of this analysis, highlighting three major groups of sectors: "science-based innovation"; "resource-based innovation"; and "low-intensity innovation." Two other sectors are outliers within the analysis: road-building, and steam and water utilities. We exclude these sectors from the three-class taxonomy discussed below.

Our descriptive classifications for these three groups of sectors are based on factor analysis (principal components), a statistical technique that reduces a dataset with many variables (30 in our case) into one with fewer dimensions (seven in our case). We separately analyze the variables in the left and right panels of Table 5.1 (i.e. innovation expenditures variables on the one hand, and information sources and effects on the other hand), using the main correlations among variables within each group to find linear combinations of variables that each corresponds to a "dimension" (factor or component) in the data.

The factor analysis highlights three factors that differentiate sectoral groupings within the expenditures-related indicators. The first of these is associated with expenditures on internal and external R&D, marketing, and design. We label this measure the "R&D and product-orientation" factor. The second factor singles out two of the variables that measure the nature of R&D: process-oriented R&D and applied R&D. We accordingly label this measure the "applied/process R&D" factor. The third and final factor singles out some of the non-R&D innovation expenditures: machinery and equipment, non-R&D external innovation services, and training. We therefore label this last measure "non-R&D based innovation".

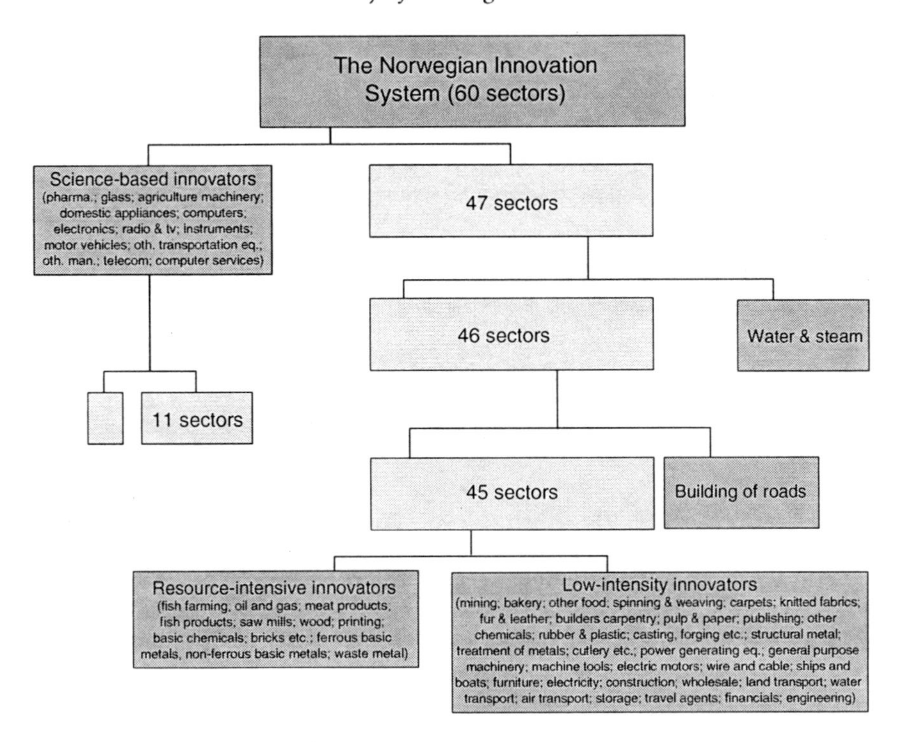

Figure 5.3. The Norwegian innovation system.

Our factor analysis of the information sources and effects variables yielded seven factors. We use only the first three of these seven factors, which account for the largest fraction of the total variance explained, in our sectoral analysis. The first of these dimensions scores high on increasing the range of products/services, and product quality as motives for innovation, and on customers and competitors as sources of information. Since a majority of these variables are related to products, we label this dimension the "product-driven innovation" factor. The second factor singles out several external information sources (consultancy enterprises, commercial R&D labs, universities, and R&D institutes) as important for innovation and differentiates sectors as well on the importance of the environment, regulatory compliance, and standards as motives for innovation. Since these information sources and motives for innovation are external to the firm, we label this dimension the "externally driven innovation" factor. Finally, the third factor measures flexibility, production capacity, and labor cost reduction as motives for innovation and suppliers as the important information source. We label this dimension

Table 5.2. Mean factor scores across the three clusters of sectors

	Science-based innovators (n=13)	Resource-based innovators (n=12)	Low-intensity innovators (n=33)
R&D and product-oriented innovation	**1.286**	−0.476	−0.289
Applied process R&D	−0.358	**0.894**	−0.155
Non-R&D-based innovation	**0.790**	−0.290	−0.129
Product-driven innovation	**1.259**	−0.443	−0.284
Externally-driven innovation	−0.177	**0.493**	−0.288
Process-driven innovation	0.225	**0.606**	−0.227

Notes: within each row, the maximum value is printed in bold; rows are dimensions of the innovation process (factor scores), columns are clusters of sectors in the Norwegian innovation system

as the "process-driven innovation" factor (note that this factor includes the "supplier-driven" innovation highlighted by Pavitt 1984).

Table 5.2 summarizes the factor scores for the three broad categories of sectors that we discussed earlier and that are depicted in Diagram 1. Although we do not report significance tests, the differences that we highlight are significant at conventional levels. Within each row, the maximum value is printed in bold. Consistent with our characterization above, the "low-intensity innovator" group scores low on all six innovation dimensions from the factor analysis.

The group of thirteen sectors that we label "science-based innovators" scores higher than the two other groups of sectors on the *R&D and product-orientation* dimension, the *non-R&D based innovation*, and the *product-driven innovation* dimension.[8] This group of sectors includes intensive performers of intrafirm and external R&D that also invest relatively heavily in non-R&D innovation activities. Although this result is interesting, it is important to note that non-R&D innovation expenditures are relatively small by comparison with R&D investment. In addition to its relatively heavy expenditures of all sorts on innovation, this group's focus on product innovation seems to set it apart in the Norwegian innovation system.

The science-based group includes pharmaceuticals, various ICT-related sectors (electronics, radio and TV, computers and computer services, telecom) and instruments, as well as several sectors not normally considered to be high-tech, such as agricultural and forestry machinery, glass and other mineral products, motor vehicles and other transport equipment. The ICT-related sectors and pharmaceuticals in this group represent the "R&D-intensive network-driven" developmental path discussed in Chapter 2. But other industries included in this group, such as motor vehicles or agricultural and forestry machinery, cannot be classified as part of the "R&D intensive" developmental path.

A second group of twelve sectors are labeled "resource intensive innovators." This label reflects the resource-intensive character of the sectors found in this group, which includes sectors associated with the "large-scale centralized path" of Norwegian economic development. The group includes the two archetypal resource-intensive Norwegian sectors, fish farming and oil and gas extraction, as well as basic metals, which includes the aluminum industry. Other resource-intensive sectors found here are meat and fish products, sawmills, wood products, basic chemicals, and building materials (bricks, stone, etc.).

Note that this group includes several industries that are included in Wicken's "large-scale centralized path" of economic development, as well as others that are not associated with this "layer" of the Norwegian innovation system, such as fish farming. Conversely, the pulp and paper sector appears in our taxonomy among the "low-intensity innovators," whereas Wicken puts this sector in the "large scale industrialization path." Such differences are hardly surprising, in view of the methodological differences between Wicken's historical analysis and the quantitative methods used in this chapter. Moreover, Wicken took the organization of production as a key organizing principle for his three developmental paths, while our main focus is on innovation.

The resource-intensive sectors in our classification system are characterized by a relatively high average score on the *applied process R&D dimension*, on the *externally driven innovation* dimension, and on the *process-driven innovation* factor. Interestingly, none of these specific characteristics is related to high innovation expenditures (although some of the individual sectors are relatively R&D-intensive). Instead, the nature of their innovation-related investments, activities, motives, and knowledge sources differentiates these sectors within the Norwegian innovation system. Process innovation plays a large role and external sources of knowledge are important in this sector. The external sources of knowledge ranked as particularly important within the sectors in the "resource-intensive" category include research institutes and universities, suggesting that the publicly funded components of the Norwegian research infrastructure are especially important for this sectoral group.

How does the innovative performance of these sectoral groups differ from one another? Table 5.3 summarizes the results of an ANOVA test for differences in the means of several innovation performance indicators across sectoral groups. The innovation performance indicators used in this statistical test include product innovation (share of innovators and share of innovative products of sales), process innovation, and non-technological innovation.

The *science-based innovators* group appears to have the highest rate of innovation, and the left part of Table 5.3 compares this group with the rest of the sample. For product innovation (the share of product innovators within the group), the average score for the high-tech group is 26 percentage points higher, and the share of total sales accounted for by innovative products is 16 percentage points higher. Process innovation and organizational innovation also are higher in the science-based group. This group scores higher than the sample average in three of the four forms of non-technological innovation, particularly organizational innovation, which is 17 percentage points higher. Management innovation is the only form of innovation that is not significantly higher in the science-based group.

Since we have already shown that the *resource based innovators* have lower innovation rates than science-based innovators, we compare them only with the low-intensity innovators in Table 5.3. The resource-based innovators have an overall (product and/or process) innovation rate that is higher than the low-intensity innovators, but this difference is barely statistically significant. The difference is largely determined by process innovation, where the resource-based group shows an average that is 14 percentage points higher than the low-intensity innovators. The resource-based innovators have a lower share of innovative products in sales than the low-intensity innovators, although this difference is not statistically significant. The differences between the resource-based innovators and the low-intensity innovators in nontechnological innovation indicators also are not statistically significant either. Interestingly, the science-based innovators have a higher rate of process innovation than the resource-based innovators.

Overall, the cluster analysis and the statistical results displayed in Table 5.3 suggest that the subset of science-based innovators in Norway is rather close to the traditional picture of high-tech industry, emphasizing R&D and product innovation. This group includes sectors associated with the "R&D intensive network-based" layer of the Norwegian innovation system discussed in Chapter 2, but includes other sectors not commonly associated with high technology, as was noted earlier. The resource-based group also contains active innovators, but they focus on process innovation. This group, which includes the sectors commonly associated with the resource-intensive segment of Norwegian industry, includes sectors corresponding to the "large-scale concentrated" industrialization path of development described in Chapter 2, but includes others that are associated with the "small-scale" path as well. Finally, the low-intensity innovators includes a number of sectors (e.g. pulp and paper, machinery) that are associated with the large-scale decentralized path of development, while including others associated with the small-scale industrialization path.

Table 5.3. Innovation performance in the three groups of sectors

	Average of science-based innovators	Difference from rest of sample	Average of the Resource based innovation group	Difference from low intensity innovators
Product and/or process innovators, as a fraction of all firms	0.61	0.26***	0.41	0.09*
Product innovators, as a fraction of all firms	0.57	0.26***	0.35	0.06
Process innovators, as a fraction of all firms	0.46	0.18***	0.37	0.14***
Innovative products, as a fraction of turnover	0.25	0.16***	0.06	−0.03
New to the market, products as a fraction of turnover	0.09	0.06***	0.02	−0.01
Strategic innovators, fraction of all firms	0.32	0.06*	0.30	0.05
Management innovators, fraction of all firms	0.19	0.01	0.20	0.03
Organizational innovators, fraction of all firms	0.49	0.17***	0.33	0.02
Marketing innovators, fraction of all firms	0.29	0.10***	0.22	0.03

Notes: One, two, and three stars denote significance at the 10%, 5% and 1% level, respectively.

The "three layers" of Norway's innovation system described in Chapter 2 thus influence intersectoral differences in innovation-related investments, activities, and outputs, but by no means provide a complete explanation for these intersectoral differences or the ways in which these intersectoral differences affect the contrasts in aggregate innovation-related activities between Norwegian and other European firms. Indeed, at least some sectors classified in Chapter 2 as part of the "small-scale decentralized" development path display innovative behavior more commonly associated with high-tech industry, underscoring the possibilities for long-established fields of Norwegian industry to shift to more knowledge-intensive trajectories. And as Chapter 9's discussion of biotechnology points out, important segments of this industry in Norway are based in sectors (fish farming, stock breeding) with deep roots in the "small-scale decentralized" path of Norwegian economic development.

INNOVATION AND INDUSTRIAL DYNAMICS

Our final empirical analysis examines market structure and market dynamics within Norway's industrial sectors, seeking to understand whether the "fingerprints" of Norwegian economic history are visible in contemporary market structure and dynamics, and how any such influence of the past affects innovation. Traditionally, economists have analyzed this issue by focusing on the relation between R&D on the one hand, and market concentration and firm size, on the other hand. This literature treated sectors as relatively homogenous entities that could be represented by simple measures of concentration and innovation.[9] But more recent reviews of this empirical literature have concluded that firm size and industry concentration are not clear-cut determinants of industrial innovation, and that the impact of firm size upon R&D seems to be conditioned by the industry context (Cohen and Levin 1989). A more complex theory of the relationships among market structure, dynamics, and innovation is needed.

We base our analysis on the premise that the empirical relationship between firm size and industry concentration upon technological change is conditioned by interindustry differences in technological opportunities and appropriability conditions (Winter 1984), suggesting that the nature of technology influences the pattern of industrial competition. This idea was pioneered in Nelson and Winter's (1982) evolutionary approach, and leads to the conclusion that innovation and market structure are jointly determined by technological conditions (in Nelson and Winter's discussion, a "technological regime") that may vary widely among sectors.

A key dimension of the technological context of a given industry is the ability of new firms to enter via innovations, as opposed to technological environments where incumbents can build on past innovative success. Nelson and Winter (1982) and Winter (1984) posit the existence of two technological regimes: an "entrepreneurial" regime associated with science based technology and relatively easy entry by innovative new firms due to a universal and noncumulative knowledge base, and a "routinized" regime that favors innovation by incumbents due to the cumulative nature of technology and the knowledge base. This work has been extended by other scholars (e.g. Breschi *et al.* 2000; Malerba and Orsenigo 1996; Marsili and Verspagen 2002; Van Dijk 2000).

This literature suggests that the three groups of sectors in the Norwegian innovation system that we have identified above may be characterized by different technological regimes and market dynamics. For example, the resource-based sector described above is likely to be characterized by a routinized innovation regime, where large firms are important, markets are relatively concentrated and entry is limited. It is more difficult to characterize the technological regime of the science-based sectors, since the empirical literature is ambiguous on this point. On the one hand, science-based sectors are often characterized by high levels of entry and relatively low entry barriers, at least for those skilled in the art (the early history of Apple is a case in point). But other science-based sectors such as pharmaceuticals are dominated by very large firms that possess large sales organizations that give them considerable market power.

We use data from the business register of the entire enterprise and business unit population in Norway to quantify market dynamics and market structure in the different sectors discussed above. This database contains information for 458,000 establishments belonging to some 415,000 enterprises, including self-employed entrepreneurs with no employees. Information at the business unit level, such as employment and turnover, was aggregated to the enterprise level in order to provide the closest possible match to the enterprises included in the combined R&D and innovation survey used for the empirical analyses of previous sections of this chapter. A novel feature of this database is that it not only includes information on entry and exit at the industry level, but also records changes within the business units of multi-establishment enterprises. Thus, we are able to provide a detailed picture of the industrial dynamics at the sectoral level and link the data on sectoral dynamics to underlying differences in the technology domain.

Our indicators of market structure and industrial dynamics can be classified as static or dynamic measures, and all refer to 2001. The static group includes an index of market concentration that measures the joint market share of the four largest firms in the industry. Higher values of this indicator mean that a sector is more highly concentrated. We also have data on firm size structure that measures the shares of industry sales accounted for by firms in various size classes (zero, 1–10, 11–50, 51–200 and >200 employees), capturing another static element of market structure.

Our dynamic indicators are related to market entry and exit. The primary measure of entry and exit is defined in the usual way: the number of new entrants or exits (both defined as firms, i.e. at the enterprise level, and based on the presence or absence of a firm in the business register in a given calendar year) as a percentage of all firms in the industry in 2001. Our establishment data also provide additional measures of market turbulence. First, we identify a specific subset of entrants and exits as those enterprises that have no employees other than the owner (i.e. self-employed entrepreneurs). We express this as a percentage of all firms in the industry (hence self-employed entry/exit rate is smaller, or in the limit equal to, the overall entry/exit rate). Although by definition, this type of entry will have very modest near-term effects on employment, or output, we include it as an indicator of the rate of small-scale

Box 5.1. Definitions of incumbent-based market dynamics variables

EVENT	DESCRIPTION
Takeover	An existing establishment is acquired by an enterprise that is not its original parent, and the former parent enterprise is closed.
Divestiture	An establishment moves from one enterprise to another, and the original owner remains active.
Expansion	The birth of a new establishment, owned by an existing enterprise.
Transformation	An existing establishment becomes an independent enterprise, and the former owner of the establishment goes out of business.

Note: in the definitions, establishments are business units that are owned by enterprises, i.e. one or multiple establishments make up an enterprise.

entry within a sector, a key element in the innovation regimes literature that we discussed above (e.g. Winter 1984). Finally, firm survival is measured as the percentage of firms in the industry that were in the industry 5 years ago.

The indicators for industrial dynamics (entrepreneurship) that we have discussed so far do not cover the activities of incumbent firms. We measure four distinct events that are related to the mobility of establishments and their relation to enterprises. These events are takeovers, divestitures, expansion, and transformation, all of which are defined in Box 5.1. We weight these events by the number of enterprises involved in an activity (e.g. when an enterprise takes over two distinct establishments, it is counted as a single takeover). We express the number of these events as a percentage of the total population within the industry in 2001.

Individual events in the box occur infrequently, and a combination of takeover, divestiture and expansion, i.e. when an enterprise combines all of these three forms in a single year, is the most frequent.[10] Based on this tendency for multiple events to occur simultaneously, we constructed a new variable that adds the individual rates for takeovers, divesting and expansion together with the combined one. We use a separate measure for transformation events, because their nature differs from the other events (e.g. one might expect a transformation to occur when a firm continues in a different form after a bankruptcy).

We use an ANOVA test to measure the statistical significance of differences in the means of these indicators among the three innovation groups. Table 5.4 provides results for all three groups relative to the rest of the sample. The summary statistics and ANOVA test results for the firm size indicators suggest

Table 5.4. Market structure and industrial dynamics in the Norwegian innovation system

	Average low intensity innovators	Difference to rest of sample	Average resource based innovators	Difference to rest of sample	Average high-tech innovators	Difference to rest of sample
0 employees	42.93	−0.37	36.55	−8.19*	45.51	3.08
1–10 employees	37.45	3.24	35.81	−0.23	33.93	−2.63
11–50 employees	13.95	0.77	16.01	3.01	12.12	−1.90
51–200 employees	4.26	−1.23	6.64	2.29	5.15	0.44
200 + employees	1.41	−2.39**	4.97	3.10**	3.28	1.01
C4	36.01	−15.13**	41.34	−1.85	63.39	26.26***
% exit by 2001	10.91	−0.23	10.03	−1.24	12.77	2.24**
% entry 2001	10.24	−0.80	10.57	−0.04	12.05	1.86
self-employed exit	5.04	0.08	3.30	−2.14**	6.79	2.28**
self-employed entry	4.26	0.08	2.66	−1.96*	5.64	1.80
Survival rate	62.90	−1.11	62.96	−0.55	63.78	0.48
Transformation	0.59	0.22*	0.49	0.00	0.28	−0.28*
Takeovers, divesting, expansion (combined)	2.20	−0.37	3.54	1.47**	1.99	−0.48

that only the two extreme size classes differ significantly across the three sectors. The population of firms in the *resource-based innovators* sector has a significantly lower share of firms with zero employees and a high share of large firms (>200 employees). These large firms account for a smaller share of the firms within the *low intensity innovators* sector. The shares of the very largest and very smallest firms in the *science-based innovators* sectors do not differ significantly from the shares of these firms in the overall sample of industries. The C4 concentration measure is lower than the average in the *low intensity innovators* (lower levels of average concentration are typically associated with lower entry barriers), and higher than average (indicating higher entry barriers) in the *science-based innovators* sectors; the resource-based sector does not display significantly higher or lower average levels of concentration. This finding for market concentration is surprising, since the scale-intensive nature of many of the sectors included among Norway's resource-based industries (e.g. aluminum, oil and gas) often is associated with higher entry barriers. One explanation for this result may be the inclusion of industries such as fish farming, which is not scale-intensive and historically not characterized by high entry barriers, in the resource-intensive sector. The finding that the science-based sector in Norwegian industry is characterized by above-average concentration is also surprising, and may reflect the lack of a dynamic ICT hardware sector, for reasons described in Chapter 10.

The *resource-based innovators* show significantly lower self-employed entry and exit than the other sectors, a result that is totally consistent with the lower share of self-employed firms in this group of industries. However, with regard to the overall entry rate, we do not find any significant differences between the three groups of sectors, which suggests that except for self-employed entrepreneurs, entry is not especially difficult in the resource-based sector. This finding is in line with our results for the static market concentration indicator (C4). The *science-based innovators* sector has a significantly higher exit rate than the overall sample, suggesting that survival in these markets was more difficult in 2001, perhaps because of the bursting of the dot-com "bubble" in that year. The science-based sector also displays higher levels of self-employed exit.

The results of the analysis of the market dynamics associated with incumbents indicate that the combined takeover, divestiture and expansion rate is significantly higher in the *resource-based innovation* group. Transformation, i.e. the "spinout" of an independent enterprise by a parent firm that ceases to exist, however, is higher in the *low intensity innovation* group, and lower in the *high-tech* group. Firms in the resource-intensive innovators group thus appear to be more active in restructuring activities. The exception to this characterization is the significantly higher levels of "transformation" events in the low innovation-intensity sector.

Overall, these findings suggest that Norway's *resource-based innovators* sector is dominated by large firms, but interestingly, this dominance is not associated with more highly concentrated markets. Self-employed entry and exit, but not the overall entry rate, appear to be lower than average in this sector. Market dynamics are dominated by transactions among incumbent firms that involve mergers, acquisitions, and restructuring, rather than entry by new firms. The results of our analysis of market dynamics in this sector thus are mixed. On the one hand, we have insider domination in market dynamics events, but overall entry and exit, as well as static measures of producer concentration are not unusually high within the Norwegian context.

Surprisingly, our analysis suggests that Norway's science-based innovators sector is characterized by relatively stable market structures and low levels of entry. Market concentration and exit rates are high relative to the average. Market dynamics are biased towards transformation, a type of structural change that often is insider-oriented, especially in small economies. The apparent lack of competition in the science-based sectors may be one lasting result of the protective measures employed by Norwegian policymakers in the unsuccessful "national champion" strategies employed in these sectors during the 1980s and early 1990s (see Chapters 4 and 10).

Our empirical comparison of these three multi-industry sectors within the Norwegian economy thus supports the conclusion that the market dynamics of these three sectors differ significantly. But the relationship between the technological regimes associated with these three sectors and observed market dynamics seem to contradict the standard views of evolutionary scholars of industrial dynamics. In particular, we do not find evidence that Norway's *science-based innovation* sector is particularly "entrepreneurial" when compared to the other parts of the Norwegian economy, nor do the *resource-based innovation* industries appear as particularly non-entrepreneurial. Perhaps the long-established industries of fish farming and metals are the locus of innovative activities that are not captured effectively in CIS or other innovation indicators, while the very small size and limited growth of Norway's science-based sector has tended to limit competition and innovation. Like the analysis of innovative behavior at the sectoral level summarized in the previous section, our measures of market structure and industry dynamics underscore the complexity of the interactions among historical development paths and current industry structure within Norway's national innovation system.

CONCLUSIONS

Our empirical analysis suggests that the three historical paths of Norwegian economic development identified in Chapter 2 have left "fingerprints" on the modern Norwegian innovation system, although these historical fingerprints might best be described as blurry. Much of Norwegian economic development during the twentieth century has relied on the growth of resource-intensive industries such as aluminum and offshore oil and gas. These two sectors in particular illustrate the "large-scale centralized" path of development highlighted in Chapter 2. But other resource-dependent sectors in Norway such as fish farming have maintained a relatively small-scale, decentralized industrial structure that is characterized by lower levels of investment in formal R&D and related activities. At the same time, however, important segments of Norway's "small-scale decentralized" industry have contributed to the growth of biotechnology and (through important innovations in marine electronics) Norway's historically significant radio and telecommunications industries. And by comparison with other high-income industrial economies, Norway in the early twenty-first century has a remarkably small industrial sector that can be described as relying on "R&D intensive, network-based" innovation.

As we noted earlier in this chapter, the prominence of resource-intensive industries in Norway distinguishes this economy and its innovation system from those of most other high-income industrial economies. Our analysis suggests that innovation in Norway's resource-intensive sectors differs from innovation in science-based sectors. The two groups of sectors display different blends of innovation-related activities, such as R&D, acquiring external knowledge in different forms, learning, the use of human resources, etc. Moreover, the anomalous nature within Europe of Norway's specialization in resource-intensive industries means that the innovation processes that characterize these sectors have received relatively little attention from scholars. Almost by default, the science-based activities that are the primary source of innovation in most highly developed economies have been taken as a benchmark for comparative analysis and evaluation of the Norwegian innovation system (arguably, including the conflation of knowledge-intensive industry with the "R&D intensive, network-based" developmental path discussed in Chapter 2). But this benchmark may be an inappropriate basis for comparison, in light of the differences we have outlined between science-based and resource-based innovation (see also Grønning, Moen, and Olsen 2008).

It is also likely that in spite of the fact that the CIS data used in this chapter seek to capture a broader array of innovation-related activities than

R&D alone, they still underestimate innovation in resource-based sectors. For example, learning-by-doing and engineering-based activities such as the design of large process plants in oil refining or basic metals are not captured by the Frascati manual definition of R&D and may not be captured by the design category in the CIS innovation expenditures question. We need more in-depth research on innovation in resource-intensive sectors in order to be able to assess the extent of any inaccuracy in the CIS picture of innovation in these sectors.

It is also possible that the definitions of innovation in the CIS are inappropriate for the resource-based innovation sectors in Norway. The CIS uses a binary variable that does not provide any information about the size or extent of the innovation, and it is possible that innovations in Norway's resource-based sector are "lumpy," i.e. a single innovation has far-reaching effects on output and productivity. Such a result means that the elasticity of productivity change with respect to innovation is higher in Norway than elsewhere in Europe, a speculative conclusion that is consistent with the data on Norway's recent productivity performance.

Although Norway's overall economic performance has remained strong during the period since 2001 (the extraordinary rise in the price of oil during 2005–08 obviously is an important but by no means the sole factor in this strong performance), Norwegian innovation output appears to be stagnant or declining slowly, as the percentage of firms reporting either a product or process innovation (or both) has declined from 33 percent in 2000 to 32 percent in 2004. Moreover, data from the most recent CIS (not reproduced here because of space limitations) suggest that the science-based and resource intensive sectors are solely responsible for this stagnation. The decline in innovation output is even greater for process innovation than for product innovation within both the resource-intensive and science-based sectors.

The innovative performance of the sectors that have been leaders in Norwegian innovation thus appear to be weakening. Given recent developments in international markets for oil and gas, these indicators of declining innovative performance within Norwegian industry are not a cause for concern over the nation's near-term prospects. But they are disquieting, and suggest that the long-term challenges faced by Norway could be substantial. Moreover, the track record of Norway's previous policies seeking to spur the growth of high-tech industry provides little basis for confidence that the financial windfalls generated by the nation's oil and gas industry can be easily translated into successful pursuit of an R&D-intensive trajectory of development.

The OECD's recent *Economic Survey of Norway* (2007) suggested (perhaps because of the lack of any plausible alternative explanation) a putative lack of entrepreneurship within Norway as a reason for low levels of measured innovation: "Entrepreneurs are thin on the ground, perhaps because the risk of failure is as great as anywhere else, whereas the rewards of success are heavily taxed. A possible reason for the unimpressive level of innovatory activity in the business sector (excepting, as always, the petroleum sector) is too-weak competition between firms. Virtually all recent empirical studies find that productivity growth and innovation activity are positively associated with the strength of competition, which other studies find to be below average in Norway" (p. 131). But our analysis of industrial structure and dynamics indicates that the most R&D-intensive and innovation-intensive parts of the Norwegian economy, i.e. the science-based sectors, show the highest market concentration and relatively low levels of entry. The unusual characteristics of market dynamics in Norway's science-based sectors may reflect the legacy of now-discarded Norwegian government "national champion" policies in sectors such as ICT, although it remains difficult to understand how the effects of policies abandoned more than a decade ago could continue to influence these sectors. Perhaps at least some of the "fingerprints" of past developments within Norway's innovation system are themselves relatively recent.

APPENDIX

Sectors used in the analysis

1 Fish-farming
2 Mining and quarrying, exc. oil and gas
3 Extraction of crude oil and natural gas, services delivered to this sector
4 Meat products
5 Fish products
6 Bakery products
7 Other food products, beverages, tobacco
8 Spinning, weaving and textiles
9 Carpets, rugs, rope, etc.
10 Knitted fabrics and products, other wearing apparel
11 Fur, leather, articles thereof
12 Saw mills
13 Wood, wooden products
14 Builders, carpentry
15 Pulp, paper, paper products
16 Publishing

17 Printing
18 Basic chemicals
19 Other chemicals
20 Pharmaceuticals
21 Rubber and plastic products
22 Glass, glass products, ceramics, other minerals products
23 Bricks, cement, concrete, stone
24 Ferrous basic metals
25 Non-ferrous basic metals
26 Casting, forging, pressing of metals, tanks, containers, boilers
27 Structural metal products
28 Treatment and coating of metals
29 Cutlery, tools, crafts, other metal products
30 Power generating equipment
31 Other general purpose machinery
32 Agriculture and forestry machinery
33 Machine tools and other special purpose machinery
34 Domestic appliances
35 Computers and office machines
36 Electric motors, generators, electrical distribution
37 Wire and cable, batteries, other electric equipment
38 Electronic valves, tubes, etc.
39 Radio and TV transmitters and receivers
40 Instruments, medical equipment
41 Motor vehicles, parts
42 Ships and boats
43 Other transportation equipment
44 Furniture
45 Other manufacturing
46 Waste metal and scrap, recycling thereof
47 Production and distribution of electricity
48 Steam, hot water, purification and distribution of water
49 Building and construction of buildings, installation, finishing
50 Building of roads, other building
51 Wholesale
52 Land transport
53 Water transport
54 Air transport
55 Transportation support, cargo storage
56 Transportation and travel agents
57 Telecom services
58 Financial services
59 Computer services
60 R&D, engineering, architecture, and design

NOTES

Funding from the Norwegian Research Council, P. M. Røwde's Foundation and the Ruhrgas Foundation is greatly acknowledged.

1. In addition to Norway, we have data for Australia, Belgium, Canada, Denmark, Finland, France, Germany, Greece, Italy, Japan, the Netherlands, Sweden, the United Kingdom, and the United States.
2. Note that the shares of individual sectors within manufacturing sum to 100 percent, in spite of the fact that the share of Norwegian GDP accounted for by manufacturing falls throughout the 1970–2003 period.
3. We do not present the detailed data on economic structure for our reference countries.
4. The CIS is carried out under the supervision of Eurostat, the statistical agency of the European Commission, and applies a harmonized questionnaire and sampling methodology across the European countries that implement it. The survey is aimed to capture a broad class of innovation activities by firms, going well beyond traditional R&D investment and patent measures.
5. These data refer to activities undertaken by firms, and therefore include intrafirm innovation-related activities that are financed by government, e.g. subsidies.
6. The CIS survey was combined with the R&D survey in Norway, and is answered by the firm manager. The combined survey was directed to a representative sample of the Norwegian enterprise population with 10 employees or more. There was a random selection of units with 10 to 49 employees. All units with 50 employees or more were included in the sampling frame. The survey was further stratified according to industrial sectors using two digits NACE and size classes. In total, the questionnaire was returned by 3,899 firms, which constitutes a 93 percent response rate, minimizing problems like sample selection.
7. The share of process R&D is equal to one minus the share of product R&D in total R&D. The sum of the shares of basic and applied R&D is equal to one minus the share of development R&D.
8. The effects that we single out as "high scores" in this paragraph are tested using various post-hoc tests in a one-way ANOVA analysis. We only report the effects that turn up significant.
9. This literature revolves around the so-called Schumpeterian Hypotheses of increasing returns to firm size and market concentration in innovation, see Kamien and Schwartz (1982); Cohen (1995); Cohen and Levin (1989); and Van Cayseele (1998) for reviews. We regard the recent contribution by Aghion *et al.* (2005) and the large literature that followed it, as a repetition of this earlier work.
10. We do not know whether this reflects the sequential occurrence of these three events in a single year for a given incumbent enterprise, or whether they occur simultaneously in a single transaction.

REFERENCES

Aghion, P., N. Bloom, R. Blundell, R. Griffith, and P. Howitt (2005) "Competition and Innovation: An Inverted-U Relationship," *Quarterly Journal of Economics* 120: 701–28.

Breschi, S., F. Malerba, and L. Orsenigo (2000) "Technological Regimes and Schumpeterian Patterns of Innovation," *Economic Journal* 110: 388–410.

Cohen, W. M. (1995) "Empirical Studies of Innovative Activity," in P. Stoneman (ed.), *Handbook of the Economics of Innovation and Technological Change*, Oxford: Blackwell.

Cohen, W. M. and R. C., Levin (1989) "Empirical Studies of Innovation and Market Structure", in R. Schmalensee and R. D Willig (eds.), *Handbook of Industrial Organization*, New York: North-Holland.

Grønning, T., S. E. Moen, and D. S. Olsen (2008) "Low Innovation Intensity, High Growth and Specialized Trajectories: Norway," in C. Edquist and L. Hommen (eds.), *Small Country Innovation Systems. Globalization, Change and Policy in Asia and Europe*, Aldershot: Edward Elgar.

Kamien, M. I. and N. L. Schwartz (1982) *Market Structure and Innovation*, Cambridge: Cambridge University Press.

Malerba, F. and L. Orsenigo (1996) "The Dynamics and Evolution of Industries," *Industrial and Corporate Change* 5: 51–87.

Marsili, O. and B. Verspagen (2002) "Technology and the Dynamics of Industrial Structures: An Empirical Mapping of Dutch Manufacturing," *Industrial and Corporate Change* 11: 791–815.

Nelson, R. R. and S. G. Winter (1982) *An Evolutionary Theory of Economic Change*, Cambridge, MA: Harvard University Press.

OECD (2007) OECD *Economic Surveys*, Norway, Paris: OECD.

Pavitt, K. (1984) "Patterns of Technical Change: Towards A Taxonomy and a Theory," *Research Policy* 13: 343–73.

Van Cayseele, P. (1998) "Market Structure and Innovation: A Survey of the Last Twenty Years," *De Economist* 146: 391–417.

Van Dijk, M. (2000) "Technological Regimes and Industrial Dynamics: The Evidence from Dutch Manufacturing," *Industrial and Corporate Change* 9: 173–94.

Winter, S. G. (1984) "Schumpeterian Competition in Alternative Technological Regimes," *Journal of Economic Behavior and Organization* 5: 287–320.

Part II

Sectoral Innovation Systems in Norway—Old and New Paths

6

Innovation and Production in the Norwegian Aluminum Industry

Svein Erik Moen

PATHS OF INNOVATION AND PRODUCTION

The emergence of the process industries in early twentieth-century Norway has been analyzed as a process of "path creation" in the Norwegian innovation system (NIS) that included the entry of new social groups, such as scientists, engineers and managers, into prominent economic roles (see Chapter 2). Aluminum production, one of the most economically and technologically significant of these process industries, began in Norway in 1908, initiated by foreign industrializts who wanted to take advantage of the country's close proximity to European markets, the nation's well-developed infrastructure, Norway's political stability, and most importantly, comparatively inexpensive hydroelectric power.

The factors that initially attracted foreign entry have been supplemented by additional institutional and knowledge-based capabilities that have maintained Norway's strength in aluminum for more than a century. The companies have built large-scale R&D laboratories at their aluminum smelters; research institutes and universities have entered the innovation system and developed strong research capabilities in aluminum, functioning as important recruitment pools for the aluminum companies and the research institute sector; R&D collaborations between firms and institutes have gradually emerged; and supporting industrial and regional policies have been set up by the Norwegian government. Earlier studies of this industry have described it as one of the major clusters in Norway, with relatively cheap energy supply and high competence (see Reve *et al.* 1992; Svendsen and Rikter-Svendsen 1992).[1]

This chapter describes the main drivers of innovation and production in aluminum from a contemporary and a historical point of view. It draws on various sources of evidence and ideas (Ragin 1994: 56–76), including interviews with present and former directors, plant managers and engineers, and representatives of industry federations. We also use public and company reports and basic statistics, as well as a large secondary literature consisting of scientific articles, books, newspapers, and magazines.

In the aluminum industry, innovation intimately links to production, and we accordingly focus on both the industry's production system and its innovation system. The industry's structure and performance are influenced by the co-evolution of actors (e.g. firms, organizations, individuals) and networks[2] (e.g. innovation and R&D collaboration), technologies and knowledge (technologies and knowledge applied for the innovations and in the production process of primary aluminum), and institutions (e.g. laws, regulations, standards, norms) (Malerba 2002). Furthermore, this co-evolution has taken place both nationally and abroad. For example, aluminum plants in Norway have historically depended on foreign companies, organizations, and institutions in order to innovate, produce and survive. Thus, the factors that have worked towards this industry's persistence in Norway over time have been embedded in an innovation system at an international level, which had, and still has, a deep impact on the path of innovation and production in Norwegian aluminum. This chapter accordingly presents an international perspective on the changing production and innovation systems for Norwegian aluminum, and discusses the nature of change in these linkages through time.

PRODUCTION, VERTICAL CHAINS, AND COST STRUCTURE

The production of aluminum involves technical processes that constitute a vertical chain extending from "upstream" to "downstream" activities. The primary aluminum smelters are vertically linked to processes that take place "upstream," e.g. mining and alumina refining where bauxite is converted into alumina by the Bayer (refining) process. During primary smelting, alumina is processed into primary metal by the Hall–Héroult process. The stages "downstream" include the production of semi-fabricated goods such as sheet, foil, wire, rod, and bar, and the end-product stage, which uses aluminum to manufacture products that range from aircrafts, automobiles, and ships to packaging. The smelting stage itself spans both upstream and downstream categories. Although most primary smelters have semi-fabrication plants attached to them, smelting and fabricating are quite distinct processes. Smelting also includes the creation of alloys, which give different properties to the metal (Stuckey 1983).

This chapter focuses on primary aluminum smelting at Norwegian plants, which, as was mentioned earlier, differs significantly from the production of alumina or aluminum products. Over 90 percent of the Norwegian production of aluminum is exported, and this accounted for nearly six percent of all Norwegian export commodities by value in 2005. Norway is the fifth largest aluminum exporter in the world (UNCTAD/WTO, 2006) and aluminum exports account for a larger share by value of Norwegian exports than fish (4, 3 percent) (Statistics Norway, 2006, SITC). There are seven aluminum smelters in Norway, operated by Hydro aluminum, Elkem aluminum ANS, and Sør-Norge aluminum (Sør-Al).[3] Presently, Elkem aluminum ANS is owned by the Norwegian company Orkla ASA and Norsk Alcoa, a subsidiary of the Aluminum Company of America (Alcoa). Sør-Al is owned by Hydro aluminum and Aluminum Limited of Canada (Alcan). Hydro aluminum operates Sunndal, Karmøy, Årdal and Høyanger, Elkem operates Lista and Mosjøen, and Sør-Al operates at Husnes. The seven Norwegian smelters produced about 1.4 million metric tons in 2005. Figure 6.1 shows the production of primary aluminum at these smelters, some of which have been active for more than sixty years.

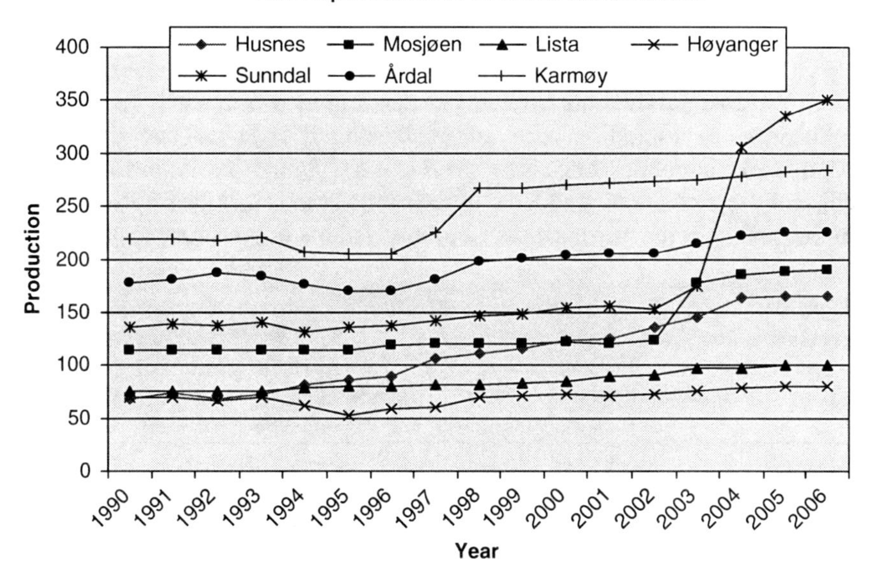

Figure 6.1. The Norwegian aluminum industry, production 1990–2005.
Source: PIL 2005.

The price of aluminum smelters' output fluctuates widely, and the competitiveness of smelters is determined primarily by their operating costs. A central driver of innovation in primary aluminum production is cost reduction through process innovation aimed at reducing energy and other costs. More recently, environmental performance has become a second important focus of innovation, which seeks to reduce or eliminate emissions. In addition, plants seek to improve the purity of the metal and to add alloys in order to obtain different qualities and properties for different uses. In Norway, aluminum oxide (alumina) and carbon prices are determined by world markets, and energy prices depend on regional and national price mechanisms (see Table 6.1). Agreements with the Norwegian state on energy prices have benefited (some would say subsidised) the aluminum industry, enabling long term investments in production capacity. Other industrialized countries have pursued similarly supportive policies for domestic aluminum firms, but since the 1970s, international aluminum companies have expanded their production investments in "energy rich" parts of the world, often in developing countries (see below for further discussion).

As the previous paragraph's discussion noted, operating costs vary from smelter to smelter and are heavily influenced by the price of energy. Norway's low energy costs are an important factor in the growth of the nation's aluminum industry. The Norwegian low-cost energy advantage now is threatened, however, by increasing domestic energy prices and the emergence of new, low cost production sites in the developing world. Other important influences on production costs are establishment scale and the vintage of plant-level technology, since older production technologies are normally less efficient. Important cost-reducing process innovations include improvements in the quality of the raw materials, and innovations in production machinery

Table 6.1. World average smelter cost structure and Hydro Aluminum's smelter cost structure in 2005 (%)

	World Average	Hydro	Markets influencing the smelter production cost at Norwegian smelters
Alumina	40	33	World market
Carbon	10	9	World market
Energy	26	28	Regional and national markets
Casthouse	4	7	Local markets
Labour	7	10	Local markets
Other	13	13	Local markets

Source: Hydro 2005a/Hydro Media 2005c/Grimsrud and Kvinge 2006.

(e.g. cells, anodes, cathodes, transmission lines for power, computer control). These and other process innovations may affect or be affected by work organization at a smelter, as well as the competence of the engineers and operators at the plants (how technology is used), which in turn are affected by the R&D and innovation networks of firms (e.g. R&D institutes and educational organizations).[4] The survival of several long-established operators in the Norwegian aluminum industry reflects their ability to develop new process innovations as well as adopting others, combined with public policies that have maintained low energy costs. These firms have also developed strategies to ensure reliable access to alumina supplies and, more recently, integration downstream into aluminum products.

GLOBAL SOURCES OF ALUMINUM INNOVATION AND PRODUCTION (1908–1945)

Aluminum has been produced commercially for roughly 150 years. Until the end of the nineteenth century, aluminum was produced exclusively in Europe and was regarded as a precious metal, because of its very high production costs. The market was small, and aluminum competed with metals like gold and silver. From 1855–85, about 40,000 kilos of aluminum were produced, mostly in England and France (Edwards *et al.* 1930: 6; Kollenborg 1962: 11). At the end of the nineteenth century, however, a radical technological shift occurred that was to lay the foundation for the Norwegian aluminum industry.

Two new processes, electrolysis and the Bayer process, revolutionized the aluminum industry in the late nineteenth century. Paul Louis Toussaint Héroult (France) and Charles Martin Hall (USA), unaware of each other's work, each invented a new electrolytic process in 1886 for aluminum smelting. Their processes dissolved aluminum oxide (alumina) in a bath of molten cryolite and passed a powerful electric current through it, resulting in a deposit of molten aluminum at the bottom of the bath. These processes were patented, and only a handful of companies acquired the patents. Innovations in mineral extraction also affected the aluminum industry by making raw materials cheaper and more accessible. One process developed by Karl Joseph Bayer (Germany) between 1887 and 1892 improved the extraction of alumina from bauxite ore (Edwards *et al.* 1930). Firms seeking to enter the production of aluminum had to gain access to the Hall or Héroult and Bayer patents, as well as acquiring bauxite deposits, in order to become self-sufficient "upstream" (Smith 1988: 99–101).

In 1889, Hall began aluminum production in collaboration with some American industrializts, who founded the Pittsburgh Reduction Company in the United States (renamed the Aluminum Company of America—Alcoa in 1907). The cheapness of the electrolysis process displaced the European smelters using the old production process. In order to maintain profit, the US company sought new production sites with cheap power (Graham 1982: 16). In the 1890s, Alcoa moved production to New Kensington, outside of Pittsburgh, which had supplies of natural gas, and to Niagara Falls, to take advantage of the hydroelectric power provided by the Niagara River Power Company. Inexpensive energy, combined with process innovations, caused aluminum prices to fall, making aluminum competitive with other metals. The process innovations developed by Hall at Alcoa and by Héroult in Europe improved the efficiency of the cells used for aluminum smelting. Innovations in the carbon electrode technology enabled producers to exploit larger cells (which are the "pots" in which aluminum is produced) and increase cell amperage (Peterson and Miller 1986), further enhancing production efficiency. Alcoa purchased equity in US automobile and electricity transmission firms, and developed a vertically integrated structure that included fabrication operations and control over bauxite deposits (Graham 1982: 16).

The European aluminum companies pursued similar strategies of innovation and integration. After the radical technological shift that took place around 1890, four companies managed to enter the aluminum industry based on the Héroult patent. Héroult cells were first operated commercially at Neuhausen in 1888 by the Swiss company aluminiumindustrie*A.G* (AIAG), where Héroult had taken his smelting process in order to take advantage of the potential for inexpensive hydroelectric power in this area. AIAG became the largest aluminum company in Europe, with smelters in Italy, Norway, Spain, and Great Britain (Graham 1982; Holloway 1988; Tresselt 1968). Héroult founded the French company, Société Electrometallurgique Française (Froges), in 1889, with the financial support of AIAG. Froges later merged with Pechiney[5] (Rinde 1996: 5; Singer *et al.* 1958: 93; Tresselt 1968: 7–8). In the UK, the British Aluminum Co. Ltd. (BACO) was formed in 1894, acquiring the British and colonial rights for the Bayer and Héroult patents for alumina extraction and aluminum reduction (Cailluet 2001).

Following the expiration of Héroult's European patent in 1906, seven new firms began to construct plant facilities, but by 1910, the larger established companies had purchased three of these. The other four remained independent and relatively small (Wallace 1937: 38–9). In response to the resulting price competition, Alcoa (US) and the largest European companies established international bauxite-aluminum cartels and acquired most of the new entrants, thus regulating price, quantities of production of aluminum and bauxite supplies. In the US, Alcoa managed to secure an additional three years of patent protection beyond the original expiration date of 1906, which, combined with the firm's acquisitions, exploration activities, purchases of riparian rights, bauxite and hydroelectric resources, established Alcoa as a monopolist within the US for half a century (Barham 1994; Smith 1988).

Until the First World War, Alcoa, Alcan,[6] AIAG, Pechiney, and BACO dominated the global aluminum industry. With the expiration of their patents on electrolysis, these firms' dominance[7] rested on their development and control of a vertically integrated system of technologies and knowledge, networks and institutions in combination with large investments to strengthen in-house research and engineering capabilities (Le Roux 2002; Smith 1988: 40–2). In sum, patents, technological monopoly, economies of scale, cartels, and vertical integration provided important barriers to entry and this had substantial effects on innovation and production in the first phase of development of the Norwegian aluminum industry.

An important precondition for the establishment of the Norwegian aluminum industry at the beginning of the twentieth century was the construction of the infrastructure for hydroelectric power generation and transmission that would be used by foreign aluminum companies (Rinde 1996: 9–10). Large investments in Norway's hydropower produced excess capacity and lower prices for hydroelectric power from 1905 (Thue 2006: 74). The resulting inexpensive electric power offset the disadvantages resulting from Norway's lack of domestic demand and related industries up and downstream, and attracted foreign aluminum firms. Although Norway had considerable competence in hydropower development and technology, the nation lacked the knowledge base and the institutions that could support entry by domestic firms into the aluminum industry (Svendsen and Rikter-Svendsen 1992: 2–11). As early as 1906, during the construction of the first Norwegian aluminum plant, an influential engineer at Norsk Hydro expressed concern that the Norwegian aluminum industry would "become controlled by a cartel who disposes the bauxite" (Andersen and Yttri 1997: 58). At that time, Hydro sought to enter the aluminum industry, but its lack of access to the raw materials led corporate advisors to recommend against entering the industry (Andersen and Yttri 1997: 58–60).

International developments proved to be key factors in the foundation of the Norwegian aluminum industry, reflecting (among other things) the control of Norway's aluminum industry by a cartel of large foreign companies. Norwegian domestic production facilities depended on these foreign companies for knowledge, technology and access to markets. Bauxite mining, alumina refining, and electrolysis were technically complex processes in which Norwegian firms had little knowledge and no experience (Rinde 1996: 15). Norwegian actors lacked organizational resources and skills, entrepreneurial competence, venture capital, access to markets and distribution channels, knowledge, and technology, all of which help explain why the development of the first Norwegian aluminum smelters took place under foreign ownership. Table 6.2 summarizes information on the aluminum companies that have operated in Norway, and their ownership, from 1908 to present.

Stongfjorden aluminum, the first aluminum plant in Norway, was initiated, built, and run by the British Aluminum Company (BACO), which acquired riparian rights in Stongfjorden from the Norwegian Ministry of Domestic

Table 6.2. Companies operating at primary aluminum plants in Norway during the three phases

Plant locations	Operating companies and their ownership*	Production start—plant closure
Stongfjorden	*Stangefjordens Elektrokjemiske fabrikker*	1908–1945
	1906–1945: British Aluminum Company (BACO)	
Vigelands Brug	*AS Vigeland Brug*	1908–1945**
	1908–1912: Anglo—Norwegian Aluminum Company	
	1912–1945: BACO	
Eydehavn	*Det Norske Nitridaktieselskap (DNN)*	1914–1975
	1912–1913: Elektrokjemisk A/S + SGN	
	1913–1918: SGN + Pechiney	
	1923–1958: Pechiney + Alcan + BACO	
	1958–1976: Alcan + BACO	
Tyssedal	*Det Norske Nitridaktieselskap (DNN)*	1916–1981
	1912–1923: Pechiney	
	1923–1958: Pechiney + Alcan + BACO	
	1958–1976: Alcan + BACO	
	1976–1981: The Norwegian State	
Høyanger	*A/S Høyangfallene Norsk Aluminum Co:*	1917–still active
	1915–1923: Norsk Aluminum Co (NACO)	
	1923–1967: Alcoa/NACO***	
	Årdal og Sunndal Verk (ÅSV)	
	1967–1986: The Norwegian State *Hydro Aluminum*	
	Høyanger	
	1986–Hydro Aluminum	
Glomfjord	*Haugvik smelte verk*	1927–1942
	1926–1931: International Aluminum Company	
	1931–1940: Allicance Aluminum Clie (cartel: Pechiney, Ugine, BACO, Alcan, AIAG, VAW, aluminiumwerke Bitterfeld.	

Table 6.2. (*Continued*)

Plant locations	Operating companies and their ownership*	Production start—plant closure
Årdal	*A/S Årdal Verk*	1948–still active
	1946–1951: the Norwegian state	
	Årdal og Sunndal Verk (ÅSV)	
	1951–1966: the Norwegian state	
	1966–1979: Alcan + the Norwegian State	
	1979–1986: the Norwegian State	
	Hydro Aluminum Årdal	
	1986– Hydro Aluminum	
Sunndal	*Årdal og Sunndal Verk (ÅSV)*	1954–still active
	1951–1966: the Norwegian State	
	1966–1979: Alcan + the Norwegian State	
	1979–1986: the Norwegian State	
	Hydro Aluminum Sunndal	
	1986– Norsk Hydro****	
Mosjøen	*Mosal/Elkem aluminum Mosjøen*	1958–still active
	1957–1963: Elkem + AIAG	
	1963– Elkem + Alcoa	
Husnes	*Sør-Norge aluminum (Sør-Al)*	1966–still active
	1962–1976: AIAG/Compadec (later Alusuisse) + Den Norske Creditbank (DnC)	
	1976–2000 Hydo + Alusuisse	
	2000– Hydro aluminum + Alcan	
Karmøy	*Alnor/Hydro al*	1967–still active
	1963–1973: Harvey + Hydro	
	1973– Hydro aluminum	
Lista	*Elkem aluminum Lista*	1971–still active
	1962– Alcoa + Elkem	

* During Second World War Germany took control over the industry and founded Nordische Aluminiumgesellschaft, in 1941 two companies—Nordisk Lettmetall and Nordag were supposed to run the existing plants, and build new projects: the "Koppenberg Plan", however, although new projects where initiated, larger capacity was not achieved during the war.

** Rebuilt after the Second World War as an aluminum refinery producing super-pure aluminum, presently owned by Hydro Aluminum and Alcan.

*** In 1928 Alcoa divested itself from its subsidiaries outside the US, including the Aluminum Company of Canada (Alcan), and transferred these companies into a single company that organized foreign plants.

**** Hydro Oil merged with Statoil in 2007, divesting itself from Hydro Aluminum.

Source: Author's construction based upon Tresselt 1968; Innvik and Kamsvåg 1996; Rinde 1996; Sogner 2003; Lie 2005; Fasting 1966; Kollenborg 1962; Tjelmeland 1987.

Affairs and began operation of its smelters in 1908 (Tresselt 1968: 8–9). In 1945 (see Table 6.2), the plant, which by then was controlled by the German occupation authorities, halted production because of a lack of bauxite supplies (Kollenborg 1962: 110). Another British company, the Anglo Norwegian

Company, initiated the second aluminum plant in Norway, Vigeland Brug, which started production in 1908. The company was based in London and used Héroult's patent. The Anglo Norwegian Company hired two Swiss engineering companies (Th. Bell and Cie Oerlikon, and Maschienenfabrik Oerlikon) to build the plant and related infrastructure, providing for most of the machinery. However, hydropower turbines were supplied by the Norwegian engineering company Moss mek. Verksted (Krogstad 1999: 71). Swiss engineers were in charge of plant management, and alumina was imported from the bauxite company Giulini, which did not participate in the international bauxite-aluminum cartel. The fall in the price of aluminum from £77 per metric ton in 1910 to £53 in 1911, combined with technical problems in the smelters, a malfunctioning sales organization, and labour unrest, led to heavy debt and, eventually, the acquisition of the plant by BACO in 1912 (Krogstad 1999: 15–85; see also Tresselt 1968: 8–9).[8]

Actieselskapet Norsk Aluminum Company (NACO) completed the third Norwegian aluminum plant, Høyanger, in 1917. Høyanger started as a Norwegian project, relying primarily on Norwegian actors and capital (Andersen and Yttri 1997: 58). NACO's strategy from the beginning was to establish a vertically integrated aluminum company. Bauxite supplies came from France, through NACO's bauxite-mining subsidiary, Société Anonyme des Bauxites et Alumines Province. A plant for the manufacturing of aluminum products (such as cans), Nordisk Aluminum (at Holmestrand), began production in 1920. However, NACO's staff had little experience in the aluminum industry, and found that it was nearly impossible to hire Norwegian staff with sufficient knowledge about alumina refining and aluminum smelting. Most of the firm's machinery was imported from the US, and its engineers and workers were hired from Europe and the US (Fasting 1966). NACO's bauxite-mining project collapsed shortly after the firm's foundation; Norway's neutrality during the First World War led French government authorities to block the firm's access to French bauxite after the outbreak of war. NACO was also crippled by problems in building and running its French alumina refinery, and eventually went bankrupt. Alcoa acquired NACO in 1923 (Fasting 1966: 109–41).

Det Norske Nitridaktieselskap (DNN) was founded by the French company Pechiney, along with other French investors and the Norwegian-owned firm, Elektrokjemisk, in 1912. Elektrokjemisk had been founded by the Norwegian entrepreneur Sam Eyde, Swedish capitalists (Wallenberg), and Knut Tillberg in 1904 in order to exploit the electric power and electrochemical innovations in hydropower and metals (Sogner 2003). DNN was originally intended to produce nitrate, but Elektrokjemisk sold its DNN share to SGN in 1913, (Kollenborg, 1962) and Pechiney built aluminum smelters instead, preventing Elektrokjemisk from entering the aluminum industry. Primary aluminum plants were built at Eydehavn, where production started in 1914, and in Tyssedal, which opened in 1916. These two plants made DNN the largest manufacturer of aluminum in Norway until 1947 (Rinde 1996: 7; Tresselt 1968: 9–10). Pechiney financed the development projects, and aluminum oxide was imported from Pechiney's plants in France (Kollenborg 1962). The British International aluminum Company Ltd. built another plant—Haugvik Smelteverk—in Glomfjord in 1926. In 1932, shares of the company were bought by a cartel consisting of Alcan Pechiney and VAW, AIAG, and BACO.

Elektrokjemisk controlled the patents on the Söderberg smelting technology, which was widely used in the metals industry and was one of the two leading technologies in the world's primary aluminum industry by the 1930s. The Söderberg technology uses a continuous anode, which is delivered to the cell in the form of a paste, and which bakes in the cell itself. By 1950, the Söderberg technology accounted for about 50 percent of the global primary aluminum production capacity (Petersen 1953: 211). Although the Söderberg technology had a major impact on the world aluminum industry from 1935 onward, Norwegian government policy provided little institutional support for Norwegian firms to enter the primary aluminum industry. The multinationals that dominated Norwegian aluminum production also controlled access to bauxite and downstream distribution.

Since Elektrokjemisk had become a specialized technology supplier by the 1930s, providing global aluminum producers and other metals companies with the Söderberg smelting technology, its entry into aluminum production would have created conflicts with the firm's customers (especially Alcoa) (Petersen 1953: 79–94; Sogner 2003: 127–8). Elektrokjemisk also did not have the finances, networks or market access to produce and sell aluminum. The early international entrants had maintained their innovative advantages while at the same time controlling production and sales. Even though the patents on electrolysis had expired in the early twentieth century, these early entrants further developed their technology, expanded production capacity, and integrated vertically, raising barriers to entry by other firms. The cartels organized by the industrial leaders further limited entry for Norwegian and other firms during the pre-1945 period. Although a significant knowledge base in primary aluminum production had developed in Norway during the 1908–45 period (e.g. Elektrokjemisk's Söderberg technology; Wulff 1992), production and distribution remained under the control of companies like BACO, Alcoa, Alcan and Pechiney. However, the explosion in demand for aluminum during the Second World War led to vast investments in the Norwegian metals industry; this greatly expanded production capacity and created the preconditions for the growth of a more autonomous Norwegian aluminum industry after 1945.

CREATING A PATH OF AUTONOMOUS DEVELOPMENT (1945–1986)

Wars often affect the evolution of industries. Its occupation of Norway meant that Germany acquired control of the Norwegian aluminum industry. Hermann Göring, head of the Luftwaffe, needed aluminum for military aircraft production, and he established the Nordische Aluminum Gesellschaft (Nordag), financed by Germany, and supported technologically by the German company IG Farben and Norsk Hydro. IG Farben and Nordag collaborated through a sister company of Nordag, AS Nordisk Lettmetall, in developing the Norwegian aluminum industry (Andersen 2005: 365–98). Germany invested 8–9 billion NOK (1997 prices) in R&D, aluminum plants, one alumina plant, and hydropower plants (Rinde 1997: 77–9).

Wartime requirements also produced a significant expansion in US aluminum production capacity and weakened the long-established monopoly within the US domestic industry, since Alcoa could not meet the allied demand for aluminum. The Federal Government financed the necessary investments in production capacity, and in 1945, the War Surplus Property Board sold the government-owned plants to independent firms, forcing Alcoa to face competition from integrated domestic producers of aluminum such as Kaiser (1946) and Reynolds (a metal company established in 1928) (Smith 1988: 192). The new companies needed more production capacity, and invested in production sites in Norway.

The wartime mobilization and demobilization policies in Germany and the US illustrate the importance of a broad array of policies, including competition policy and wartime mobilization strategies, in addition to global flows of technology and capital, in the evolution of Norway's aluminum industry. The postwar restructuring of Norway's aluminum industry almost certainly could not have occurred without significant change in the international structure of the industry. However, the involvement of the Norwegian state in the aluminum industry after 1945 also proved to be important in the industry's development: In the aftermath of the war, the Norwegian government took over all enemy property, finished the German aluminum projects at Årdal (1948) and Sunndalsøra (1954), and established the state-owned company Årdal og Sunndal Verk (ÅSV).

The postwar Norwegian government, controlled during this period by the Norwegian Labour Party, pursued an economic strategy of supporting the development of heavy industry by exploiting domestic hydroelectric power resources. Policy toward Norway's aluminum industry sought to increase domestic production, add more value in the Norwegian industry by expanding downstream fabrication; and expand exports (Byrkjeland

and Langeland 2000: 27–35). The plant at Sunndalsøra also benefited from funding supplied by the US Economic Cooperation Administration (ECA), which administered the Marshall Plan after the outbreak of the Korean War. ECA funding was motivated by the US government's concern over maintaining access to aluminum supplies during wartime mobilization (Ingulstad 2006).

Alumina supply has always been a major concern of primary aluminum producers. Beginning in 1947, ÅSV acquired long-term supplies of alumina through agreements with Alcan and Alcoa, which in return received a share of the primary aluminum that ÅSV produced for their fabrication activities in Europe. ÅSV also received loans from these companies to increase its smelting capacity.

During the period of strong postwar demand for primary aluminum that lasted until 1958, ÅSV increased production and aluminum sales significantly. Profits were reinvested in expanded production capacity and better cell technology and infrastructure to further increase smelting capacity. The company did not build up competence in areas other than electrolysis, however, nor did they buy shares in the bauxite industry, pursue downstream vertical integration, or collaborate with other companies to build oxide plants. There were possibilities for vertical downstream integration in collaboration with the American aluminum company Harvey, and for upstream integration into bauxite in Yugoslavia (Rinde 1996: 36–44). However, ÅSV did not find the bauxite mines to be an attractive investment (Gøte 2001: 20–1). Instead, ÅSV focused on "concentrating labour and capital on electrolysis", for which Norway had "exceptional natural conditions"; these were activities that had given the company economic success, and the firm's state ownership made its strategic concentration on Norway politically important (Rinde 1996: 43).

Beginning in 1953, ÅSV entered into research collaborations with researchers at the Norwegian School of Science and Technology (NTH), which was supported by the Research Council for Scientific and Industrial Research (NTNF). One of the projects at NTH, "the theory of aluminum electrolysis," was initiated in 1953 and proceeded until the late 1980s (Gulowsen 2000: 144). Much of this early collaborative R&D focused on improving efficiency at ÅSV's plants, which was rather low by international standards. In 1965, the managers at ÅSV found that in the period 1955–1962, the productivity of ÅSV was 33–45 percent of the productivity level of US plants (Myrvang 2000: 77).

From 1945 until the late 1970s, the world aluminum industry was dominated by oligopolies represented by the "Big Six": Alcoa, Alcan, Reynolds, and Kaiser in the US; and Pechiney and Alusuisse in Europe (Holloway 1988).

Norwegian government policies and public research support proved to be essential for the entry of Norwegian firms into the aluminum industry during the 1960s. ÅSV, Elektrokjemisk, and Norsk Hydro and other large Norwegian companies in the process industries benefited from governmental industrial policies.

The Norwegian government set up a committee for industrial finance in 1959 to attract foreign capital into Norway, and launched a programme (1959–1960) to rapidly increase the production of aluminum (Johannesen *et al.* 2005: 233–56). Governmental policies gave incentives (e.g. tax breaks and long-term power supply) for investment by domestic firms, but also supported expansion by foreign firms into the Norwegian aluminum industry. Some of the policy tools used to achieve this goal provided great advantages to Elektrokjemisk. The company had large surpluses from its ferrosilicon and mining operations, and new tax regulations enabled the firm to reinvest in aluminum production in the early 1960s. Relatively cheap and stable electricity contracts were also offered to Norwegian companies, such as Elektrokjemisk, ÅSV and Hydro (Sogner 2003: 141–73). Similar governmental policies also influenced the Sør-Al's decision to build the aluminum plant at Husnes (1965). In this case, however, the foreign company AIAG/COMPADEC, owner of the proprietary technology, prohibited Norwegian companies and the Norwegian state from owning stock in the new company. AIAG thereby managed to retain control of their technology, and knowledge about the aluminum market in house (Tjelmeland 1987: 24–39).

Elektrokjemisk's entry into aluminum production depended on more than Norwegian government policies. As already mentioned, US antitrust policy forced Alcoa to let other competitors like Reynolds and Kaiser enter the market. However, these new companies were not interested in supporting the Söderberg technology and the patent-pool that Elektrokjemisk had established as a technology supplier to the aluminum industry, and sought to develop their own smelting technology (Sogner 2003: 141–73). Even Alcoa wished to develop its own technology rather than relying on Elektrokjemisk's Söderberg technology. These developments limited further development of the Söderberg technology, and in the long run led to its replacement by other technologies. The Söderberg technology currently accounts for only 27 percent of global primary production, mainly at old plants (IAI 2006). Elektrokjemisk had no choice but to maintain its commitment to the Söderberg system in entering the aluminum industry, because of the technology's importance in corporate licensing revenues (Sogner 2003: 141–73). Elektrokjemisk's efforts to enter aluminum production relied heavily on the technological system that Alcoa offered for upstream and downstream activities regarding innovation, production and sales. Elektrokjemisk built two aluminum plants, at Mosjøen (1958) and Lista (1971) in a joint venture with Alcoa (Sogner 2003: 190–5).

The third Norwegian actor, Norsk Hydro, had sought to become an aluminum producer since the early twentieth century, but its lack of bauxite supply and lack of access to technology precluded its entry (Johannesen *et al.* 2005: 233–56). After 1945, however, the Norwegian government's industrial policies and the presence of new international actors seeking new markets facilitated Hydro's entry. Nonetheless, its limited technological capabilities in aluminum meant that Hydro could only provide hydroelectric power, which meant that entry into aluminum required collaboration with foreign companies, just as had been the case in the early twentieth century. Harvey aluminum, a US firm, provided such an opportunity. Harvey had entered the primary aluminum industry in the late 1950s and needed more smelting capacity. Norway's inexpensive electricity supply as well as its proximity to the European market made production in Norway an attractive option. In 1963, in a joint venture, Hydro and Harvey aluminum built a vertically integrated plant for aluminum smelting, rolling and extrusion. Norwegian financing was provided by Hydro and the Norwegian state (Johannesen *et al.* 2005: 233–56). This facility enabled Hydro to build up new competences in many stages of the aluminum value chain. However, in-house R&D on aluminum within Hydro was out of the question, since the firm did not have an experienced aluminum division.

Hydro's insufficient knowledge and experience in relation to the aluminum smelters became apparent in the mid-1970s, after Hydro purchased Harvey Aluminum's share of the joint venture and the plant began to experience production and efficiency problems (Johannesen *et al.* 2005: 402–9). In 1977, the company established an R&D centre at the Karmøy aluminum plant, consisting of a "metallurgical development group" with large scale laboratories and research groups focusing on extrusion, casting and rolling. Karmøy also began to collaborate more with other Hydro corporate operations that had an established R&D centre at Porsgrunn (Andersen and Yttri 1997: 246–51).

Until the mid 1960s, Norwegian R&D investment was low within the aluminum industry itself, as well as within the university/research institute sector.[9] Moreover, collaborative R&D between the aluminum companies and other organizations in the Norwegian innovation system was nearly absent because of the companies' secrecy and unwillingness to fund basic research. The production of aluminum in Norway until the mid-1960s has even been characterized as "pre-scientific" (Gulowsen 2000: 140–59). This characterization may be oversimplified, but it points to the fact that neither

Norwegian R&D organizations nor universities had any substantial R&D or innovation collaboration with the aluminum producers. This gradually changed during the 1950s and 1960s, as graduates from NTH were recruited into the industry and university–industry linkages were developed, partly due to personal networks. Furthermore, R&D investment in aluminum in Norway began to grow in the mid-1960s, when ÅSV started to collaborate more with NTH, the Centre for Industrial Research (SI), the Institute for Energy Technology (IFE), and the Foundation for Scientific and Industrial Research (SINTEF) on plant automation and electrolysis processes, resulting in increased productivity (Gulowsen 2000; Wulff 1992).

AUTONOMY ACHIEVED (1986–)

The global aluminum industry's structure is less concentrated than it was in the early twentieth century, and numerous actors have obtained access to technology and knowledge, raw materials and markets. Vertical integration has been a key organizational feature and a survival strategy for most aluminum companies since the origins of the "modern" industry in the late nineteenth century. Companies that have not vertically integrated have been subject to takeovers (Peck 1961; Stuckey 1983; Wallace 1937). There can be many motives and drivers behind vertical integration (both upstream and downstream), such as securing access to raw materials, improving access to markets, enabling horizontal expansion, or enabling innovation (Armour and Teece 1980; Cassiman 2006; Teece 1996). Interruptions in supplies of alumina or power are very costly. Integration upstream enables primary producers to acquire increased leverage over the alumina supply, which is especially important during times of high demand for raw materials.

Most Norwegian aluminum companies, however, were not fully integrated vertically until the mid-1980s. Although they have not always been vertically integrated, the Norwegian primary producers have all been part of larger conglomerates operating within other industries, in contrast to the very large, vertically integrated specialist producers exemplified by Alcoa and Alcan (Chandler 1962: 337–40). Until very recently, Hydro Aluminum was a subsidiary of a firm with a large oil and energy division, one aluminum division and one division composed of "other businesses".[10] Hydro Aluminum also has developed strategic alternatives to vertical integration, using long-term contracts and equity investments in other firms to "lock in" access to the output of independent alumina suppliers. Hydro presently holds equity interests in three alumina refineries in Brazil, Jamaica and Germany, and the company has long-term contracts with other bauxite-aluminum multinationals such as Comalco. Elkem ASA has business operations in energy, silicon metal, foundry products, microsilica, carbon, calcium carbide and solar energy, in addition to its aluminum smelters.

In both Hydro Aluminum and Elkem Aluminum, aluminum production and innovation has been affected by this diversified structure. For example, Hydro's revenues from energy and oil (Lie 2005: 172–202) have financed much of Hydro Aluminum's recent expansion. The survival of Elkem Aluminum, on the other hand, involved the divestiture of some unrelated divisions.[11] In the late 1980s, Elkem produced relatively simple products and competed on volume. The collapse of the Soviet Union in 1989–91 led to large exports of aluminum into Western Europe and the USA, causing aluminum prices to fall dramatically. Elkem was nearly bankrupt in 1992, and needed to change its strategy in order to survive. One company director explained that they had to consolidate and sell unprofitable business divisions, shifting the company's focus to becoming the best low-cost producer of aluminum. The success of this strategy by the late 1990s enabled Elkem to acquire the downstream company Sapa in 2003.[12]

Acquisitions and mergers have been key survival strategies for primary producers of aluminum. Hydro Aluminum AS was established in 1986 through the merger between Norsk Hydro and the Norwegian state-owned Årdal og Sunndal Verk (ÅSV) with the strong support of the Norwegian government. Hydro acquired the German Vereinigte Aluminiumwerke AG (VAW) in 2002, making Hydro Aluminum the fifth largest producer of primary aluminum in the world (Roskill Information Services 2003). In addition to increasing market share, mergers and acquisitions enable companies to gain access to new technology and knowledge through R&D centres at the plants of former competitors and their networks in the innovation system, such as universities and research institutes.[13] For example, after Hydro's acquisition of VAW, the company obtained a dominant position in the German aluminum industry, and may have benefited from stronger linkages to the German innovation system. According to one company executive, "researchers and engineers in the former VAW system are working together with the Hydro-people".[14] Similarly, when Alcan and Pechiney merged in 2003, Alcan acquired Pechiney's smelting technologies AP 18 and AP 30, seen as the best technologies available for license on the commercial market in terms of their cell amperage, efficiency of cells (measured in current efficiency) and number of cells (EEC 2003). Analysts at the time suggested that Alcan would benefit from its improved ability to apply the Pechiney AP 30 and the more modern AP 50 cell technology, which is not available via licensing (Metal Center News, August 2003). According to Alcan (2006: 9), the AP technology package is now the lowest-cost process technology in smelting.

Bauxite is still refined into aluminum oxide (alumina) and then electrolytically reduced into metallic aluminum; as one engineer said, "What we really are doing, is fine-tuning the processes that were invented over 100 years ago".[15] Thus, the Hall-Héroult process is a mature technology, but R&D on productivity and environmental improvements remains active. Reducing energy requirements has always been a major driver for innovation,[16] and the energy required for production has steadily been reduced. In 1899, smelters used over 50 kWh to make a kilogramme of aluminum from alumina (IAI 2006). The best smelters today require only 13kWh, while the Norwegian smelters use 13–14 kWh/kg, which is better than the world average (Grimsrud and Kvinge 2006).

Governments across the world are increasingly active regulators of aluminum companies' operations because of environmental concerns that in turn have been heightened by the Kyoto agreements on CO_2 emissions, and much of the R&D in the Norwegian aluminum industry focuses on environmentally friendly process technologies. Norwegian smelters are among the leaders in environmentally friendly technologies and have reduced their PFC emission (C02, CF4, C2F6) by 62 percent in the period 1990–2005, while increasing production by 61 percent in the same period (Norsk Industri 2006). Reducing PFC emissions from primary production relies mainly on improved alumina feeding techniques through computer controls, trained operators, improved computer control to optimize cell performance, and monitoring cell operating parameters (US Environmental Agency). Hydro aluminum has found that its technological leadership in this field can be an advantage when acquiring plants and penetrating new markets, where national environmental legislations are of concern.[17]

Norwegian smelters equipped with the old Söderberg technology face shutdowns because of the OSPAR convention (Protection of the Marine Environment of the North-East Atlantic) that guides international protection of the marine environment of the North-East Atlantic. The plants at risk of shutdown are Karmøy, Årdal and Høyanger, all of which are run by Hydro. However, at some of these plants (e.g. Årdal), production is nevertheless expected to rise because of increased cell amperage (Hydro 2003, 2007).[18] Lista, owned by Elkem, also has Söderberg production lines, but has managed to reduce their emissions significantly by improving the Söderberg technology (Teknisk Ukeblad 2006).

The production costs that determine the competitiveness of the smelters can be reduced by developing better cell technology. It has been claimed that

Hydro has developed a cell system that is one of the "best in the world" (NTB 2004). The latest investment in new smelters and related technology by Hydro at Sunndal (2003) was about 5.6 billion NOK, the largest investment in the Norwegian onshore industry in the last 20 years (Hydro 2004). Prior to this overhaul, the smelter used HAL230 cells, considered by Alcan to be the second best technology that is licensed (EEC 2003). Hydro aluminum claims that its smelter is now one of the most efficient (and largest) in the world, with 350 new cells that use Hydro's proprietary technology, HAL250. The HAL250 technology will not be licensed to other firms, according to Hydro (EEC 2003); only the "next-best" technology is licensed to other companies.

Norwegian producers' R&D activities take place both in Norway and abroad. Company directors stress that the companies still have important R&D collaborations with the Norwegian University of Science and Technology (NTNU), and SINTEF. One company research director argued that NTNU has few academic equals in aluminum-related competence. The recruitment of engineers by Norwegian aluminum companies has for a long time relied on students from NTNU, and the companies work closely with the University, even influencing the education and research activities of the university.[19] Hydro Aluminum has R&D centres at its aluminum plants in Norway and around the world. R&D on electrolysis in Norway is mainly carried out at Sunndal and Årdal, while the company conducts R&D on alloys at Karmøy. Customers rarely use pure aluminum, and alloys are therefore important in meeting the need of clients in the manufacture of semi-finished goods (Le Roux 2002: 727). Typically, customers (e.g. the automotive industry) require specific material properties, and manufacturers of aluminum develop alloys to meet these performance requirements. Iron, silicon, zinc, copper, and magnesium are among the materials combined with aluminum to produce metals with various properties for applications in semi-fabricated products and end-user products. These downstream fabrication activities are key drivers in developing new alloys. Thus, interaction along the value chain is important for innovation activities.

Corporate R&D investment by the diversified firm Hydro amounted to 716 mil. NOK in 2005, of which 226 mil. was spent on energy and oil, and 456 mil. NOK on aluminum. About 50 percent of their R&D budget on aluminum is spent abroad (Hydro 2005b), most of it in Germany, due to the acquisition of VAW in 2002.[20] Hydro Aluminum also has R&D collaborations with technical universities in Germany and the US. Hydro has developed its own smelting technologies, but Elkem is more dependent on its American partner Alcoa. The Prebake technology that has gradually replaced the Söderberg technology was developed within Alcoa. However, most of the improvements on the Söderberg technology are generated by Elkem, which has a research centre at Kristiansand that focuses on the Söderberg technology and carbon materials.[21]

The Norwegian aluminum companies have enjoyed long-term support from the government. Management at Norsk Hydro refers to "the Hydro Model", in which the Norwegian state is one of the major owners of the company, but plays a minimal role in management. Since its founding in 1905, Hydro has benefited from its Norwegian state ownership (Lie 2005), but the formerly strong ties between Hydro and the government are now looser. For example, the liberalization of Norway's domestic electricity market, combined with low levels of investment in new capacity, have increased electricity prices and made long-term electricity supplies harder to obtain. In periods of high demand and high aluminum prices, producers with relatively high energy costs can cope with this. When aluminum prices drop, however, high energy-cost plants may shut down. Hydro's plant in Stade, Germany, was closed in 2006 because it could not cope with the substantial increase in German power prices during recent years. The increased difficulty of obtaining long-term power supplies, combined with the growing feasibility and attractiveness of investment in new capacity in developing countries where power is less expensive, mean that Norwegian smelters will have to be extremely energy efficient in order to survive.

What do these concerns imply for the future of the Norwegian aluminum industry? Analysts expect the global market for aluminum to double to over 60 million tonnes by 2020, reflecting growth in demand and, in particular, surging markets in China, India, and Russia (Alcoa 2005). Companies are not likely to invest in new Norwegian capacity, since power prices are predicted to remain high and most of the Norwegian long-term power contracts expire in 2010. Renewal of these contracts, even if feasible, will almost certainly involve significant increases in electricity costs. For example, at Sør-Al the re-pricing of power to current levels will increase the plant's annual production costs by nearly 40 million USD, which may force the smelter to shut down (Industri-avisen 2006). The president and chief executive officer of Hydro predicts that most aluminum plants in Europe will close during the next 10–15 years due to high energy prices and the global restructuring of the aluminum industry (E 24 2005).

Thus, the future prospects for the Norwegian primary aluminum industry appear to be grim, and current expectations concerning these future prospects are already influencing innovation and production strategies. Hydro Aluminum is focusing future investments within its aluminum business on upstream activities, and will shut down costly primary plants and replace them by new competitive capacity in "energy-rich areas" outside of Norway (Hydro 2005b). Moreover, since most R&D staff and their large-scale laboratories need to be located close to the smelters, the researchers and engineers working within Hydro may have to relocate, exit, or focus on other aluminum-related R&D.

CONCLUSION

The factors that influence and drive the path of innovation and production in sub-national industries may exist at many levels, e.g. within the firm itself, at the sector, the national, or at the international level. Analysing one single level is not always the most appropriate approach. The case of the Norwegian aluminum industry illustrates this empirically. In particular, the case adds important insights to the innovation system literature concerning the effects on a sub-national "sectoral" innovation system of developments in both the national and global innovation and production systems.

As Table 6.3 sums up, the innovation and production system of the Norwegian primary aluminum industry has changed over the last 100 years. Worth mentioning here are the changing characteristics of the actors and networks, technologies and knowledge and the institutions during the three phases. In the first phase, from 1908 to 1945, the locus of innovation resided primarily abroad. Production facilities were created and run by foreign companies that wanted to take advantage of the relatively cheap hydro power that was available in Norway's waterfalls. Other than the anode innovation of Elektrokjemisk, developed in collaboration with Alcoa, most of the technology and knowledge of Norway's aluminum industry during this period was developed in-house by these companies. These foreign companies also exerted considerable market power through vertical integration and cartels. As explained in Chapter 2, the Norwegian aluminum industry nonetheless represented a radical new path in the Norwegian NIS, as it brought new capital, technologies, knowledge and actors into the system.

In the second phase (1945–86), the Norwegian aluminum industry created a more autonomous path in innovation and production. The Norwegian government created supporting institutions that in combination with domestic firms' strategies laid the foundations for entry into primary aluminum production by the national champions Hydro and Elektrokjemisk. Global networks and access to foreign technology remained essential—the joint ventures with Harvey, Alcan, Alcoa, and European companies secured alumina supply, technology transfer, collaboration, and market access. US antitrust policy was crucial in weakening the Alcoa monopoly, enabling other US and non-US firms to enter, thereby creating numerous new opportunities for international collaboration and technology access by Norwegian firms. Marshall Plan financing for ÅSV also facilitated the entry of a state-owned Norwegian national champion in the early 1950s, and illustrates the important influence exercised by global institutions on the industry's development within Norway. Beginning in the mid-1960s, Norwegian aluminum companies improved their knowledge base through increased in-house R&D activities and interaction and collaboration with other Norwegian organizations such as NTH, SINTEF, and IFE.

The third phase (1986–) is primarily associated with the creation of Norwegian vertically integrated aluminum companies and horizontal expansion, typified by the merger of ÅSV and Norsk Hydro (1986). This merger also strengthened the combined firm's technology and access to offshore sources of knowledge. Hydro had knowledge and experience with the production and sales of semi-fabricated products, while ÅSV had developed their own electrolysis smelting technology, which Hydro further developed into one of the best available production technologies. The new company was able to strengthen its technological and market position, and avoided a takeover by larger multinationals. Elkem has continued to develop its smelting technology collaboration with Alcoa and has expanded downstream through the acquisition of SAPA.

Networks of cooperative relations have underpinned firms' investments in innovation and modern, environmentally friendly production capacity. This has supported the Norwegian aluminum industry. For example, the government, professional associations, and other organizations, in public and semi-private R&D and education programmes, have supported the firms' innovation and production activities. During the postwar period, in particular, this reinforced the industry's central position in the Norwegian innovation system. Moreover, powerful global forces, mediated by foreign states and large multinationals, have enhanced the influence of domestic policies on the industry's evolution. The ability of Norway's post-war "industrial policy" in this sector to operate in concert with, rather than in opposition to, larger global trends within the industry, is noteworthy.

Castellacci (2007) has pointed out that sectors need support from the NSI in order to become industrial leaders. This observation certainly applies to the Norwegian aluminum industry as well as the Norwegian oil industry (see Chapter 7). In addition, however, this study shows that innovation and production in the aluminum industry depend as well on global flows of technologies and knowledge, global networks, and global institutions. National and global forces shape the innovation system of the Norwegian aluminum industry, which means that an international perspective on the innovation

Table 6.3. A stylized overview of the key building blocks in the innovation and production system of Norwegian primary aluminum from 1908–2007

	Technologies and knowledge	Key actors and networks	Institutions
1908–1945: global sectoral sources dominate	- The Söderberg anode (world dominance until the 1950s). - The wider smelting technological system is developed and controlled through technological monopolies by MNE's.	- Elektrokjemsk (technology supplier). - Innovation collaboration between Elektrokjemisk and Alcoa. - Alcoa/Alcan, AIAG, Pechiney and BACO initiate the Norwegian projects.	- Monopolies (US) and oligopolies (Europe). - Cartels controlled by MNE's on production at the various sites and the sales of aluminum and bauxite worldwide.
1945–1986: Creating a path of autonomous development	- ÅSV develops its own cell technology. - Harvey Aluminum provides Hydro with technology. - The large MNE's loose their technological monopoly. - Automatization of the technological system at the plants.	- Nationalization of the industry: Hydro, Elkem and SørAl. - The Norwegian State. - Joint ventures between Norwegian firms and the foreign industrial leaders. - Norwegian R&D at production sites. - R&D linkages to the between domestic firms and at NTH, SINTEF, SI and IFE.	- The Monopoly in the US broken. - There where oligopolies in the US and in Europe. - Norwegian firms gained national support. - The Marshall Plan + ECA.
1986– Autonomy achieved	- Norwegian R&D groups and divisions abroad. - Global orientation and control with regard to technology development, knowledge base, and production. - Increased focus on environmental technologies.	- Norwegian companies expand and integrate vertically. - Stronger linkages to foreign universities and R&D organizations.	- Liberalization and less national support with regard to tax incentives, electricity agreements. - Higher standards regarding environmental control nationally and Globally (Kyoto and OSPAR).

Source: Compiled by the author.

and production systems is essential to understanding the sector's historical evolution and prospects.

The competitiveness of Norway's aluminum plants today, which is crucial to long-term industry survival, depends on a few fundamental factors. These include: improvements in the cost competitiveness of Norwegian smelters by improved energy efficiency; reductions in the capital costs per production tonne; environmental considerations; and improvements in quality and functionality for specific applications of the metal. Although the Norwegian aluminum industry possesses state of the art technology and has the knowledge to run plants efficiently, one might question this industry's potential to acquire inexpensive energy (relative to other competitors), which is necessary in order to sustain competitiveness.

ALUMINUM GLOSSARY

Source: Metallurg Aluminum Online Glossary: http://www.metallurgaluminum.com/glossarymain.html

Alloy: A mixture containing more than one element.

Anode: Positive electrode made of carbon in electrolytic cells used in the smelting process.

Bauxite: A soil type and ore derived from the weathering of granite (sialic parent rock), and depleted of nearly all soluble elements. Aluminum oxides are virtually all that remain after calcium, potassium, sodium and silicon have been leached out. Bauxite is the principle ore of aluminum, and is predominantly found in tropical and sub-tropical regions.

Cathode: Negative electrode in electrolytic cells used in the smelting process.

Casting: Pouring molten metal into moulds at any stage of the fabrication and manufacturing process.

Cells (or "pots"): A reduction cell in which aluminum is produced.

Cryolite: Salts-based flux (e.g. potassium aluminum fluoride).

Extruding: The act of deforming solid aluminum, usually in the form of a cylindrical billet, by pushing it from an enclosed container through a die to form a product of consistent cross section.

Ingot: Metal cast in a shape that is convenient for handling, storage, shipping and remelting. Ingots may be small enough to be stacked and handled by one person, or in the case of a larger T-ingot, may be designed to be handled by a forklift.

Process control: The act of controlling the key parameters of a process to try to ensure product consistency.

Smelting: Process of producing molten aluminum by the reduction of alumina.

NOTES

This chapter is a slightly modified version of TIK working paper on Innovation studies No. 20070604, which I submitted to the project "Innovation, Path-dependency and Policy" (IPP) carried out at the Centre for Technology, Innovation Culture (TIK), University of Oslo with the support of the Norwegian Research Council (Contract no. 154877). However, the Centre, University and Research Council are not responsible for the content of the chapter, which is the sole responsibility of me. I would like to thank Jan Fagerberg, David C. Mowery, Bart Verspagen and all the researchers involved in the IPP project, including all the project's external commentators for valuable comments during many interesting seminars and workshops. In addition, I would like to thank Kjetil Gjølme Andersen, Helge Hveem at the University of Oslo, and Asbjørn Karlsen and Mats Ingulstad and their colleges involved in the Comparative aluminum Research Programme at NTNU for important insights. Finally, I would like to thank Hydro aluminum and Elkem aluminum for taking time and effort in giving interviews, and the Norwegian Research Council for financing my research.

1. It has also been claimed that Norwegian firms in the process type sectors such as aluminum are subject to a 'systemic lock-in' in their NIS (Narula 2002). The sources of innovation come mainly from the national innovation system, consisting of strong and supporting Norwegian organizations and institutions, but the overall economy remains committed (and, arguably, overcommitted) to relatively mature, long-established industrial sectors.
2. "A sector is composed of heterogenous agents that are organizations or individuals (e.g. consumers, entrepreneurs, scientists). organizations may be firms (e.g. user, producers, and input suppliers) or non-firms (e.g. universities, financial institutions, government agencies, trade-unions, or technical associations), and include subunits or larger organizations (e.g. R&D or production departments) and groups of organizations (e.g. industry associations)." (Malerba 2005: 385).
3. Vigeland Metal Refinedy (Vennesla plant) is not included in this analysis, since the plant produces a somewhat different product—superpure aluminum (99.99 percent). Today it produces about 6,500 tonnes of super-pure aluminum. The plant imports its primary metal from Russia, and is owned by Norsk Hydro (50 percent) and Alcoa (50 percent).
4. Interview with plant manager December 2005 at Mosjøen, and interview with company director at Elkem September 2005, Oslo.

5. Compagnie de Produits et Electrométallurgiques was founded in 1855 and produced aluminum through chemical processes for about 30 years, applying the Deville-process that was developed in 1860. This company became known as Pechiney after the merger with SMEF, and until 1886 was the dominant aluminum company in France. But the new electrolysis process made the Deville technology uncompetitive.
6. As a part of Alcoa, the Canadian operation was incorporated in 1902 as Northern Aluminum Company, but in 1925 its name was changed to Aluminum Company of Canada, Limited. In 1928, a federal antitrust suit forced Alcoa to divest its principal subsidiaries outside the United States, including Aluminum Company of Canada, Limited, which was reorganized as an independent Canadian company that focused on the development of the industry in Canada and internationally. The directors and management were independent of Alcoa from 1928 onwards; a final adjudication of legal proceedings in 1950 forced the divestiture of overlapping equity ownership between the two firms (AluNET International 1999).
7. "...industries in which being ahead of one's competitors in product or process technology, or in production and marketing, gives firms an advantage in the world markets" (Mowery and Nelson 1999: 2).
8. After the Second World War the plant (Venesla) was reconstructed to produce super-pure aluminum.
9. With the exception of the R&D investments and collaboration that produced the Söderberg system and the Pedersen process for extracting alumina out of Norwegian clay, there had been little R&D and little collaboration among metals-processing companies and other actors within the Norwegian NSI (see Wulff 1992).
10. In 2007 Hydro merged with the Norwegian oil company Statoil producing a firm named Hydroil.
11. Interview with a company director at Elkem, September 2005.
12. Interview with a company director at Elkem, September 2005.
13. Based on interviews with company directors in Elkem, September 2005, and Hydro aluminum, October 2006.
14. Interview with a research director at Norsk Hydro, January 2006, Oslo.
15. Interview with plant engineer at Elkem Mosjøen, December 2005, Oslo.
16. Interview with a research director at Norsk Hydro, January 2006, Oslo.
17. Interview with a research director at Hydro Aluminum, Primary Metal, October 2006, Oslo.
18. Interview with Thomas Knutzen at Hydro, press contact for Hydro Aluminum, 13 April 2007.
19. Interview with a research director at Norsk Hydro, January 2006, Oslo.
20. Based on Hydro annual report 2005, and interview with a research director at Norsk Hydro January 2006, Oslo.
21. Interview with company director at Elkem, September 2005, Oslo.

REFERENCES

Alcoa (2005) Annual Report.

AluNET International (1999) Retrieved 26.11.2006, from www.alunet.net/main.as

Armour, H. O., and Teece, D. J. (1980). "Vertical Integration and Technological Innovation," *The Review of Economics and Statistics*, 62(3), 470–4.

Andersen, G. K. (2005) *Flaggskip i fremmed eie: Hydro 1905–1945*, Oslo: Pax Forlag A/S.

—— and Yttri, G. (1997) *Et forsøk verdt: Forskning og utvikling i Norsk Hydro Gjennom 50 år*, Oslo, Universitetsforlaget.

Armour, H. and Teece, D. (1980) "Vertical Integration and Technological Innovation," *Review of Economics and Statistics*, 62(3) 470–4.

Barham, B. (1994) "Strategic capacity investments and the Alcoa-Alcan monopoly, 1888–1945," in B. Barham, S. G. Bunker, and D. O'Hearn (eds.), *States, Firms and Raw Materials: The World Economy and Ecology of Aluminum* (69–111). London: University of Wisconsin Press.

Byrkjeland, M. and Langeland, O. (2000) Statlig Eierskap i Norge 1945–2000, Oslo, *Fafo-notat* 2000:22.

Cailluet, L. (2001) "The British Aluminum Industry, 1945–80s: Chronicles of a Death Foretold?" *Accounting, Business and Financial History* 11 ((1)1).

Cassiman, B. (ed.) (2006) *Mergers and Acquisitions: The Innovation Impact*, Cheltenham: Edward Elgar Publishing.

Castellacci, F. (2007) *Innovation in Norway in a European Perspective*, TIK Working Papers on Innovation Studies, Centre for Technology, Innovation and Culture, Oslo.

Chandler, A. D. (1962, 1990) *Strategy and Structure: Chapters in the History of the American Industrial Enterprise*, Cambridge, MA: MIT Press.

Edwards, D. J., Frary, F. C. and Jefferies, Z. (1930) *The Aluminum Industry, Aluminum and its Production*, New York: McGraw-Hill Book Company, Inc.

EEC (2003) Commission of the European Communities Regulation (EEC) No 4064/89 Merger Procedure, Case No COMP/M.3225—ALCAN /PECHINEY (II), Article 6(2) NON-OPPOSITION (29.09.2003).

E 24 Næringsliv (2005) "*Spår aluminiumdød*," Retrieved 30.11.2006, from http://e24.no

Fasting, K. (1966) *Norsk aluminum gjennom 50 år, Aktieselskapet Norsk aluminum Company A/S Nordisk aluminiumsindustri, forhistorie og historikk, 1915–1965*, Oslo.

Graham, R. (1982) *The Aluminum Industry and the Third World: Multinational Corporations and Underdevelopment*, London: Zed Press.

Grimsrud, B. and Kvinge, T. (2006) Har aluminiumsindustrien en framtid i Norge? Oslo, *Fafo-rapport 539*, Fafo.

Gulowsen, J. (2000) *Bro mellom vitenskap og teknologi: SINTEF 1950–2000*, Trondheim, Tapir: Akademisk Forlag.

Gøte, O. C. (2001) Det lette metal, Oslo: Chr. Schibsteds Forlag A/S.

Holloway, S. K. (1988) *The Aluminum Multinationals and the Bauxite Cartels*, New York: St. Martin's Press.

Hydro (2003) "Derfor blir Søderberg-anleggene stengt," Retrieved 13.18.2006, from www.hydro.com

——(2004) "Full production at Sunndal," Retrieved 16.10.2007, from www.hydro.com

——(2005a) "Upstream repositioning 2005–2010," Torstein Dale Sjøtveit—Head of Primary Metal, aluminum. Hydro Media.

——(2005b) Annual Report.

——(2005c) Aluminum, Retrieved 26.11.2007, from www.hydro.com/upload/Documents/Presentations/Capital%20Markets%20Day/2005/cmd2005_nilsen.pdf

——(2007) Fortsatt høy produksjon i teknologikommunen Årdal. www.hydro.com

IAI (2006) The International aluminum Institute, Retrieved 23.11.2007, from www.world-aluminum.org/

Industriavisen (2006) "Gode aluminiumspriser, men energiprisen bekymrer" Retrieved 14.11.2006, interview with Jan Yttredal, Plant Manager at Sør-Al. Retrieved 14.11.2006, from http://industriavisen.no/

Ingulstad, M. (2006) USA og ÅSV: *Amerikansk strategisk råvarepolitikk, Marshallplanen og finansieringen av Sunndal Verk* (Master's thesis, NTNU, 2006).

Innvik, P. E. and Kamsvåg, J. L. (1996) *Verket: Sunndal Verks historie gjennom 40 år.* Sunndal, Hydro Aluminum.

Johannesen *et al.* (2005) *Nasjonal kontroll og industriell fornyelse, Hydro 1945–1977*, Oslo: Pax Forlag A7A.

Kollenborg, E. (1962) *Det Norske Nitridaktieselskap 1912–1962*, Oslo, Rambæks Trykkeri.

Krogstad, A. (1999) *Vigeland I Vennesla, Bruket—Bedriften—Gården*, Venesla, Vigeland Brug/Vigeland Metal Refinery A/S.

Le Roux, M. (2002) Innovation relationships between Pechiney and Alcoa, a complex competition for a technological monopoly from the 1890s to the late 1930s. In H. Bonin, C. Bouneau, L. Cailluet, A. Fernandez, and S. Marzagalli (eds.), *Transnational companies (19th—20th Centuries)*, Paris: Editions PLAGE.

Lie, E. (2005) *Oljerikdommer og internasjonal ekspansjon, Hydro 1977–2005*, Oslo: Pax Forlag A/S.

Malerba, F. (2002) "Sectoral Systems of Innovation and Production,' *Research Policy*, 31(2), 247–264.

——(2005) "Sectoral Systems: How and Why Innovation Differs Across Sectors." in J. Fagerberg, D. C. Mowery, and R. R. Nelson (eds.), *The Oxford Handbook of Innovation* (380–406). Oxford: Oxford University Press.

Metal Center News. (August 2003) "Alcan-Pechiney Merger. Interview with Lloyd O'Carrol, Vice President and Chief Economist and Equity Research Analyst at BB&T Capital Markets.

Mowery, D. C. and Nelson, R.R. (eds.) (1999) *Sources of Industrial Leadership: Studies of Seven Industries*, Cambridge: Cambridge University Press.

Myrvang, C. (2000) "Teknikker i transformasjon: Ledelse, organizasjon og teknologi ved Årdal Verk og ÅSV fra 1940- til 1970-åra," *Skriftserie Nr. 2/2000*, Centre for Technology Innovation and Culture, University of Oslo.

Narula, R. (2002) "Innovation Systems and Inertia in R&D Location: Norwegian Firms and the Role of Systemic Lock-in," *Research Policy* 31, 795–816.

Norsk Industri (2006) Retrieved 13.07.2007, from www.norskindustri.no/

NTB (Norsk Telegram Byrå) (2004) 04.11.2004, Interview with Truls Gautesen, Director of Hydro Aluminum Metals.

Peck, M. J. (1961) *Competition in the Aluminum Industry 1945–1958*, Cambridge, MA: Harvard University Press.

Petersen, E. (1953) *Elektrokjemisk AS 1904–1954*, Elektrokjemisk, Oslo 1953.

Peterson, W. S. and Miller, R. E (eds.) (1986) *Hall-Héroult Centennial: First Century of Aluminum Process Technology 1886–1996*, Pennsylvania: TMS (The Minerals, Metals, and Materials Society).

PIL (2005) The Federation of Norwegian Process Industries. Data on Production of Aluminum at Norway's Aluminum Plants.

Ragin, C. C. (1994) *Constructing Social Research*, Thousand Oaks, London, New Delhi: Pine Forge Press.

Reve, T., Lensberg, T., and K. Grønnhaug (1992) *Et konkurransedyktig Norge*, Tano: Bergen.

Rinde, H. (1996) *Utenlandske interesser i Norsk aluminiumsindustri 1908–1990*, Arbeidsnotat 1996/10. Handelshøyskolen BI.

—— (1997) "Den lange ventetida," in R. P. Adam, D. Gjestland, & and A. Hompland (eds.), *Årdal, verket og bygda* (58–82), Det Norske Samlaget, Oslo.

Roskill Information Services (2003) *The Economics of aluminum* (8th edn.), Roskill Information Services.

Singer, C., Holmyard, E. J., Hall, A. R., and Williams, T. I. (eds.) (1958) *A history of technology* (vol. 5), Oxford: Clarendon Press.

Smith, G. D. (1988) *From Monopoly to Competition, the Transformation of Alcoa, 1888–1988*, Cambridge: Cambridge University Press.

Sogner, K. (2003) *Skaperkraft, Elkem gjennom 100 år, 1904–2004*, Oslo, Messel Forlag.

Statistics Norway (2006) International Standards for Classifying Commodities (SITC).

Stuckey, J. A. (1983) *Vertical Integration and Joint Ventures in the Aluminum Industry*, Cambridge, MA and London: Harvard University Press.

Svendsen, B. and Rikter-Svendsen, K. (1992) Et konkurransedyktig Norge, Aluminiumsindustrien. *SNF-rapport Nr. 60/1992*. Stiftelsen for samfunns-og næringslivsforskning, Norges Handelshøyskole.

Teece, D. J. (1996) "Firm Organization, Industrial Structure, and Technological Innovation," *Journal of Economic Behaviour and organization* (31), 193–224.

Teknisk Ukeblad (2006) "Miljøvennlige aluminiumsverk", Retrieved 27.04.2006, from www.tu.no

Thue, L. (2006) *Statens Kraft 1890–1947*, Oslo: Universitetsforlaget.

Tjelmeland, S. (ed.) (1987) *Sør-Norge Aluminum A/S, 1968–1987*, Husnes, Sør-Norge Aluminum A/S.

Tresselt, D. (1968) Strategi og kontroll i norsk aluminiumsindustri: Fra teknisk funksjonshemning til risikoaversjon, *Norges Handelshøyskoles særtrykk-serie* Nr. 72.

UNCTAD/WTO (2006) International Trade Statistics.

Wallace, D. H. (1937) *Market Control in the Aluminum Industry*, Cambridge, MA: Harvard University Press.

Wulff, E. (1992) Aluminiumsmiljøet i Norge i dag. Trondheim, *SINTEF report* Nr. STF05A92010.

7

The Development of the Norwegian Petroleum Innovation System: A Historical Overview

Ole Andreas Engen

This chapter addresses the development of the Norwegian Petroleum Innovation System. The foundation of this system was laid through the establishment by government and industry of a particular organization for offshore oil production that allowed Norwegian participants to convert domestic industrial competence bases into competencies for offshore oil production. The adaptation of the international oil industry into a Norwegian context was encouraged by a political strategy of integrating domestic firms and enterprises into the large development projects that were necessary for the exploitation of Norway's offshore oil and gas deposits. International oil companies, along with a wide range of suppliers and local communities accepted this strategy because it facilitated their exploitation of the Norwegian continental shelves and contributed to job creation, the creation of new industrial competences and regional and national development.

The petroleum industry in Norway, which at the outset was largely foreign-controlled, was thus transformed into a Norwegian Petroleum Innovation System through alliances within the Norwegian industrial environment involving Norwegian oil companies, the R&D sector, public administrative institutions and Norwegian politicians (Andersen 1993; Nelson 1993; Engen 2002). Moreover, through different historical phases the petroleum innovation system became more closely integrated into the Norwegian national innovation system and actually turned into a cornerstone of that system (see Chapter 3). In this chapter, we ask whether Norwegian politicians, the Norwegian administration, and Norwegian industrialists contributed to the establishment of this particular industry, and how a domestic knowledge and competence base was created to enable the industry's adaptation and integration.

INSTITUTIONAL CONDITIONS AND ABSORPTIVE CAPACITY

An initial assumption for adapting a new innovation system such as the petroleum industry is that during the course of inward technology transfer, domestic actors (firms, institutions, and individuals) acquire sufficient competence in the application of the technology and organizations such that domestic personnel are able to undertake new innovations and improvements. This competence is normally referred to as "absorptive capacity" (Cohen and Levinthal 1990), and often results from a particular strategy for the selection and utilization of technology. In the first development phase of Norway's oil and gas industry, the international companies established the premises for transfer, but both public institutions and private actors in Norway negotiated in a manner that laid the foundations for "national absorptive capacities" upon which it was possible to build at a later stage in the industry's development. The capacities comprised fragments of expertise that functioned in concert and helped create national institutional systems for receiving and adapting a petroleum innovation system.

Institutional capacity

The first official Norwegian measures for institutional absorptive capacity comprised the granting of concessions—and taxation laws (Hanisch and Nerheim 1993).[1] The concession laws became the government's primary control instrument in determining which companies should be granted permission to operate in the Norwegian sector and where the operations should be concentrated, namely which blocks were open for tender. These laws were legally binding for all implicated parties and thereby clarified the relationship between the petroleum sector and the Norwegian State. The laws guaranteed the companies' rights while simultaneously expressing the sovereignty of the state over the Norwegian area of the continental shelf.

In Norway, these types of "contracts" were not unfamiliar. The main principle was largely similar to the concession system formulated some fifty years previously in association with the development of hydroelectricity in Norway (see Chapter 2). In the development of both the electricity and petroleum innovation systems, the relationship to foreign capital was central. Foreign capital has always been regarded with scepticism, but has become recognized in varying degrees as necessary to the industrialization process in Norway. Both at the turn of the century and in the 1960s, senior figures in Norway's government and industry sought to attract foreign capital to Norway. But this capital was to be controlled, to serve the interests of Norwegian society.

In both of the regulatory regimes mentioned, state policy created incentives for integrating new innovation systems into Norwegian society. The incentives were derived partly from the concession system itself, which required that Norwegian subcontractors were to be employed, but were also associated with the authority granted to Norwegian government agencies to distribute rights among those companies and bodies that they considered would best take into consideration Norwegian interests.

It is reasonable to assume that previous practice in the electricity concessions influenced the procedures for oil and gas. But representatives of the Norwegian administration did not independently determine the outcome of the negotiations over oil and gas concessions. During the course of the negotiations it was the oil companies who set the pace and proposed a system of licenses or concessions. In this regard the regulations were formulated according to the desires of the oil companies. But alternatives existed both for the Norwegian negotiators and the international oil companies. Certain companies desired to obtain the sole rights to resources on the Norwegian continental shelf. Such a solution had been chosen by the Danes, giving AP Moeller sole rights on the Danish continental shelf (Andersen 1993; Hahn Pedersen 1997). This, however, implies that the authorities lose an essential part of their controlling functions, and risk losing a potential fortune. Norwegian government officials instead granted concessions to several foreign firms, and did so only through a series of negotiations that enabled revision in the terms of the concessions to accommodate Norway's interests in building up domestic capacities in offshore oil and gas.

Industrial capacity

During the first stage of establishment, actors in the private sector began to exploit the possibilities for Norwegian industry created by offshore oil and gas exploration and production. These comprised three groups: the engineering industries, shipping and the only national firm of international scale—Norsk Hydro. Norway's shipbuilding industry experienced a boom period that lasted from the late 1960s into the early 1970s. During this period, the industry's design and construction techniques were transformed, emphasizing sectional construction rather than the building of a complete unit from the keel up. In sectional construction, the separate sections are constructed independently and then welded together as a unit. This technique implied greater specialization at each of the yards in design and construction, enabling individual shipbuilders to develop specialized expertise and maintaining the geographically dispersed pattern of the industry.[2] This form of production organization also increased the need for reporting and administrative controls to monitor the individual parts of the process. In aggregate one may refer to a development of a joint competence within individual firms and the industry as a whole. This contributed to the emergence of new groups of shipbuilders who individually were able to execute specialist tasks and together could undertake complex construction operations such as building oil rigs and platforms.

When the international oil companies became active in the North Sea in the 1960s, parts of the Norwegian shipbuilding industry were prepared. As early as 1964 the Aker-group had planned the establishment of three companies: one concerned with drilling, one associated with supply services, and a partnership company for the construction of oil platforms. The latter resulted in the construction of the drilling rig, Ocean Viking, where work was split among several yards with the main base in Nyland Vest. An important partner in the Viking project was Rosenberg Shipyard in Stavanger, and the resulting division of labor enabled deliveries as early as 1967.[3] This early accord and the experience which the Aker Group acquired were important for subsequent platform-construction projects, notably the Aker H-3 oil platform. The development of cooperation and partnership relationships opened up the possibility for executing large construction contracts that utilized the expertise accrued by the local yards and engineering workshops (Hanisch and Nerheim 1993).

The second group that became involved with the petroleum industry in the early years of the sector's development comprised Norwegian shipowners, consisting of two groups. The first group covered those who were already familiar with the workings of the international oil industry—in 1965, 20 percent of the world's tanker tonnage was registered in Norway. In spite of the fact that the oil companies themselves transported half of the crude oil from the Near East and North Africa, it was the independent shipping companies that transported most of the oil from the Persian Gulf. In other words, Norwegian shipowners were a part of the international petroleum system and understood the rules of the game. The other group represented Norwegian financial institutions with considerable interests in Norwegian shipyards and the shipbuilding industry, a group that had significant political influence at both the local and national levels.

The Norwegian international company, Norsk Hydro, played a role in petroleum exploitation in Norway from the beginning, initially relying on its international links. At this time the company had a number of French shareholders, which proved important in connection with approaches from the French oil industry regarding participation in a project aimed at oil exploration and extraction in the Norwegian sector. Hydro's interest in French capital led to the formation of the Petronord group and for Hydro, oil became an increasingly central feature. As early as the mid-1960s, Hydro had begun to build up an internal staff with expertise in offshore oil production, and a number of plans were formulated which had the aim of establishing an independent oil industry in Norway. Compared with the UK, Norway lacked strong actors with the ability to play an independent role in the oil system; Hydro was the single Norwegian firm capable of such a role in Norway's embryonic oil industry. Whether the company actually had this ability in the early phase has never been conclusively proved. But there was not sufficient political backing for Hydro's stance.[4] Instead, the Norwegian government decided to establish a new fully integrated state oil company, Statoil, relegating Norsk Hydro to a secondary role in the early years of the industry's development. Only in the later phases of the oil industry's development in Norway did Hydro become one of the important actors alongside the state owned company, Statoil.

Norwegian industry thus included a number of firms seeking to establish a position in the field. New associations were made through channels of existing technology. Some actors attempted to establish themselves in such manner that this would be to the advantage of the industries already established while simultaneously utilizing the international network. Nevertheless, it is important to point out that in this phase there was no mention of an independent Norwegian investment in technology, and special competence had to be imported. The central issue was to establish appropriate connections where the Norwegians could participate and thereby gain profitable experiences.

INTERNATIONAL ECONOMIC AND POLITICAL CONDITIONS

Compared with oil production off the Texan coast or in the Middle East, the development projects in the North Sea were initially an expensive and risky venture for the international oil companies. Nevertheless, there was no lack of will to invest. The individual companies manifested strong interest to carry out surveys in what became the British and Norwegian sectors. This interest increased pressure on the companies to search for new territories.

In the years following the Second World War, the oil industry was dominated by seven companies: Exxon, Gulf, Standard Oil of California, Texaco, Mobil, BP, and Shell (the Seven Sisters). In the global context these operated more or less as small, mobile states with considerable independence of and influence over national governments. Nevertheless, these firms' market power was reduced during the 1960s and 1970s. The market share of the so-called Seven Sisters in global oil production declined from 98.3 percent in 1950 to 89.0 percent in 1957, and 76.1 percent in 1969. During the same period, the share of other smaller companies increased correspondingly from 1.8 percent to 23.9 percent (Engen 1997). The loss of market share by the Seven Sisters reflected the rapidly increasing demand for and consumption of energy in Europe and the USA. The Seven Sisters alone were not able to meet this demand and thereby created niches in the market for new and smaller companies. The declining power of the Seven Sisters was also due to the nationalization ideology that had sprung up in the Arab world. The oil companies' host nations began to exert pressure, demanding higher payments for concessions. The Seven Sisters were generally not particularly interested in this form of renegotiation, but the independent companies were more willing to accept new economic conditions, and therefore obtained access to the new oil reservoirs.

In 1960, OPEC—the Organization of Oil Exporting Countries—was established. The aim of OPEC was to encourage changes in the taxation systems, to introduce production quotas and to support measures for the nationalization of concessions already granted. During the 1960s, changes were largely limited to the taxation systems. But in the 1970s, OPEC emerged as a significant influence on production and markets through its use of a broader range of devices. From March to October 1973, OPEC announced sharp production curbs and strong tax hikes for member countries, which brought about a quadrupling of the spot price for crude oil (Hanisch and Nerheim 1993). The nominal price level for oil around 1970 was actually under $2 per barrel. At the beginning of 1973 the price had risen to $11.50 per barrel. The rise continued from 1975 to a level of about $15 in 1978/79, and then jumped to $38 in 1980. In real terms this is approximately the same price level as that of 2007.

The 1960s signified the start of a process that threatened the dominance of the international companies in parts of the world that they had hitherto considered as their sole domain. This new and challenging international political climate was of decisive significance for the companies' investment programs. Both the independent and the major companies in this period were interested in regions that satisfied their demands for political stability, and the North Sea emerged as one of great promise. The finds in Groningen (the Netherlands), the possibilities for locating closer to the energy markets, plus the fact that the area was characterized by permanent political stability, were clear "pull factors." When the region was opened for oil exploration, the oil companies emphasized the significance of establishing a framework that would regulate production and guarantee a certain security for their investments. On the whole, however, the broader changes in the international environment strengthened the bargaining position of the Norwegian government and Norwegian firms in the early development of the oil industry.

NATIONAL INSTITUTIONAL RESPONSES

The economic and political foundations of Norway's oil industry were created in the 1970s with the establishment of Statoil, the Norwegian Petroleum Directorate (NPD) and the Ministry of Oil and Energy (MOE). Simultaneously, the state slowed the issue of new exploration and production licenses, in order to build up Norwegian competence in the industry. The opening of the fourth concession round in 1978 coincided with a doubling of the oil price and gave additional impetus both for oil companies and public officials to unite on common goals regarding competence development and choices of technology (Hanisch and Nerheim 1993). The technology that eventually would dominate on the Norwegian continental shelf was large integrated platform constructions, a topic that we discuss in greater detail below.

From its establishment, Statoil became the main instrument of the state in developing Norwegian petroleum competence and shaping a particular technical and organizational trajectory, relying on the concession system to strengthen its dominance.[5] Agreements regarding training and transfers of knowledge and technology from other companies were negotiated, and Statoil itself took the role of intermediary in delegating tasks to Norwegian industry. Industrially, the company contributed greatly to technical and organizational adaptation by utilizing traditional industrial networks and functioning as an agent transferring and adapting international petroleum techniques and competence (Engeland 1995).

The outlook for long-term production and positive investment returns also induced the international oil companies to be flexible in meeting the demands of public officials regarding their development strategies, choice of suppliers, and measures to transfer competence to Norwegian companies and research institutes.[6] This form of governmental organizing for a national petroleum industry is fully compatible with an "infant industry policy," and it contributed to the creation of a petroleum industry cluster in Norway.[7] Representatives of the Norwegian government considered these terms as absolutely necessary for securing the greatest possible share of oil profits for the Norwegian society.

The UK developed a regulatory framework for exploitation of its North Sea oil and gas deposits that resembled that of Norway in many respects.

The UK government initially implemented a petroleum tax system and a system of handing out licences, and established institutional bodies to oversee offshore exploration and production. But developments in the UK domestic oil industry followed a different path from that of Norway after the 1970s. Only the Norwegians pursued the "infant industry policy" during the eighties, while the UK chose to deregulate and de-emphasize active governmental participation. In general, we may say that UK oil policy in the 1980s consisted of little more than a passive concession practise that relied on a tax policy that was compatible with the objectives of the Thatcher government (Andersen 1993).

As the Norwegian oil companies and main suppliers gradually built up their competence in the shelter of a protectionist policy, the public administrative apparatus for oversight of the oil and gas industry matured. The development of the Norwegian Petroleum Directorate (NPD), established in 1973, shows how the role changed in accordance with the evolution of Norwegian industrial competence. The formation of NPD followed traditional administrative practice (Olsen 1989), but the Directorate engaged personnel with little experience in the petroleum industry, frequently resulting in an inflexible approach to regulation of the industry. This rigidity began to change in the late 1970s and 1980s, however, as NPD built up independent competence to handle complicated technical problems and complex development projects.

THE DEVELOPMENT AND INTEGRATION OF THE PETROLEUM INNOVATION SYSTEM—FIVE PHASES

Would the technical and organizational development and the associated competence creation in the Norwegian oil industry have taken another direction in a different political and economic environment? The answer is probably yes, but it is highly uncertain what would have happened and which concepts and competence bases would have come into being. In any case, given the situation in the international petroleum market at that time, the international companies chose not to take a very tough position in negotiating with Norwegian officials. The international companies accepted the fact that Norwegian participation in the exploitation of this portion of the North Sea would grow and that Statoil had achieved considerable independent competence. They also accepted directives from the Norwegian administration on the implementation of regulations, although these directives restricted technological choices.

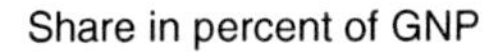

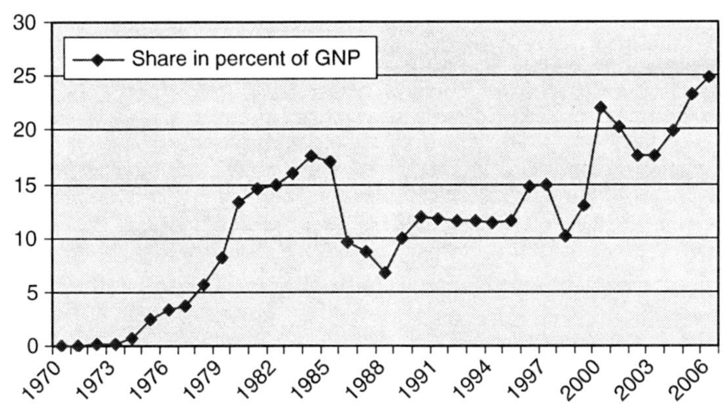

Figure 7.1. The Norwegian oil and gas industry, share of GNP.
Source: Statistics Norway.

Figure 7.1 and Figure 7.2 show the impact of the petroleum sector on the Norwegian economy in a historical perspective, focusing on the growth in the share of GNP and the total employment accounted for by the industry from the early 1970s through the early twenty-first century. From the beginning of the seventies until 1985, output and employment grew rapidly. Subsequently, the sector experienced periods of instability—both after 1986 and in the late 1990s. Every period of decline in output or employment is associated with declines in the international price of oil that trigger cutbacks in domestic investment in the sector. Not surprisingly, instability in oil and gas has affected growth and employment in the overall Norwegian economy.

In spite (or because) of these economic fluctuations, the industry has over 40 years experienced major organizational and technological change and has shown a remarkable capacity to handle changing environments. From its beginnings in the 1960s up to 2007, the offshore petroleum industry has passed through several stages of economic development, regulatory framework and organizational/technological evolution. It is useful to divide this developmental history into five phases.

The first two phases, "Entrepreneurial" and "Early Consolidation", describe the establishment of the innovation system and the shaping of a particular Norwegian way of organizing this. This period witnessed the development of a regulatory regime and the establishment of Statoil and NPD. Moreover, it included an active infant industry policy and the emergence of Aker and Kvaerner as main suppliers. The third phase, "Maturation", denotes a transitional period during which the innovation system was exposed to less

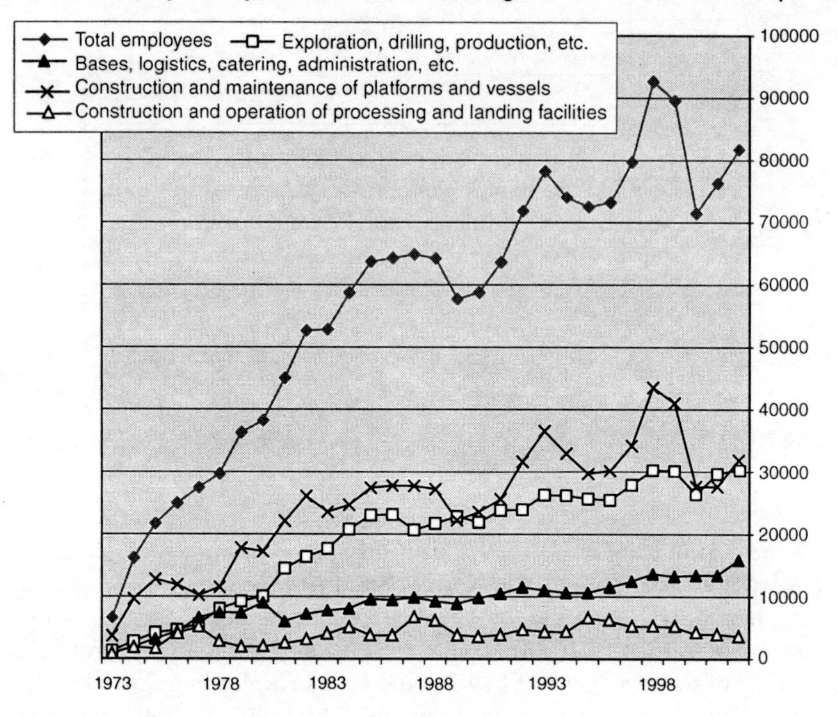

Figure 7.2. Rate of employment, 1973–2002
Source: Statistics Norway.

favorable economic conditions. Nevertheless, this phase was characterized by improvements in the capabilities of the Norwegian R&D system, increased public spending and larger funding for petroleum-related R&D within Norwegian universities. Finally, the last two phases, "Reorganization" and "Second consolidation," describe the erosion of the infant industry policy and the rebuilding of an innovation system better adapted to international competitive requirements and liberalized energy markets. This latest development resulted in part from the NORSOK program, a cooperative project between the Norwegian government, the oil companies and the supplier industry that sought to reduce the average cost of exploration and production on the Norwegian shelf by approximately 50 percent. The next section describes these five phases in greater detail and analyses how particular characteristics of the innovation system have affected technological choices on the Norwegian continental shelves.

Entrepreneurial phase 1970–1976

The main characteristic of the entrepreneurial phase is that neither the international oil companies nor the Norwegians understood the requirements and implications of oil development in the North Sea. Phillips brought with them technological experiences from other regions, e.g. onshore drilling in the Middle East and offshore experiences in the Gulf of Mexico, that were of limited relevance for the North Sea. Norwegian policymakers were unsure of how the regulatory regime (concession system, tax system) would work in practice when the oilfields started to produce.

The development of the petroleum fields "Ekofisk" and "Frigg" in the 1970s thus became important arenas for "learning by doing" for the international oil companies, the Norwegian authorities and Norwegian industry (Olsen & Engen 1997). While the international oil companies preferred light platform designs and sub-sea solutions with little involvement from Norwegian subcontractors, the Norwegian authorities insisted on designs that implied the highest possible participation of Norwegian industry. The Ekofisk- and Frigg-fields represented a breakthrough for Norwegian contractors in their introduction of concrete technology for offshore platforms that was well known in Norway and relatively easy to adapt to the oil industry. These oilfield production projects were of great importance in the transfer of technology to Norwegian contractors as well as the development by these firms of new platform technologies.

Nevertheless, Ekofisk was an expensive project with large cost overruns. But the high price of petroleum provided very good financial results in spite of 100 percent cost increases for parts of the project, and the favorable results allowed Phillips to accept the "Norwegianization" policy. In spite of a sceptical attitude towards the Norwegian government, the company adjusted to the regulations, illustrated by its construction of a huge concrete storage tank in the Ekofisk field in response to a suggestion from the Norwegian contractor firm Høyer Ellefsen. For Phillips, it was politically advantageous to engage a Norwegian firm for the construction job. Institutionally speaking, the storage tank illustrates the effects on technology choice and project costs of the incentives stemming from the concession system. The storage tank was built in Norway and involved Norwegian suppliers and engineering competence associated with the hydroelectric power system of Norway (water power plants and dams). For Phillips, building the storage tank gave the company credit when applying for new extensions of the Ekofisk area. The large potential profits from Ekofisk also increased the US firm's willingness to reach an agreement. Ekofisk must be considered the first offshore project in which a Norwegian petroleum innovation system affected technology choice and design.

The Frigg field was developed with a total of seven anchored platforms, three made of concrete and four of steel. The total cost was NOK 26 billion (1997). When the field was opened for development in April 1972, no stipulations were laid down by the Norwegian government on choice of technologies for production or transportation of natural gas to land. Frigg was at that time the world's largest offshore gas field and began production in 1977. As with Ekofisk, there was a great cost explosion, but nevertheless the Frigg field was considered profitable from the start. The economic risk associated with meeting the design and technology requirements of the Norwegian government was therefore judged by the foreign oil firms to be small. All internal concept studies showed that the project was economically sound and after the quadrupling of the prices in 1973 any remaining doubts about its financial prospects disappeared.

Expectations among Norwegian officials for the Frigg field were also high, not least because of an increased Norwegian share in the field. A 50/50 division of anchored steel and concrete construction suited Norwegian officials. The Norwegian portion of the supply business (28 percent as of 1974) exceeded that of Ekofisk (20 percent in 1973) (Engen and Olsen 1997). Nevertheless, the international oil companies dominated technology and management decisions during this period. The Norwegian Innovation Petroleum System did not yet exist, not least because there was no Norwegian R&D involved and there was little or no involvement by Norway's research institutes and universities.

Consolidation 1977–1980

The development of the Condeep platforms on the Norwegian shelf brought Norwegian actors into the center of the petroleum innovation system. The Condeeps were large gravity platforms placed on the sea bed (see Box 7.1 for detailed discussion), and the construction of these production systems relied on numerous Norwegian sub-contractors—among them Norwegian Contractors (NC) which was owned by Aker. To manage the complicated set-up, oil companies, suppliers, sub-contractors and authorities had to build up huge bureaucracies in order to control each other. This extremely bureaucratic way of organizing the workplace was to some extent unknown in Norway, although the international oil companies had long relied on extensive management control systems in their international operations, in order to protect themselves against opportunistic behavior by contractors. In the North Sea, these elaborate management structures resulted in very high operating costs. During this period, the influential Norwegian stakeholders in the oil

Figure 7.3. The largest Condeep ever built. The towing of the Troll A platform, 1999 (photo: Norwegian Petroleum Museum/StatoilHydro)

industry expanded to include politicians, governmental institutions, oil companies, suppliers and labor unions, creating a complex network of interests and objectives that further shaped this phase of evolution of the innovation system.

There are three integrated platforms with independent processing works on each field at Statfjord, the major project undertaken during 1977–1980. This was where Statoil's competence was developed sufficiently to enable the company to operate independently. For Statoil it was important that their modest competence gained from large offshore production plants be utilized to the fullest. At that time, Statoil's only experience was with Condeep technology, and Statoil and Mobil together constituted an absolute majority in the licensee

Box 7.1. The Condeeps

Condeep (abbr. *concrete deep water structure*) refers to a type of gravity base structure for oil platforms. A Condeep usually consists of a base of concrete oil storage tanks from which one, three or four concrete shafts rise.The original Condeep always rests on the sea floor, and the shafts rise to about 30 meters above the sea level. The platform deck itself is not a part of the construction. Following the success of the concrete oil storage tank on the Ekofisk field, Norwegian Contractors introduced the Condeep production platform concept in 1973. This gravity base structure for platform was unique in that it was built from reinforced concrete instead of steel, which was the norm up to that point. The platform was made for the heavy weather conditions and the great water depths found in the North Sea. In addition to their use on the Norwegian Shelf, there are three condeeps on the UK shelf, Beryl A, B, and Brent A. Condeeps are used nowhere else. The picture shows the last and biggest condeep ever built. The Troll platform was towed over 200 km from Vats in the northern part of Rogaland to the Troll field 80 km north-west of Bergen. Troll A has an overall height of 472 meters and weighs 656,000 tons and has the distinction of being the tallest structure ever moved by mankind.

committee.[8] During the construction of Statfjord, the working partners were subjected to more substantial demands from Norwegian authorities.[9] Political decisions and interpretations of regulations guided technological choices to a larger extent than before.

The large development projects of this period required approaches to management that were new to Norway. During the seventies, the traditional shipyards were transformed into suppliers of offshore production platforms (Olsen and Engen 1997; Andersen 1986). Large engineering firms were responsible for designs and detailed specifications, and frequently undertook surveillance and control of the very same construction work. Manufacturing firms were given the task of producing according to ready-made drawings—whether they worked or not. Operators built up their own shadow organizations that in turn monitored the engineering and manufacturing firms. In this way, every firm had its own clearly delineated and overlapping areas of work and responsibility. But the sheer complexity and concern over opportunism that characterized the resulting system produced enormous organizations, extensive bureaucratic procedures, and significant cost increases.

Engineering companies held an important position in the design and construction phase. Through the strength of an initiative by the Norwegian company Aker, NPC (Norwegian Petroleum Consultants) was founded in 1973 for the purpose of securing national engineering control of Statfjord A (Engeland 1995). The Norwegian main suppliers Aker and Kvaerner also built up equivalent departments, requiring the recruitment of many more engineers than were available at the time. When all operators were required to choose Norwegians, a costly bottleneck developed in the industrial system, producing considerable wage increases for Norwegian engineering staff on these projects.

Although the policies in place during this phase of the industry's development sought to create a rational and efficient management system, the outcome was a cumbersome, bureaucratic, labor-intensive management model. Norwegian participation thus had economic consequences and higher costs, something that is commonly observed in infant industry policies. It was costly to develop new technical concepts and transfer competence to Statoil, and costly to train the Norwegians. Nevertheless, this developing innovation system generated national income and jobs, and began to bring Norwegian firms and institutions into a more active and autonomous role within the petroleum industry. Statoil began its transformation into an integrated petroleum company and Aker and Kvaerner established themselves as main suppliers. This implied among other things that the main suppliers increased their emphasis on internal sources of innovation, developing and improving their technical and engineering solutions. Nevertheless, there was little scientific or technological support from public sources. The Norwegian universities and R&D institutions remained absent from the petroleum scene.

Maturation 1981–1988

The development and commissioning of two big oilfields (Statfjord and Gullfaks) by Norwegian firms during the 1980s strengthened the organizational and technological networks linking national actors, who specialized in the only exploitation technology they knew at that stage (integrated gravity platforms). Strong political interventions gave priority to uniform and continuous investments in projects where Norwegian actors had acquired competence. Gradually, Norwegian actors acquired "know-why" competence within most sub-systems, making it easier to design procedures for safer operations.

In 1979 the Norwegian government took a critical step towards building up domestic research capacity with the establishment of the "Goodwill agreements". Under these agreements, the international petroleum companies were given so-called "goodwill points" by contracting with Norwegian firms and research institutes for oil and gas related R&D and by developing Norwegian research institutions. The policy included a well-articulated system for evaluating foreign firms' contributions to domestic capacity building.

Financial support for R&D was rewarded and transfer of know-how along with financial support was rated even more highly. Finally, the companies' overall contributions were reflected in their standing in the next concession round. For operators on the Norwegian shelf, the tax system also promoted R&D spending by classifying R&D-related costs as immediately deductible. Since the corporate tax rate had in 1975 increased to 78 percent, this policy implied that the government covered 78 percent of these costs. In sum, these regulations gave the institutional research sector in Norway a significant boost. Sintef (Trondheim), Christian Michelsens Research (Bergen) and Rogaland Research (Stavanger) quickly became major public R&D performers in the petroleum innovation system. Applied geology, well drilling technology and principles for enhanced and improved oil recovery became central research areas for all of these institutions (see Chapter 3).

The profitability of Statoil was dramatically improved by the oil price rise in 1980, and the firm accordingly maintained a high level of development activity and costs. Statfjord C, which began production in 1985, was an approximate copy of Statfjord B with an equivalent production capacity. In contrast with the two previous platforms, however, Statfjord C maintained good cost discipline and avoided large overruns.[10] As early as 1983 the operator could report that total costs had been reduced by approximately NOK 600 million and this, along with the high oil price, provided a very favourable profit picture for Statoil for the rest of the eighties.[11]

The Norwegian technical and organizational experiences on Statfjord were further utilized on Gullfaks, and the platforms are clear examples of the monumental technical concepts characterising Norwegian development projects offshore. Even more important, this was the first development on the Norwegian continental shelf in which both the operator and the main suppliers were Norwegian. Statoil was the operator, receiving 85 percent of the output, with the remainder being divided between Hydro and Saga. The first construction consisted of two integrated platforms, Gullfaks A and Gullfaks B. The Gullfaks projects avoided the sorts of cost overruns that typified Ekofisk, Frigg, or Statfjord, and it was therefore not necessary to adjust the budget upwards for either Gullfaks A or B.

The decision to develop phase I of Gullfaks, however, was not received with enthusiasm by the international oil firms. A sceptical attitude had been expressed by the Norwegian Conservative Party (Høyre) towards Statoil's growing role in Norwegian oil policy (Rommetvedt 1991), which resulted in Statoil being deprived of some control of the income flow from the shelf.[12] Nevertheless, Norwegian firms continued to benefit from offshore development and production, and the positive financial development of Gullfaks led to a second phase of development. Statoil exerted significant pressure on the

Parliament to hasten the completion of the field, and the decision to do so was largely the result of the company's effective marshalling of arguments concerning employment and regional policy (Thomsen 1991).

Another important explanation for the accelerated development of Gullfaks C was the delay in the construction of the Sleipner field, originally scheduled to begin development in the mid 1980s.[13] The main reason for the delay was Britain's refusal to enter into an agreement in 1985 to purchase gas from the Sleipner field. The delay of Sleipner could have caused a temporary downturn in the Norwegian suppliers' portfolios, something neither Norwegian politicians nor the supply industry wanted. It was therefore important to fill the vacuum.

Gullfaks C revealed a strong connection between public officials and operators, as well as showing how local and national industry and employment interests dominated the petroleum industry at that time. In retrospect, Gullfaks C appeared to be a purely political project in which considerations of employment onshore dominated concerns about technical and economic factors. The development of Gullfaks C was also in line with the new petroleum policy adopted in 1983. Previously, a production ceiling had limited output to 90 billion oil ton-equivalents. But by 1985, the focus of policy had shifted to ensure a stable pace of production investment, so as to stabilize the market for Norwegian suppliers.[14] The foundation for development of Norwegian competences in offshore exploration and production had now been established, and policy focused on reaping profits from the developments and further increasing Norwegian participation.

With the development of Oseberg North, Norsk Hydro emerged as an important Norwegian oil company. This represented a significant shift in the government's policy towards Hydro. The Oseberg field had been discovered in 1979, and Phase 2 developed the northern part of the field. Oseberg North utilized integrated production-drilling platforms fabricated from steel, and its production technologies have been characterized as more cost-effective than those used in Statfjord or Gullfaks (Kristiansen 1997). The Oseberg North project extended the concepts and organization model that characterized "Condeep", but also incorporated important incremental innovations that contributed to lower unit prices and more efficient operation and production.

Norwegian firms and subcontractors, together with Statoil and Hydro and the smaller Norwegian Oil company Saga Petroleum, increased their dominance on the Norwegian shelf during this period. The technical concepts and organizational models used in Norwegian oil and gas exploration in the late 1980s and subsequently reflected Norwegian policymakers' interest in expanding the nation's participation in the international petroleum industrial complex. Participation by domestic firms in large projects such as Oseberg North expanded independent Norwegian competence in offshore oil and gas production. Norwegian offshore oil and gas policy during this period continued to rely on the principles of domestic collaboration that developed during the 1960s and 1970s, as Norwegian suppliers cooperated rather than competed for contracts on large projects. Finally, the technology agreements with foreign firms and increased public funding for R&D in petroleum exploration and production (see Figure 7.4) linked the Norwegian research system, including the research institutes, to the international petroleum innovation system. For all of these reasons, we denote this phase as the "maturation" of the Norwegian Petroleum Innovation System. But Norwegian oil and gas production during the "maturation" period continued to be characterized by costs that were much higher than those of other oilfields. Only after the commercial shock of the 1990s did the domestic industry undergo the structural transformation that has made it internationally competitive.

Reorganization 1989–1996

From 1981 to 1985 the petroleum price sank in real terms, bottoming out at \$9 per barrel in 1986. From 1987 to the mid-nineties, the price varied nominally within a range of \$17 to \$25 per barrel (Engen and Olsen 1997). The 1995–2005 period was also characterized by fluctuations, but in general—if we include the last three years—prices have climbed steadily.

The decline in the oil price from \$40 to \$9 per barrel imposed a shock on the whole innovation system (Olsen and Sejersted, 1997; Engen, 2002). The main actors within the Norwegian Petroleum Innovation System were strongly encouraged to co-operate in order to develop new cost-saving technologies by forming NORSOK (abbreviation for Norsk Sokkels Konkurranseposisjon), an industrial program for development of new technologies and standards, organizational development and new contractual relations, regulations, and new initiatives for cooperation and negotiations between oil companies and their suppliers. The main objective was to reduce average costs by as much as 50 percent. The program was inspired by the similar British initiative, CRINE (Cost Reductions in a New Era). NORSOK, which began in 1993, introduced a process that gave the actors greater freedom in planning and implementing alternative technological solutions to achieve ambitious overall goals for increased efficiencies, standardization, and reduced costs.

During the 1980s, the development paths of the Norwegian and British continental shelf diverged. The Norwegian Innovation Petroleum System was integrated into the Norwegian economy by institutional arrangements such as the Goodwill agreements, and the great development projects (Statfjord and Gullfaks) were used by the government to reduce unemployment and build up petroleum-industry competence in the regions of Norway. In the UK, the Grampian region and Aberdeen experienced a similar development as Stavanger in Norway (Gjelsvik *et al.* 2006), but the North Sea oil and gas industry accounted for a much smaller share of total UK employment (about 100,000 in general employed in the eighties) and GNP than was true of Norway.

The average size of British fields was smaller and did not require technologies that were as costly or complex as Condeep. Compared to the British offshore oil sector or fields developed in the Gulf of Mexico, the time from discovery to production is almost three times longer in Norway. These long development cycles, among other factors, make development costs far higher in Norway's oilfields than in comparable developments elsewhere in the world. Despite new technical concepts and increased competence among all actors, the Norwegian oil system did not yet possess sufficient competitive capabilities to place it among the international elite of the energy producers. At the beginning of the 1990s these inefficiencies were seen by policymakers and managers in Norway's oil industry as serious problems that threatened the entire Norwegian economy.

When the Norwegians faced the fact that the old "technological regime" had to undergo a cost-efficient "makeover", it was natural to look to Britain and try to establish a program that was similar to CRINE. To a certain extent, NORSOK represented an institutional break with the protectionist praxis of the concession system and the technology policy of the Goodwill agreements. NORSOK was a collaborative project among the government, oil companies, main suppliers, and labor unions to replace the old technological regime with new, more cost-efficient routines and procedures. This implied a political shift from an active and interventionist oil policy to a more passive one that sought to link various actors rather than just dictating terms. In practice this meant that both oil companies and main suppliers enjoyed greater freedom when choosing technological concepts, subsuppliers, location of bases, headquarters and so forth.

During this phase, the R&D institutions in Norway connected to the petroleum innovation system increased their role. R&D funding from public and private sources grew rapidly throughout the 1980s and came to account for a significant fraction of total Norwegian R&D spending. During the late 1990s, oil and gas companies alone funded 12 percent of Norway's total R&D expenditures. The government also assumed a greater direct role in supporting R&D for the industry, assigning the Norwegian Research Council a prominent role in 1990 for supporting R&D in offshore production technologies. The public research programs in oil and gas during this phase included one program that was especially important in developing efficient technologies for the Norwegian Shelf.

The Ruth research program (Reservoir Utilization through advanced Technological Help) was initiated in 1991 as a collaboration among the Norwegian Research Council, the NDP and several oil companies and research institutions. The program produced significant results in a short time period, attracting industry attention and new business participants. The two most important technologies that came out of Ruth were gas injection and combined water and gas technologies, whereby water and gas are pumped into the reservoir to improve recovery. When the program came to an end the average planned oil recovery rate from existing fields had increased from 34 percent to 41 percent. These results made it possible to utilize existing technologies more efficiently for field development, thus making smaller and marginal fields more attractive to develop.

Figure 7.4 shows the fluctuations in public spending in oil and gas exploration and production, illustrating an interesting pattern. After the price drop in 1986, public R&D almost doubled until 1988, before falling during the next two years and recovering during the early phases of the NORSOK program.

The evolution of the new technical concepts employed in Norway's emergent oil and gas industry also reflected the operation of complex economic and political influences. The Condeep production platforms that dominated production activity through the 1980s were accompanied by large bureaucratic organizations. The dominance of Condeep reflected the fact that this

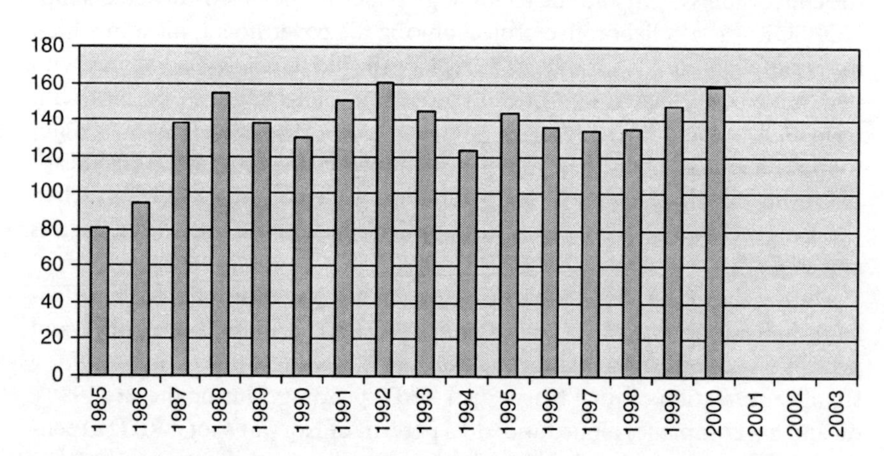

Figure 7.4. Public R&D expenditures in the Norwegian National Budget 1985–2004. Oil & Gas exploration and production. Current prices.

Source: Statistics Norway.

particular technology had generated its own momentum in the eighties. Thirty to forty thousand jobs were connected to the concept, Norwegian suppliers had learned to do things in that particular way rather than developing more general construction competences, and the Condeep design satisfied the desire of the Norwegian government to secure a high level of employment. Its dominance indicated the integration between the international and the Norwegian petroleum system around a production design and competence base that was developed to satisfy political, social, and technical criteria simultaneously.

By the 1990s, the shift in policy as well as the success of programs to develop industry-relevant competences among Norwegian firms meant that the national origin of suppliers and producers no longer dictated technological choices offshore in Norway. Large integrated platforms such as Troll and Draugen were built by Shell, and both Statoil and Hydro are involved in developments in which subsea installations are used. To some extent we may say that the similarity in domestic and foreign firms' technological choices was one effect of the Goodwill agreements. Shell was one of the first international oil companies to utilize the Goodwill agreements effectively, e.g. by building the Ullrigg Well Centre that drew on RF-Rogaland Research in Stavanger. The centre became one of the most important test facilities in Norway. The Goodwill agreements linked Shell closely with the Norwegian oil industry and influenced the firm's acceptance of Norwegian technology policy and practice.

Although technological choices for the Troll development were based partly on the infrastructure that had been built, independent integrated installations were chosen for individual fields. Hydro had the responsibility for the construction of phases II and III of the Troll development and for the gas injection on Troll-Oseberg, exploiting results from the Ruth program. Statoil held the main rights to Troll with about 75 percent of the total; Shell and Hydro accounted for 8 percent and 7 percent respectively of the field's output.

The Troll project was seen in petroleum circles as technologically advanced, since it utilized new technology for monitoring and controlling the seabed via computers. The organization of planning and construction however, did not break with established routines. Troll's construction showed that the Condeep design was durable. In spite of the fact that the installations were scaled down several times in response to changed conditions, one of the installations in the Troll field is the world's largest offshore platform.

The oil-price drop in 1986 induced more serious consideration of new production-platform designs. On the Troll project and fields development afterwards new innovations in drilling and exploitation were implemented, demonstrating that knowledge of and competence for alternative construction concepts had been developed before the price drop in Norwegian as well as foreign companies. In addition, one of the most important immediate effects of the price fall was not a change in companies' organization of development projects, but political and public administrative acceptance of simpler designs (Olsen and Engen 1997). Finally, the effects of increased R&D began to appear by the late 1980s in the improved feasibility of new designs.

Few fields were discovered in the period between 1986 and 1995. Indeed, development of numerous fields was considered and rejected during this period. During 1986–96, between 50 and 60 fields were tested in the North Sea (and a total of about 120 on the Norwegian shelf) and assessed to have little or no profit potential. The reason for the lack of interest on the part of the oil companies in developing these fields was that they were on the whole small, located far from existing infrastructure, and/or contained mainly gas. Even with new concepts, the high costs of development under the established organization of work prevented exploitation of these fields. The underlying mismatch between the Norwegian approach to offshore development and the profitability of oil production thus remained a pressing problem for public officials, companies, and suppliers early in the 1990s, the results of NORSOK notwithstanding.

Second consolidation 1997–?

During the second half of the nineties, economic insecurity plagued the Norwegian oil industry, as oil prices seemed to have stabilized at a lower level. The largest fields had been discovered and developed, or their final development was planned in the near future. The problem facing the Norwegian industry, however, did not reflect a lack of new discoveries, but rather the tendency for new discoveries to be smaller and less accessible than previous fields. In addition, the discoveries contained more natural gas than oil. Nevertheless the investment rate continued to rise as previously planned developments in the Sleipner, Troll, and parts of Oseberg sectors were implemented.

Two conditions nonetheless provided grounds for optimism concerning the offshore oil and gas industry in Norway. First, new technologies and simpler construction methods were being deployed, enabling operating-cost reductions in the neighborhood of 30 percent. The deployment of these new techniques required political acceptance of rationalization and job cuts in the operating organizations, and both public officials and labor unions accepted the necessity for such measures in order to maintain profitability. The other condition was the opening of exploration and development of oil and gas in the Norwegian Sea.

The new discoveries, along with continued concern over the high costs of exploration and production in Norway, meant that public funding of petroleum research increased during the late 1990s, and two important programs, "Demo 2000" and "Offshore 2010," were launched. Suppliers and SME's have assumed much greater responsibilities for R&D performance in both programs than previously was the case. The main funding for both programs, however, still comes from the Norwegian Research Council and the oil companies. The main objects of Offshore 2010 are subsea production, drilling technologies and well/fluid transportation technologies. DEMO 2000 seeks to develop new oil and gas fields through new technology, improved execution within project budgets and development of new industrial products.

In 2001, the private share of R&D expenditures in Norway's oil and gas industry exceeded the public share for the first time. The publicly funded programs appear to have catalysed additional private R&D investment in this sector, in accordance with the NORSOK objectives. The petroleum research programs in this phase illustrate the continuing collaborative efforts in the Norwegian Petroleum Innovation System and the close links between the public R&D infrastructure of Norway and the petroleum sector.

During the year 2000, floating installations, production vessels and subsea solutions were installed in the Njord, Norne Varg, Visund, and Åsgard fields. New organizational models were employed and the supplier firms Aker, Kvaerner, and UMOE accomplished extensive inter-organizational readjustments that aimed at reducing administrative overhead. Higher oil prices sparked an unprecedented boom in the petroleum industry during this period, and in 1999 the number of employees exceeded 90,000. The terrifying scenarios developed only a few years earlier now seemed irrelevant (see Figure 7.2).

New inter-organizational models were gradually implemented to accompany the new technological regime. The old Condeep-based contractual system, based on dividing contracts among many suppliers, was no longer rational when the core expertise was gathered in one organization. The post-Condeep technical concepts required simplified organizational models that were themselves one of the main effects of the NORSOK program, which ended in 2001. Other characteristics of the new technological regime manifested themselves in greater use of unmanned installations and computer-based technical solutions that required further development of subsea technology and drilling techniques.

Box 7.2. Industrial spinoffs of the Norwegian oil and gas industry

At least two important sectors have benefited from the development of Norway's oil industry. The breakthrough of under water system and advanced methods for seismic surveillance gave new opportunities for petroleum-oriented IT companies, e.g. Simrad, Kongsberg Offshore System (see Chapter 10). A second major industrial beneficiary of Norway's oil and gas industry has relied on the development of LNG technology (Liquid Natural Gas), which has sparked demand for advanced ship designs and processing equipment. The LNG technology (Ships and terminals) has increased its significance during the last years. Aker Kvaerner is an important actor in the LNG market and MARINTEK in Trondhjem has become one the major R&D suppliers to the Norwegian LNG producers.

This phase of the Norwegian industry's development was accompanied by significant industrial consolidation. First Statoil and Hydro consolidated their role by dividing the smaller Norwegian oil company, Saga, between them. Two large oil fields, Snorre and Tordis, thus became part of the portfolio of the largest Norwegian oil companies. Together Statoil and Hydro achieved total control of the Norwegian Shelf. Similar processes took place by the main supplier. The amalgamation of Aker and Kvaerner in 2001 meant that only one Norwegian supplier remained that could handle large petroleum contracts. The Norwegian Petroleum Innovation System is now dominated by a small number of powerful Norwegian actors. On the one hand we may say that the emergence of these firms is the result of different industrial strategies applied in the Norwegian oil industry in order to make the oil companies more competitive and to increase the technological capacities of the suppliers. On the other hand, this is a general trend that reflects the closer integration of the Norwegian petroleum industry—both suppliers and oil companies—with the international energy system.

However, controversies remained between the oil companies and the main suppliers about risk sharing within new development projects. These tensions reflected the fact that suppliers now had primary responsibility (and therefore were exposed to greater risks) throughout the entire construction process. Moreover, the greater responsibilities of suppliers for R&D produced new questions about the ownership of the intellectual property and patents resulting from their activities. Economic optimism nevertheless created a foundation for new relations between the main suppliers and the oil companies. Both oil companies and suppliers were eager to show they had common interests in future development projects—not least when it came to projects in new areas such as Azerbaijan, Angola, and Nigeria.

The internationalization of the Norwegian petroleum industry marks a new era. The first step in this direction was the BP—Statoil alliance in 1990, which focused on exploration and production in the former Soviet Union, Angola, Nigeria, Vietnam, and China. Although the alliance was dissolved in 1999, Statoil intensified their efforts to establish themselves on the international arena with the acquisition of the Irish oil company Aran in 1995. Hydro Oil & Gas chose the same strategy and joined Statoil when they in 2006 tried to gain access to oilfields in Venezuela and capture a large share in the Russian field, Stockman. Both of these initiatives failed, but in 2007 Statoil and Hydro, Oil & Gas merged as part of a strategy to compete in the international arena. The very term, "the Norwegian Petroleum Innovation System," may no longer be accurate, since the major producing companies and the main supplier in Norway, Aker Kvaerner, now seek to operate on a global scale (Ryggvik & Engen 2005).

The establishment in 2007 of StatoilHydro indicates that the oil companies, major suppliers and public officials have succeeded in entering the international energy industry. The Norwegian government has to a larger extent than previously taken on the role of agent and advocate for efficiency and internationalization. The oil companies and main suppliers compare their performance against international energy firms, and Norwegian politicians and administrative representatives appear to have accepted the premise that Norwegian petroleum competence is an export item that must be competitive on international markets. Norway's future petroleum industry thus may seek greater autonomy from the political system that gave birth to it and the Norwegian Petroleum Innovation System.

CONCLUSION

The Norwegian Petroleum Innovation System has achieved the twin (and not necessarily compatible) goals of developing technologies to enable cost-effective exploitation of offshore oil and gas deposits in one of the most challenging environments in the global oil industry, while simultaneously enabling Norwegian firms to enter and become globally competitive production and supplier firms. The accomplishment of both goals in less than 40 years is a remarkable achievement.

The initial phase of development of Norway's oil and gas industry in the sixties was characterized by an absorptive capacity that was limited to receiving new technology. The building of Norwegian competence in the seventies and eighties was shaped by public policies that sought to expand Norwegian participation in technology development and supply as well as production. In this phase, the development of Norwegian organizations and ambitions for public control were important. This strategy also sought to use the growing oil sector as an instrument both for job creation and economic policy. Accordingly, the international oil industry represented a general challenge for the existing Norwegian industrial structure, while also playing the role of teacher and transmitter of new technology. New organizational routines were implemented and extended into the Norwegian industrial environment.

With the development of the Condeep production-platform technology, it became possible to speak of an independent Norwegian petroleum industry. The development of Statoil, Hydro, and the main supplier industry signified that petroleum activity in Norway was entering a new phase. Large resources were also used in education and research. Publicly funded petroleum education and research were introduced at several levels within the R&D system of Norway. In the last phases we may say that the adjustment was concluded, as the absorptive capacity that existed in the initial phases had supported the development of a "participant competence" or what we have denoted as the Norwegian Petroleum Innovation System.

Over time, arguments for Norwegian jobs also carried less weight with public officials, as older "developmental strategies" fell out of favor (Engen 2002), and oil companies placed greater emphasis on profit criteria. With the emergence of a constellation of factors—the projected total resource foundation did not grow, newly discovered fields were generally smaller, and the price level sank—it became legitimate to search for completely new technologies that were both cheaper and less labor intensive. During the 1990s, new approaches and more cost efficient technologies and organizational structures overcame many of the cost-based obstacles to exploitation of new fields, even as the price of oil reached unprecedented heights. These organizational and technological adjustments were concluded by the beginning of the twenty-first century. The Norwegian oil and gas actors finally perceived themselves ready to fully participate in the international system of energy producers.

NOTES

1. The history of the first oil concession laws is treated in detail in Norsk Oljehistorie (The Norwegian Oil History Vol 1) by Tore Jørgen Hanisch and Gunnar Nerheim.
2. In Stavanger, Haugesund, Stord, and Verdal, small cities/rural areas along the western coast of Norway, shipyards became important contractors to the oil industry.

3. At this point of time, however, Rosenberg shipyard had not considered oil rigs and long into the 1970s it continued to give priority to the construction of gas tankers.
4. One explanation is that a large part of the shares at this time were in foreign hands. The authorities possibly felt that this situation was too risky to invest capital which in any case would be controlled by foreigners. It was better, therefore, to rely solely on foreign companies. The repurchase of Hydro shares by the Borten government in 1970 meant that the state acquired 51 percent of the company.
5. It should be noted that Hydro and the former Norwegian oil company Saga also benefited from the "Norwegianization policy", but in a much less dominating way than Statoil. Both of the former had to stand at the end of the queue during this period. See Ryggvik 2000.
6. In connection with the fourth concession round in 1978, foreign companies were called on to cooperate with Norwegian industry in research and development. That was the beginning of the so-called technology agreement. The introduction of the technology agreement should be seen in the context of the desire for Norwegianization and for developing an active Norwegian participation in petroleum activities (Blichner 1995).
7. Reve, Lensberg, and Grønhaug (1992). However, this study gives greater significance to the state's intervening role than Michael Porter himself does in "The Competitive Advantages" (Porter 1985).
8. When Mobil argued for a strategy that suited Norwegian interests, Statoil voted consistently with Mobil against the other companies in the concession group.
9. Organizing in the form of module construction was important here. Under pressure from Statoil and the Norwegian government, a letter of intent was signed which ensured that this building principle would be carried out consistently on Statfjord B and C. On Statfjord A, Aker argued against Mobil that it would be wise to build a large number of modules on the platform itself, something that increased the Aker group's portion of the supply business appreciably. That fitted in well with Statoil's overall strategy.
10. It was claimed that the large cost overruns of the seventies and the criticisms which followed resulted in the oil companies improving their management system for developing fields, and in up-scaling the cost estimates in their budgets rather than changing technology. Thus the final investment costs looked much more acceptable in comparison with the original budgets.
11. White paper no. 24 (1983–84).
12. White paper no. 35 (1984–85). Parliament decided in 1984 on a new organizational form for its direct economic engagement SDFI (in Norwegian, SDØE—statens direkte økonomiske engasjement). This principle implied, among other things, that part of the income, taxes, and investment that previously had been transferred to Statoil would now be channelled directly to Statoil.
13. White paper no. 40 (1982–83).

14. White Paper no. 35 (1984–85). This report was based entirely on the conclusions from the Tempo committee (Nou 1983: 27).

REFERENCES

Andersen, H. W. (1986) *Fra det britiske til det amerikanske produksjonsideal. Forandringer i teknologi og arbeid ved Aker. Mek. Verksted og i norsk skipsbyggingsindustri 1935–1970* (From the British to the American Production ideal. Changes in Technology and Work in Aker and in the Norwegian shipbuilding Industry 1935–1970). Trondheim.

Andersen, S. S. (1993) *The Struggle over the North Sea. Governmental Strategies in Denmark, Britain and Norway*, Scandinavia University Press: Oslo.

Blichner, L. Chr. (1995) *Radical Change and Experimental Learning*, LOS-rapport 9511: Bergen.

Cohen and Levinthal (1990) *Absorptive Capacity*, Administrative Quarterly Journal.

Engeland, S. (1995) *Ingeniørfrabrikk på norsk. Oppbygginga av norsk petroleumsrelatert engineeringkompetanse*, Hovedoppgave i Historie. Universitetet i Oslo: Oslo.

Engen, O. A. (1997) En evne til å ta imot in Sejersted, Francis and Olsen, Odd Einar (red.) *Oljevirksomheten som teknologiutviklingsprosjekt*, Oslo: Ad Notam.

——(2002) Rhetoric and Realities. *The NORSOK programme and Technical and organisational Change in The Norwegian Petroleum Industrial Complex*, Dissertation submitted for the degree of dr. polit. University of Bergen: Bergen.

Engen, O. A. and Olsen, O. E. (1997) "Et teknologisk system i endring—fra norsk stil til internasjonale ambisjoner" i Olsen & Sejersted (red). *Oljen som teknologiutviklingsprosjekt*, adNotam Gyldendal: Oslo.

Gjelsvik, M., Hatenaki, S., Lester, R. and Westnes P. (2006) *From Black Gold to Human Gold*, MiT—IPC-IRIS: Stavanger.

Hahn Pedersen, M. (1997) *Ap Moeller og den Danske Olie* (AP Moeller and the Danish Oil), Schultz, Denmark.

Hanisch, T. J. and Nerheim, G. (1993) *Norsk Oljehistorie Bind* I, Leseforlaget: Oslo.

Kristiansen, T. S. (1997) *Teknologiske valg under utbyggingen av Osebergfeltet. Hovedoppgave i Historie*, Universitetet i Oslo: Oslo.

Nelson, R. (ed.) (1993) *National Innovation Systems. A Comparative Analysis*, Oxford University Press: New York/Oxford.

Olsen, J. P. (1989) *Petroleum og Politikk*, Tano: Oslo.

Olsen, O. E. and Engen, O. A. (1997) "Konservativ nyskaping i offshore oljeproduksjon—Olsen and Sejersted. *Oljen som teknologiutviklingsprosjekt*, AdNotam Gyldendal: Oslo.

Olsen, O. E. and Sejersted, F. (1997). Oljen som teknologiutviklingsprosjekt, adNotam Gyldendal: Oslo.

Porter, M. E. (1985) *Competitive Advantage*, Free Press: New York.

Reve, T., Lensberg, T., and Grønhaug, K. (1992) *Et koukurransedyktig Norge*, Tano: Oslo.

Rommetvedt, H. (1991) *Statoils rolle*, RF—Rogaland Research.24/91: Stavanger.

Ryggvik, H. and Engen, O. A. (2005) *Den skjulte dagsorden. Rammer for en alternativ oljepolitikk*, SAFE—rapport: Stavanger. The report is available in English at <http://www.corribsos.com/uploads/Statoil_report.pdf>.

Thomsen, E. (1991) *Vedtaket om å bygge ut Gullfaks fase II. En beslutningsteoretisk analyse av forholdet mellom Statoil og Storting, regjering og Olje—og Energidepartementet*, Hovedoppgave i Statsvitenskap, Universitetet i Oslo: Oslo.

8

The Innovation System of Norwegian Aquacultured Salmonids

Heidi Wiig Aslesen

"Norwegian salmon" is probably Norway's best-known export product. It is the main driving force behind the development of the Norwegian aquaculture industry. The value of exported fish and fish products from Norway was about 32 billion NOK in 2005, representing the third largest export category (by value) in Norway that year. The export value of aquacultured salmonids[1] accounts for approximately 50 percent of this total. As wild capture decreases in the years to come (FAO 2002) the growing demand for seafood must be produced through aquaculture production. During the last twenty years, Norway has increased its market share within the global supply of salmon. Since the early 1970s, Norwegian production of farmed fish has doubled every three years (Berge 2000).

Today, the aquaculture industry has reached a size and maturity where business can no longer be run efficiently merely by employing the methods that were successful in the past. Moreover, during the last couple of years Norway's world market share in salmon has diminished, in large part due to competition from other countries like Chile, which has a strong and expanding aquaculture industry (some of which is Norwegian-owned). The industry is confronted with a series of new challenges. The strong demand experienced by the industry since its inception is no longer outpacing supply. The firms are increasingly exposed to price pressures and Norwegian firms face competition from producers with much lower operating costs. At the same time, customer demands are becoming more differentiated, and not easily addressed by all aquaculture firms.

This chapter focuses on the innovation system in aquaculture for salmon and trout in Norway. The aim of the chapter is to describe the sector in a national and global context, and to highlight how different aquaculture firms manage innovation. The chapter discusses the external relations and interactive learning processes that are involved in innovation and describes the sectoral innovation system. I devote particular attention to current approaches by Norwegian aquaculture firms to managing innovation in a time where the global industry is changing significantly. By looking at how various actors use Norway's aquaculture innovation system in different ways, the chapter provides input to policy makers who seek to strengthen the innovation system in a sector with the potential to become even more knowledge intensive and innovative than today.

The chapter addresses the following questions:

1. How has salmonid aquaculture developed within Norway and what characterizes the sector in today's global economy?
2. What innovation strategies do aquaculture firms currently employ and what do these strategies tell us about the functioning of the sectoral innovation system?

The outline of the chapter is as follows: section 2 discusses the evolution of the aquaculture industry and the development of important innovations in the sector. Section 3 gives a short description of important regulations in aquaculture and the sector's interconnections with regional policy. Section 4 describes the current structure of Norway's aquaculture industry and places it in a national and global context. Section 5 discusses the operations of different types of aquaculture firms and how these firms relate to external actors in their innovation efforts. The final section summarizes key findings and discusses innovation barriers in aquaculture.

THE EVOLUTION OF THE NORWEGIAN AQUACULTURE INDUSTRY

This section presents the evolution of the aquaculture industry by looking at important actors and innovations in the industry's history.

The early years of aquaculture

Commercial farming of salmon and trout in Norway is a relatively young sector, dating back to the 1970s. Nevertheless, experimentation and testing related to feeding, breeding and technology for fish farming has a long history among entrepreneurs and scientists at different localities in Norway.

In the early 1930s, a researcher named Rollefsen at the Institute for Marine Research in Bergen carried out experiments to find out how captured saltwater fish could survive. One of the central questions was ensuring that the fish (in this case, cod) survived their larval stage (Schwach 2000: 223). Rollefsen eventually managed to feed the cod larvae with a little crawfish, a breakthrough that set the foundation for *marine* aquaculture (as opposed to farming of anadromous species such as salmon and trout). Rollefsen documented his experiments in both hatching and feeding, and played an important role as knowledge provider when the interest in farming of *salmon* escalated at the institute around the late 1960s.

The growth of fish farming was a bottom-up, decentralized process that was centered in communities dominated by fisheries. The first impulses to aquaculture in Norway came from Denmark in the 1950s, where farming of trout in dams and lakes had begun (Schwach 2000: 338). At the same time, several pioneers in Norway experimented with fish farming independently of each other. Much of the farming activities were linked to wild salmon and trout in rivers and lakes in Norway, where voluntary associations worked to support the natural production of wild salmon. Much of the early development of commercial aquaculture occurred in coastal regions populated by people sharing a common culture and knowledge and experience from fisheries. These were communities where the informal exchange of knowledge and other resources was part of the values and norms of everyday life (Ørstavik 2004).

In 1968, several of these fish-farming pioneers approached the Institute for Marine Research, seeking solutions to problems with diseases among hatchery-produced fish. One of the younger researchers at the institute, Dag Møller, had completed a master's degree on salmon and had a special interest in population genetic studies. After spending a year in Canada, Møller returned to the institute in 1970, and expanded his research on aquaculture of salmon.

Another important breakthrough came about in the late 1960s, when a salmon farmer conducted successful tests of salmon farming in seawater net-pens. The success with the seawater net-pens ended years of trial and error in proper location specifications and technology choices and lead to an increased demand for specified coastal space (Aarset 1998). Strong market demand for salmon and the well-established existing distribution and sales systems meant that salmon farmers could obtain very good prices in Norway as well as internationally (Ørstavik 2004).

Another scientist who contributed to the development of Norwegian fish farming was professor Skjervold at the University of Life Science at Ås. Skjervold, a specialist in animal husbandry, conducted experiments on the genetic development of fish as a species and pioneered in the systematic breeding of salmon (see below for further discussion).

Even though early scientific experiments and trials had important consequences for the industry's development, farming of salmon and trout cannot be viewed as a "science driven" industry. Instead, fish farming was an entrepreneurial experience-based activity carried out by fishermen along the coastline who had capital and knowledge of fish farming as a sideline to their primary activity in fisheries. Norwegian fish farming thus drew on management, technology and experience-based knowledge from the fisheries sector under the ownership of small-scale entrepreneurs in Norway's far-flung coastal region. In this respect, Norwegian aquaculture presents a sharp contrast with Scottish fish farming, which historically has been dominated by large enterprises and capital from other regions (Jakobsen *et al.* 2003).

Knowledge of fish farming was accumulated among individuals in dispersed coastal communities through a prolonged learning process. During the 1970s, a combination of experience-based knowledge and research-based knowledge were used to solve the different biological, physical and chemical problems that arose.

When salmon aquaculture emerged, its products entered the supply chain of wild-caught fish, and as a consequence utilized many of the prevailing market institutions, product requirements and traditions of the fisheries sector. But technological innovations in aquaculture enabled greater control over the production process, and control of this activity by fishermen gradually declined (Tveterås and Kvaløy 2004). From the late 1970s the Norwegian government supported small-scale vertical and horizontal integration within salmon farming, mandating a producer cartel similar to those in fisheries and agriculture in many countries (*ibid.*). These regulations and institutions only survived until the beginning of the 1990s, when the government removed the laws that protected the sales cartel and prohibited horizontal integration (*ibid.*).

Important innovations combining knowledge bases

Innovation is often a result of interactive learning processes among actors, and for aquaculture firms, breed, feed, vaccination, and technology suppliers are important sources of new knowledge. The supply industry has developed new and better feed, advanced feeding systems, surveillance equipment, health- and veterinary services, etc. In the following sections important innovations in the history of aquaculture are discussed as illustrations of the operation of the aquaculture innovation system. They all illustrate how the fish farmers who pioneered the industry initially employed a practice-based approach to innovation in fish farming. Over time, however, aquaculture entrepreneurs have become more dependent on scientific research as well as government policy for innovation. The last paragraph has a short description on aquaculture-related R&D activity today.

Breeding and genetics

The early research challenges associated with fish breeding concerned selection of natural species that were best suited for farming. Subsequent efforts concentrated on scientific breeding programs that focused on the development of breeds that differed from those found naturally (Aslesen *et al.* 2002). One of the most important breakthroughs in breeding of salmon and trout for fish farming was the work of Professor Skjervold in the 1970s, who initiated a breeding scheme for salmon that drew on principles developed for cattle breeding, known as Norsk Rødt Fe (NRF). This was the starting point of a global breeding scheme for salmon.

Today, much of the same knowledge is used on other/new fish and shellfish species. Norwegian institutions and companies have developed and established the most advanced breeding systems for fish and shellfish worldwide, and Norwegian breeding companies are established in the most important salmon farming countries (Olafsen *et al.* 2006). The Norwegian research institution AKVAFORSK—The Institute of Aquaculture Research, has also developed a breeding systems for tilapia (the Philippines) and shrimps (Thailand).

Feed

In the initial stages of fish farming, knowledge from fur-bearing animals (mink) was used by marine farmers to create what is called wet-feed. During the 1980s, dry-feed was developed that was much easier to handle and of higher quality. Since the 1980s, Norway has been a center for research on feed for salmon aquaculture. Important factors to consider when making fish-feed include how to make feed economical, but still nutritious for growth and health, while preventing the remains from polluting the environment, etc. (Aslesen *et al.* 2002). The Norwegian fish-feed industry has carried out intense research efforts in different institutions and venues in order to improve the feed. Today, Norwegian research institutes and universities doing research on feed have a high international standing.

Beginning in 1996, salmon production was regulated by means of feed quotas, in order to control the growth of the industry. The amount of feed that each fish farmer could purchase annually was set by the Ministry of Fisheries. The aim of the feed quotas was to limit the growth in salmon exports and thereby maintain a floor on prices. The feed quotas have been effective in reducing production, since fish-feed is not easily substituted, but the quotas also sparked innovations, such that the type of feed used in Norway contains more fat than in other fish producing countries, producing higher-value salmon. The feed quotas were abolished in 2005 and the effects of this both on total production and the type of feed used are still uncertain.

The Norwegian fish-feed industry has become a global actor, and recent consolidation has produced three dominant companies: Skretting AS, Ewos AS, and Biomar AS, all of which have research staff in Norway. Salmon feed is the main product of these firms and their products and services are based on Norwegian expertise. The feed companies are present in all salmon-producing countries and other aquaculture countries worldwide, and they are all working to develop dry-feed for new species (Olafsen *et al.* 2006). During the last few years Nutreco, which is a supplier of feed for animals as well as fish, has made major acquisitions in Scotland and Norway. Through the acquisition of Hydro Seafood, Nutreco also has become Norway's largest fish farmer, strengthening links between supplier-firm R&D and aquaculture innovation.

Nearly 90 percent of the global production of fish oil and more than 50 percent of the fishmeal produced is consumed by the aqua-feed industry: It is expected that reduced supplies of fish oil resulting from overfishing will limit the growth of the global aquaculture industry (Olafsen *et al.* 2006). The Norwegian research community and feed companies have considerable expertise in the development of fish-feed, and are working to develop high-energy diets for salmon that will utilize feed more efficiently and effectively (*ibid.*).

Health

Sickness and diseases have long been a key problem for fish farmers. As was noted earlier, one of the first areas of involvement by scientists in modern aquaculture was the search for effective remedies against parasites, bacterial and viral infections, deformities, etc. (Aslesen *et al.* 2002). During the 1980s, Norwegian fish farmers were continuously challenged by diseases and their use of antibiotics rose, highlighting the need for further research and new solutions. In the early years of fish farming the research institutes had few resources for research on fish mortality in net cages. There was also limited contact among research institutions concerning these problems, and tensions between the research institutes and universities representing agriculture and fisheries over responsibility for such research further retarded its progress.

As mortality in fish farming rose during the 1980s, however, the industry seized the initiative to combat the diseases, while emphasizing the need for the different research communities to collaborate in order to find solutions to these severe problems. Through the "Frisk Fisk" (Healthy Fish) research program constituted in 1983 and financed by NFFR (The Norwegian Fisheries research, since 1993 part of the Research Council of Norway), industry initiatives increased collaboration within the industry and among the different research institutes.

An evaluation of aquaculture research in 1990–91 for the Research Council of Norway concluded that: "In the past, aquaculture diseases studies in Norway have been notorious for the lack of agreement and cooperation between different groups. It was a pleasure to see that this was now a thing of the past and collaborations between the individual scientists and between institutes was so positive and mutually rewarding" (Møller 1996). The Health Fish program supported the development of efficient vaccination procedures and paved the way for collaboration between important actors in the innovation system of aquaculture. Today, vaccination programs and other fish health systems are well established in Norway. Important contributions to these innovations have come from the National Veterinary Institute and the Norwegian College of Fishery Science.

Aquaculture technology

Challenges in the development of aquaculture farming technology and equipment concerned the selection of specific species most suitable for farming, finding locations, deciding what density of population that ought to be maintained, when and how to feed, etc. With regard to equipment, closing nets had to be constructed and anchored adequately, to make them resist strong winds, to keep salmon from escaping, etc. (Aslesen *et al.* 2002).

The first net cages used in aquaculture were based on homemade technology using wood as material, and fish farming was carried out near to the coastline. In order to carry out farming in deeper seawater, stronger net cages were needed, along with automatic feeding, feed control etc.

Traditionally, the technology[2] suppliers to the aquaculture industry were small and medium sized companies, many of them family-owned. In parallel with the consolidation of aquaculture firms (on the demand side), consolidation has also occurred among suppliers. In 2005, the Norwegian Equipment Producers Association (NLTH) counted 29 firms as important equipment producers to the aquaculture industry, of which nine have sales of less than 15 million NOK, and twenty have sales greater than 15 million NOK. Data from the Norwegian Equipment Producers Association (NLTH) shows that the total sales in 2005 was *c.*800 million NOK domestically and *c.*250 million for export (24 percent of total). Although the Norwegian aquaculture technology suppliers are becoming more active in export markets, their technology development activities are still based in Norway.

Education and the knowledge infrastructure

Compared with other salmon producing countries, Norway was quick to build educational establishments supporting the industry. In 1994, reforms were carried out (Reform 94) establishing specialized education in aquaculture and fisheries at secondary level. There is also specialized education at the university level (i.e. at Norwegian University of Life Science, Norwegian College of Fisheries Science, University of Bergen, and Norwegian University of Science and Technology (NTNU). Most of these educational institutions have close relationships with public research institutes that support the industry. For example, the Norwegian College of Fisheries Science collaborates with the Norwegian Institute of Fisheries and Aquaculture Research and the Norwegian University of Life Science works with AKVAFORSK—The Institute of Aquaculture Research. Many of the lecturers at these higher educational institutions are also employed as researchers in the institute sector.

The aquaculture industry's invests little in training. *Attitudes* towards qualified labor in the seafood industry were studied by Reve and Jakobsen (2001), who found that 30 percent of the aquaculture firms in their study lacked a strategy for skills upgrading.

According to Berge (2002), the research milieus relied heavily on the knowledge from the fisheries sector until 1985, after which date the combined results of more experience-based private knowledge and private R&D investments began to expand the aquaculture-specific knowledge base of the industry. Among the most important contributions of this growing industry-specific knowledge base is the development of fish health systems, breeding schemes and fish feed. The enormous growth of Norwegian aquaculture during the 1990's might well have been impossible without the research efforts from the 20 public, semi-public, and private research institutes in Norway that perform R&D in aquaculture. These are clustered in four main areas of Norway, namely Northern Norway (Tromsø and Bodø), Mid Norway (Trondheim), Western Norway (Bergen and Stavanger) and Eastern Norway (Ås and Oslo). The bulk of their research activity is focused on the salmon species.

According to Sundnes *et al.* (2005) the marine Research and Development institutions (including universities and university colleges) in Norway had a total R&D spending of 1.7 billion in 2003. In 2003, 1,600 researchers, scientific personnel and specialist employees were engaged in marine R&D in Norway working either within the institute sector, at universities and colleges or within the industry (Sundnes *et al.* 2005). R&D expenditure in aquaculture was

684 million in 2003, of which 29 percent was performed in the industrial sector, 53 percent in the institute sector and 18 percent in university and higher educational institutions.

GOVERNMENT REGULATION OF AQUACULTURE AND REGIONAL POLICY

Government regulation has played an important and growing role in the development of Norwegian aquaculture (Jakobsen *et al.* 2003). The aquaculture industry also has been affected by Norwegian regional policy, which has sought to sustain communities in sparsely populated areas in Norway, creating an aquaculture infrastructure along the coastline that is internationally unique (Jakobsen *et al.* 2003). The importance of regional policy reflects the longstanding political and social influence of geographically peripheral areas in Norway (see Chapter 2).

The regulation of aquaculture during the 1970s has been described as a "Regional policy hegemony" (Berge 2002), and these early policies strongly affected the subsequent development of fish farming. The influence of these early policies, however, has not prevented considerable changes in policy and its regional effects (Jakobsen *et al.* 2003). As political attention towards the industry has grown, the development of regulatory and regional policy has been affected.

Norwegian regulation of aquaculture is built on the 1973 Aquaculture Act, which is administered by the Ministry of Fisheries. Central to this law is the obligation to acquire a license for operating a fish farm. The system required the registration of new entries and most early applicants were granted permits, but in 1977, the authority banned new permits pending revision of a new Aquaculture Act (Jakobsen and Aarset 2007). The second Aquaculture Act (1981) sought to maintain an industrial structure based on small enterprises, an ownership structure based on local ownership, and a geographically dispersed industry. The law had a strong regional-policy element in its support for small-scale operations that would disperse fish farming activities and to support employment in sparsely populated areas. In the third Aquaculture Act (1985), the basic principles of the 1981 Act were reaffirmed.

The aquaculture sector has become an important industry for the coastal areas of Norway. Production of trout and salmon has traditionally been concentrated to Western Norway and Trøndelag (mid-Norway), areas that were assigned a large number of concessions in the early years of the industry's regulation by government. The later concession rounds have favored the sparsely populated Northern parts of Norway. The regional distribution of employment shows that in 2005 Nordland was the county employing the most people in aquaculture, followed by Hordaland and Møre and Romsdal.

In the late 1980s, the Norwegian aquaculture industry faced falling prices for farmed salmon because of oversupply and increased international competition. As a consequence, profit margins came under pressure and bankruptcies increased. This led in 1991 to a liberalization of ownership regulations which allowed firms to own multiple licenses and relaxed requirements that firm owners must reside in the same region as their enterprise. A wave of mergers and acquisitions followed that led to a sweeping restructuring of the industry. The market value of licenses increased as well. In 2001 the Acquaculture Act was changed to allow the Ministry to require payment for permits for the breeding of salmon and trout, four million NOK for assignments in Nord-Troms and Finnmark and five million NOK for the rest of the country. The higher fees will limit entry by small-scale entrepreneurs and therefore potentially may reduce local control of aquaculture, undermining a longstanding goal of Norwegian regional policy.

EMPLOYMENT AND OUTPUT IN NORWEGIAN AQUACULTURE

Employment in *salmon and trout* aquaculture has decreased steadily the last decade from 3,503 in 1995 to 2,208 in 2005. On the other hand, sales of salmon and trout have increased steadily, to 645,387 tonnes in 2005, three times the sales in 1994 (Figure 8.1).

The adoption of new technologies and learning within the aquaculture industry has contributed to the rapid productivity growth reflected in the employment and output trends cited above. Production costs declined steadily until the year 2000, when reductions slowed because the most obvious sources of farm-level cost reduction had been exhausted (Tveterås and Kvaløy 2004: 11). Today, firms within the industry are focusing their cost-reduction efforts on activities that are further downstream in the value chain, particularly processing and distribution to customers.

Even though the number of employees working with salmon and trout in Norway is relatively low, the aquaculture industry's contribution to gross national product (GNP) was 2.6 billion NOK in 2004. Overall, the marine sector (fishing, catch and aquaculture) contributed to 1.05 percent of total GNP in 2004 and represents the only primary industry other than oil and

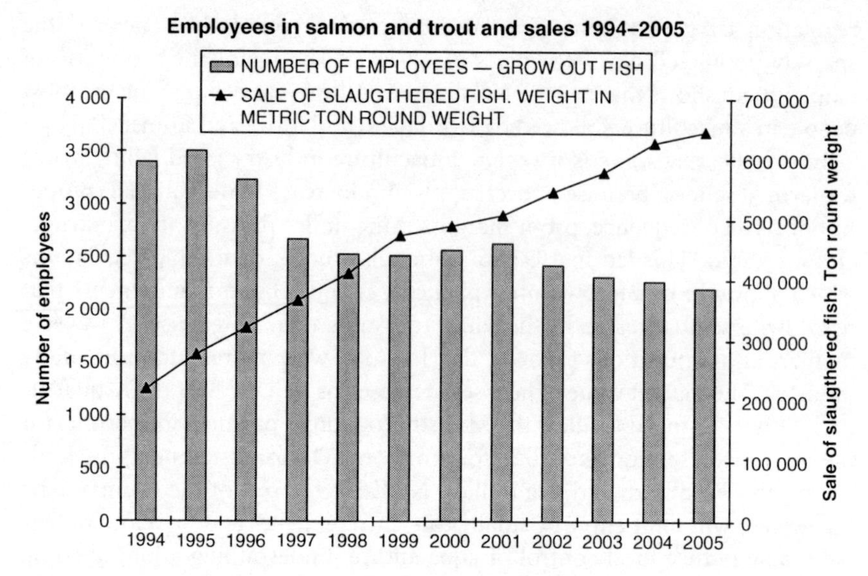

Figure 8.1. Employment of persons working with salmon and trout in Norway, 1994–2005, and sales of salmon and trout in the same period.
Source: Directorate of Fisheries /NFF. Pr 30.11.2006.

gas that has maintained its share of Norway's GNP (Sandberg *et al.* 2005). By comparison, oil and gas extraction accounted for 20 percent of Norwegian GNP in 2004.

Apart from the aquaculture industry's direct contributions to Norwegian employment and output, aquaculture is linked with supplier industries that provide feed, vaccination, technical equipment, etc. These linkages with other industries mean that the aquaculture industry all together supported total employment of 13,200 employees in 2003 (Sandberg *et al.* 2005) and the contribution to Norwegian GNP from other sectors amounted to *c.* 6.6 billion NOK in 2004.

In 2005 the value of exported fish and fish products from Norway was 32 billion NOK of which 17 billion came from wild catch and 15 billion from salmonid aquaculture (Olafsen *et al.* 2006).[3] Norwegian aquaculture exports consist primarily of "primary" fish products; i.e. fresh round or filleted fish. In 2002, 78 percent of Norwegian exports were head-on gutted fish whereas 80 percent of Chilean exports were fillets (Tveterås 2004). These differences in the composition of exports reflect a tradition in exporting "primary" fish products, partly due to trade barriers (high tariffs) for processed products and high labor costs in Norway.

Partly because of its high level of exports, the aquaculture industry is very cyclical, the export ratios varying within the year and between years. Aquaculture-related supplier industries also export their products. Olafsen *et al.* (2006) estimated that the export value of *marine expertise* such as deliverables from "Suppliers", "Research, Development and Educational institutions" and "Other knowledge-intensive services" amounted to 3.8 billion NOK in 2005, representing 10 percent of total marine export value. According to Olafsen *et al.* (2006) this segment could increase its share of marine exports to 25 percent by 2025.

THE EVOLUTION OF FIRM STRUCTURE AND INNOVATION STRATEGIES IN NORWAY'S AQUACULTURE INDUSTRY

The Norwegian aquaculture industry has undergone significant structural change since 1990, reflecting consolidation among firms that have increased in size, as well as firms' pursuit of horizontal and vertical integration in their operations. Another important source of change in Norwegian aquaculture is the growth of international activities by leading Norwegian firms, which paradoxically have strengthened the production and technological capabilities of non-Norwegian aquaculture industries. Finally, the role of innovation, which has always been central to the development of this industry, has undergone considerable change, reflecting the growth and decline of firms that utilize different approaches to innovation management. This section discusses the changing domestic and global structure of Norwegian aquaculture firms and considers the different approaches to innovation management pursued by different types of firms within the industry.

Horizontal and vertical integration in Norwegian aquaculture

The Norwegian salmon industry was originally an owner-operated industry with hundreds of small single-farm firms. The changes in the Aquaculture Act of 1991, however, triggered considerable sales of existing licenses and a growing number of licenses for salmon and trout are today owned by limited-liability companies. In 2001 97.3 percent of fish food licenses were held by such entities, a substantial increase from 86.3 percent in 1992. The share of salmon and trout licenses owned by sole proprietorships has seen the opposite trend. In 2001 only 1.2 percent of the licenses were held by such entities, a drop from the 1992 share of 6.7 percent. The 10 largest companies' share of total

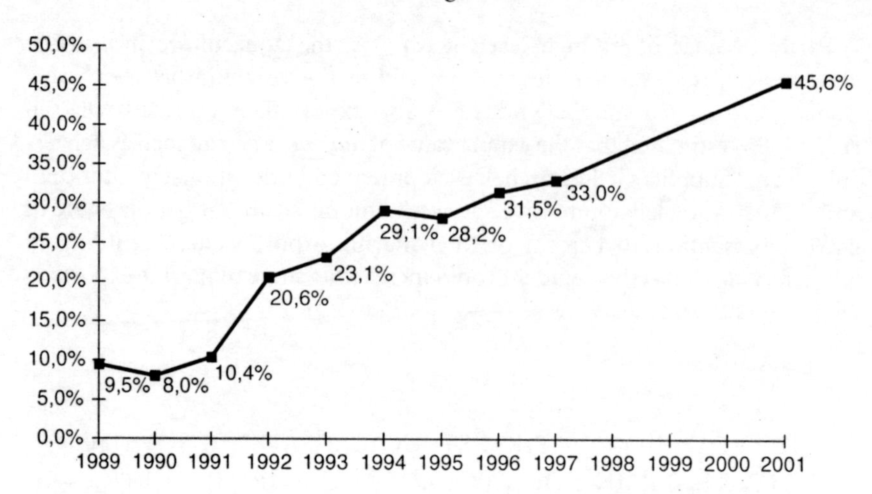

Figure 8.2. The ten largest aquaculture companies' share of total production, 1989–2001(%).

Source: Jakobsen *et al.* 2003.

production of farmed salmon in Norway also has increased from 8 percent in 1990 to 46 percent in 2001 (Figure 8.2). Growth in these firms' share is especially steep from 1991 to 1992, a period with many bankruptcies in fish farming (Jakobsen *et al.* 2003).

Increased horizontal integration in the salmon farming industry reflects the considerable production and price risk that characterizes the industry. The production period from release of salmon fingerlings to harvest is typically 12–18 months, a period during which the fish are exposed to diseases, temperature changes and extreme weather conditions (Tveterås and Kvaløy 2004: 12). The price of salmon also can change significantly during this lengthy "growing season."

Vertical integration of firms supplying salmon to Europe also has increased since the 1990s, reflecting growth of large companies with direct ownership of production hatcheries, fish processing and export operations (*ibid.* 4). Such vertical integration is relatively uncommon in other fields of large-scale commercial agriculture; but salmon farming has some idiosyncrasies that enhance the benefits of vertical integration. Fish are more perishable and unpredictable, which means that larger investments and higher degree of coordination in the supply chain is necessary to preserve product quality and increase shelf life (a lack of futures markets also exacerbates producer risk). It also is more costly to monitor external inputs in the production process and in a supply chain that spans international boundaries (*ibid.* 8). Figure 8.3 provides an overview of the core activities linked to fish farming, including

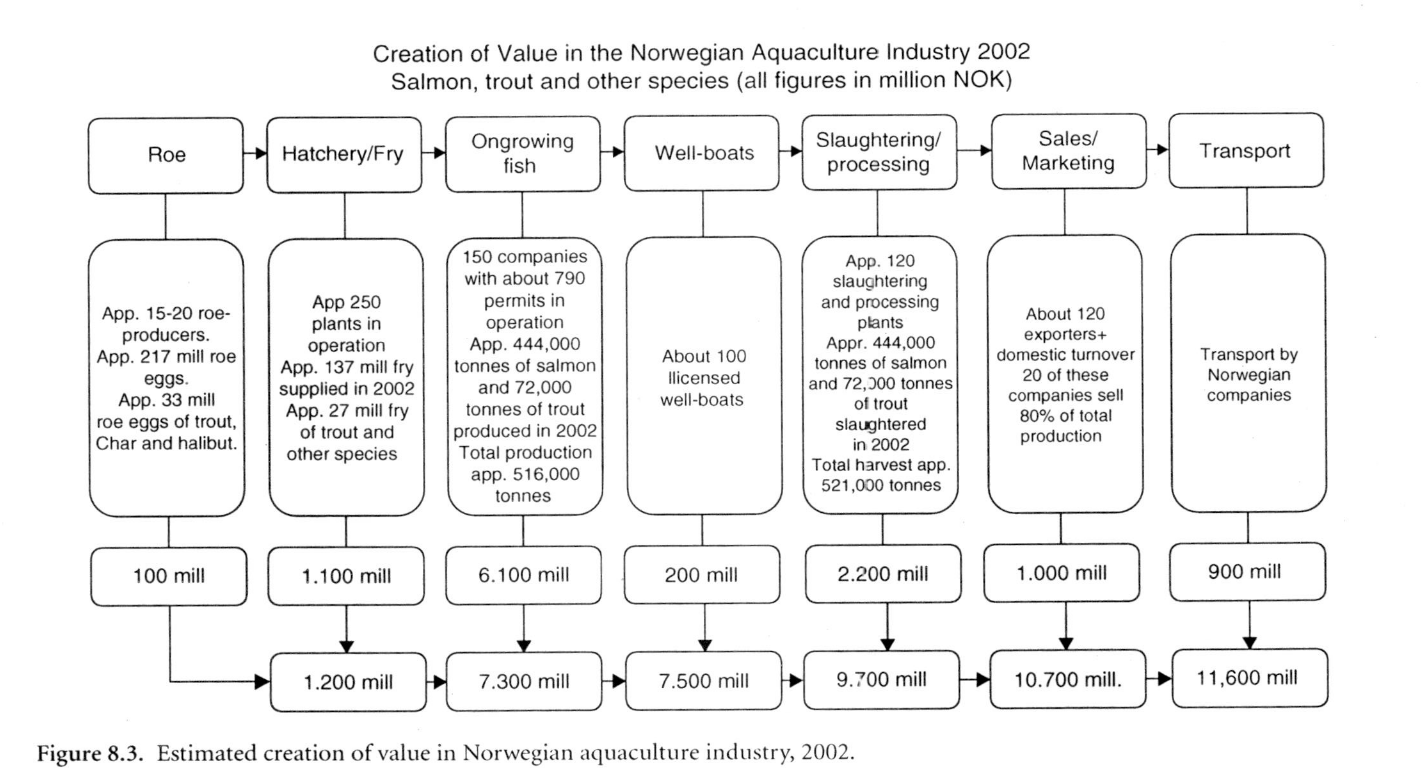

Figure 8.3. Estimated creation of value in Norwegian aquaculture industry, 2002.

Source: Kontali Analyse AS.

estimates on the contribution to total value added of each step of the value chain.

Today, there are between 15–20 roe producers and approximately 250 fry plants in operation in Norway, accounting for 1.200 million NOK in sales. However, the greatest revenues in aquaculture (6.100 million NOK in 2002) are associated with growing fish. 150 companies work in this segment of the value chain, which involves the raising and feeding of fish until slaughter, usually at a size of 3–6 kg. Well-boats are used to transport smolt from the hatchery to the farms, and fully-grown live salmon from farms to the slaughterhouse. All the salmon are slaughtered in specialized fish processing plants. In Norway, there are 120 exporters involved in the export of salmon and trout. Exporters are either part of a vertically integrated company that is involved in fish farming, such as Pan Fish Sales AS, or they may act as exporters for several companies, such as NRS (Norwegian Royal Salmon), or they can be independent. The exporters represent the link in the value chain that has most frequent contact with the market, including close relationships to foreign companies that import salmon and trout from Norway. At the same time they have relationships with a broad set of actors in Norwegian aquaculture.

The product portfolio in the export business has changed since the late 1990. Large-scale horizontal integration among the salmon producers has led several of these firms to integrate "forward" into the sales and distribution functions that were previously provided by intermediaries (Tveterås and Kvaløy 2004: 19). A diminishing share of the primary products now flows through traditional distribution channels and wholesale dealers. Simultaneously, the downstream end of the value chain now includes larger firms that optimize distribution of seafood through collaboration with retailers who reduce costs by building long-term relations with one or a few suppliers of fish. These suppliers in turn must meet stringent demands concerning volume and timing of deliveries, raw material attributes, product range and differentiation, production process and transactions costs. According to Tveterås and Kvaløy (2004), this vertical coordination of the supply chain—from salmon aquaculture production to the supermarkets—is a relatively recent phenomenon in Norway.

The largest fish farmers in Norway grew rapidly between 1990 and 2000, especially companies like Pan Fish and Fjord Seafood. Other firms that grew rapidly during this period, however, such as Hydro Seafood/Marine Harvest Norway, have stagnated more recently. According to Tveterås and Kvaløy (2004: 23) there is little evidence that the larger firms are more cost efficient than smaller firms, since biological production such as salmon farming requires motivated workers and managers with an economic stake (e.g. an ownership share) in production outcomes (Tveterås 1999). The economic effects of the consolidation processes in the industry are still to be seen.

Norwegian salmonid aquaculture in a global context

The main non-Norwegian producers of aquacultured Atlantic salmon are Chile, UK, USA/Canada, Faeroe Islands, Ireland, and Iceland (see Figure 8.4). Aquaculture of salmonids in these other countries has in varying degrees relied on Norwegian expertise, finance, and technology (Olafsen *et al.* 2006). The total production of farmed salmon from these countries was 1,130,700 tonne wfe[4] in 2003 (Winther *et al.* 2005) (see Table 8.1). Nearly one-half of this production, 587,000 tons, was sold in the EU.

Norway's world market share has decreased from 49 percent in 2000 to 46 percent in 2004, while Chile's market share grew from 19 percent in 2000 to 29 percent in 2004 (Winther *et al.* 2005). The most important market for Norwegian exports of farmed salmon is the European Union, where the Norwegian market share increased slightly in 2004. Chile's share of the EU market grew from 4 percent in 2000 to 7 percent in 2004 (Myrland 2003; Nielsen 2003; referred in Winther *et al.* 2005). The main explanations for Chile's increasing market share are the country's adoption of best practice

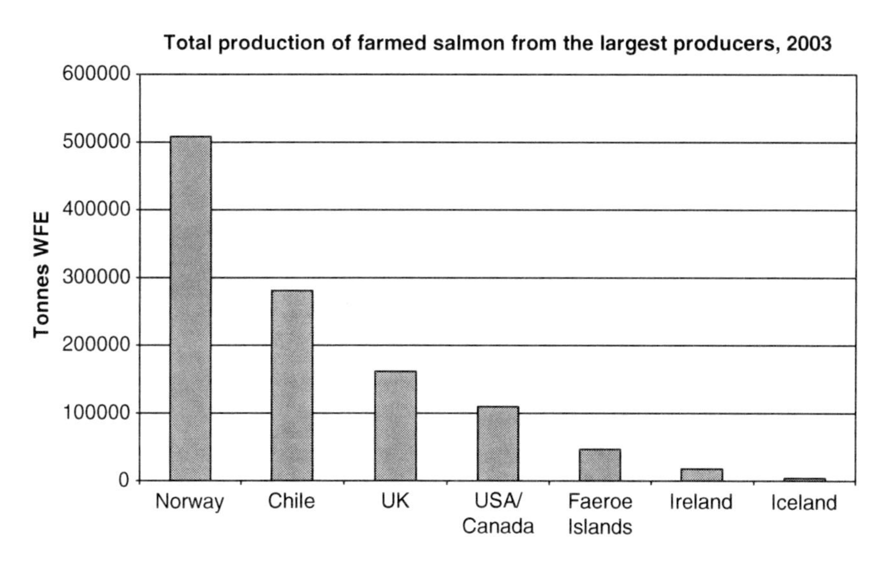

Figure 8.4. Total production of farmed salmon from the largest producers, 2003.
Source: Adapted from Winther *et al.* 2005 based on statistics from Kontali Analyse A/S.

Table 8.1. Companies ranked by global production of salmon and rainbow trout, 2003 (in metric tonnes)*

Company	Headquarter	Total production 2003	Norway	UK	Chile	Canada	USA	Other countries
Nutreco	Netherlands	178,500	70,000	32,000	59,000	12,500		5,000
Pan Fish	Norway	86,100	31,100	20,500		9,800	12,000	12,700
Fjord Seafood	Norway	72,500	35,000	7,000	28,000		2,500	
Stolt Sea Farm	Norway	70,500	15,000	6,000	24,000	25,000	500	
Cermaq	Norway	48,500		8,000	32,500	8,000		
Aquachile	Chile	48,000			48,000			
Pesquera Camanchaca	Chile	37,000			37,000			
Cultivos Marinos Chiloe	Chile	34,500			34,500			
Salmones Multiexport	Chile	34,000			34,000			
Pesquera Los Fiordos	Chile	33,000			33,000			

* The total production of farmed Atlantic salmon was, according to Kontali, 1,130,700 tonnes wfe in 2003 (Winther *et al.* 2005). Although trout is included in the table, comparing these figures gives an indication that the ten largest companies produced nearly half of the supplied Atlantic salmon in 2003.

Source: Tveterås and Kvaløy (2004).

technologies, exploitation of economies of scale, and lower labor and feed input costs (Bjørndal *et al.* 2004 referred in Winther *et al.* 2005).

Norwegian fish farmers have become more internationalized over the past few years, investing in fish farms in other countries, especially Scotland, the Faeroe Islands, USA, and also in Chile. Some of the diminishing share of Norway's world production reflects the growing "offshore" operations of Norwegian firms.

Table 8.1 below shows the ten largest global companies engaged in farming of salmon and rainbow trout in 2003. Nutreco, headquartered in the Netherlands, was the largest global producer of salmon and rainbow trout in 2003. Nutreco also was the largest single producer in Norway in 2003, and has large operations in Chile, UK and "other countries". Pan Fish, Fjord Seafood, Stolt Seafarm, and Cermaq, all headquartered in Norway, are the four next largest producers, all having large production operations in Norway and in several other countries. According to Tveterås and Kvaløy (2004), Pan Fish produced almost the same amount of farmed salmon in the UK as the UK-based company Scottish Seafarmers, and Stolt Sea Farm accounted for larger production of farmed salmon and rainbow trout in Canada than the largest Canadian firm, George Weston/Connors. Chile is also an important home for global aquaculture firms, with five of the ten largest firms headquartered in Chile. These companies have no farming facilities in other countries.

The responses to interviews with Norwegian aquaculture firms that produce in foreign locations (Aslesen *et al.* 2002) suggest that Norwegian companies derive operating-cost and marketing advantages from these offshore operations. The changing Norwegian government policy on concessions meant that the costs of expansion of aquaculture activities in Norway during the past decade have fluctuated, and at various times have made foreign expansion more attractive. Establishing production operations in other countries also provides proximity and flexibility in the most important markets for salmon. Operating in leading foreign markets (in i.e. Scotland–EU, Canada–North America) also reduces the risk that Norwegian firms could face trade barriers. Foreign operations also can reduce the risks associated with concentrating production operations in any single site (e.g. if a site experiences a serious disease outbreak). Facilities abroad also lower logistics costs and increase flexibility in production, improving Norwegian firms' delivery performance in these markets.

Innovation strategies within aquaculture firms

A central concern in most studies of innovation systems is how learning is achieved through interaction with others—so-called interactive learning. The knowledge base of the firm is one of the most important background variables to understand innovation strategies and interactive learning. By understanding the structure of the knowledge base, we may also gain insight into the types of interactive learning processes that are observed empirically.

The analysis is based on 25 in-depth interviews with actors in the aquaculture industry (Aslesen *et al.* 2002; Aslesen 2004). The interviews were based on semi-structured guides developed in the two projects, focusing especially on innovation activities, innovation collaboration, networks, and innovation strategies. The main aim of the interviews were to understand how new knowledge is created, used and spread in the innovation system. Based on these interviews we found it reasonable to distinguish between two different strategic approaches among aquaculture firms:

1. A large number of firms base their activities, and the development and gradual improvement of these activities, almost exclusively on practical knowledge. Learning happens by experimenting, and the knowledge base underlying innovation activities is to a large degree tacit. Knowledge is gained through experience from operations and through solutions based on accessible practical and tacit knowledge in the field of aquaculture. We denote such a knowledge base a *practical knowledge base.*
2. Another significant number of firms base their innovation efforts on interaction with and contributing to a sectoral system of scientific knowledge. The learning in these firms is based on interactive development and use of new scientific or technological knowledge and can as such be denoted as *formal knowledge* involving both scientific and technological knowledge.

This field research, suggests that aquaculture firms are not placed in a continuum between the two alternative approaches to innovation, but instead tend to choose either one or the other approach. Innovation and knowledge strategy varies with firm size, but our interviews indicated that this relationship is complex. The aquaculture industry consists of a large number of small firms, and a much smaller number of big firms. But the smaller firms pursue very different approaches to organizing innovation-related activities. A surprising number of firms rely on ad-hoc solutions to structural problems. A number of other firms, usually older firms, have a more structured management structure and more elaborated, functionally differentiated organizations.

This line of argument provides us with the following classification of aquaculture firms and their different approaches to managing innovation, which is discussed in greater detail immediately below (see Table 8.2).

Table 8.2. Stylized variants of aquaculture firms

Knowledge base Organzation	Practical	Technological/scientific
Entrepreneurial, ad-hoc	(1) "The family firm"	(3) "Research based entrepreneurs"
Structured management system	(2) "The coastal enterprise"	(4) "Science based process industry"

(1) "The family firm"

"The family firm" is usually a small aquaculture company run by the owners, often a single family that may include second-generation family members in the running of the company. The different functions within the firm often overlap and rely on knowledge obtained from practical experiences. The aquaculture technicians, the management and the operational personnel are often one and the same person ("a Jack of all trades"). If there are more people involved, they rely on face-to-face contact and dialogue. This facilitates operational improvements as well as other innovations. The "family firm" rarely controls more than its own little piece of the whole value chain. Management focuses on costs and is financially conservative, because of the unpredictability (influenced by biology, market conditions, and regulations) characterizing production. Decisions are taken swiftly and the "written word" is of little value. Management philosophy is often that "*money not made today will never be made*."

Problems do arise that the firms cannot solve by themselves. These actors primarily search for solutions to such problems by drawing on the knowledge intensive-service activities available from the vertical networks of suppliers of feed, equipment (also ICT), and medication, as well as their competitors (similar firms) (see Box 8.1). The family firm will take part in development projects with other firms and institutions in their vertical network and receive valuable knowledge of practical- and experience-based nature (as opposed to scientific). Knowledge related to adjustments and incremental development activities (especially technologically related knowledge) is a very important source of innovation in the "family firm."

Another important influence on innovation within the family firm is the open and accessible knowledge that exists in the Norwegian aquaculture cluster as a result of fish farmers meeting informally and sharing their knowledge on "best practices". "Everybody knows everyone", and as such the channels for the free flow of knowledge operate effectively, albeit more efficiently within regional clusters of family firms. Reflecting their regionally concentrated

Box 8.1. Example of an innovation in the "family firm"

Incremental innovation and adjustments rely on regular interactions between the firm and especially the technology suppliers. For example the use of new steel constructions at different localities and new anchorage systems has been an important source of improvement in fish farming. The firm has also acted as a test site for equipment developed by a local entrepreneur that gathers dead fish and feed spillovers and provides information that supports optimal feeding of the fish. As such the firm has engaged in an innovation project without bearing the financial burden or the entirety of the risk.

character (among other things), however, these knowledge nodes do not support diversity in knowledge sources, and channels to the scientific community are both rare and inefficient.

Together with their suppliers, some aquacultures firms have joined user-oriented projects managed by the Research Council of Norway (RCN) and through these projects have received input from Regional Technology Offices. Supplier firms in particular often serve as important bridge builders between the family-run aquaculture firms and the knowledge infrastructure.

(2) "The coastal enterprise"

"The coastal enterprise" has developed beyond the ad-hoc organization of the family firm, and displays a more mature and formalized organizational structure and operating procedures. The firm has developed a more professional and functionally differentiated organization, with permanent management. The "coastal enterprise" seeks to exploit the advantages of horizontal integration (mergers with other fish farms) and vertical coordination down the value chain that were discussed elsewhere in this chapter. The "coastal enterprise" represents a shift from the organization adopted from traditional fisheries (exemplified by the "family firm"), towards a more "industrial" value chain, although the "coastal enterprise" organizations still have a long way to go.

Several of the "coastal enterprises" pursue a strikingly "anti-innovation" strategy. We are told explicitly that they do not, and do not plan to, carry out any research and development, and they systematically avoid being in early adopters of new technologies and solutions. Most innovation thus results from activities that are similar to those in the family firms, basically involving copying and through trial and error. The "anti-intellectual" characteristic of this culture seems to have been only reinforced by the increasing pressures towards increased efficiency and adjustment to lower prices that have led

Box 8.2. Example of an innovation in a "coastal enterprise"

A firm had taken part in a development project with a local supplier of technology. The collaboration ended in an innovation that makes it easier to change the fishing net in the net cages and as such gives a better environment in the net cages. The use of the so-called "Environment drum" means that fish farmers do not need to impregnate the fishing net as frequent as before. This has both an economic and environmental benefit. Firms interviewed indicated that one of the aims of the project was to make the drum fully automated. This product is today commercialized, suggesting a successful innovation.

managers in many of these firms to characterize administrative capacity as a low-priority "luxury".

In spite of these anti-intellectual and anti-innovation strategies, the coastal enterprises must engage in innovation. Pressures towards innovation in the "coastal enterprise" stem from the operational side, and, as in the family firm, often result from interaction with suppliers of equipment (see Box 8.2). Innovations take place by employing new equipment, new feeds, etc., and at times, the risks associated with possible failure are shared with the suppliers. The interviews indicated that these firms' innovation activities also relied on finance from RCN's user-driven projects. Innovation Norway was used by several of the managers of these firms who were interviewed as an important contributor to innovation, both for direct investment support and for "softer" inputs of importance to innovation, such as different types of business courses and training. One of the interviewed firms had also received public money for the marketing of their products abroad.

Important parts of the knowledge infrastructure are still not integrated in the "coastal enterprises" innovation efforts, raising challenges to these organizations' long-run viability. The growing requirements within the industry for vertical coordination are likely to increase pressure on these firms for long-term, strategic innovation projects (i.e. buyers demand for high quality attributes, raw material in feeds) of a type that has been rare.

(3) "Research based entrepreneurs"—profiting from being in front

The focus of this class of aquaculture firm has moved away from operational aspects to scientific and technological knowledge development. These firms rely on knowledge generation, coupling diverse forms of knowledge and including knowledge developments at the frontiers of scientific research. Operational aspects of aquaculture may actually be irrelevant to such firms,

Box 8.3. Example of an innovation in a "research based entrepreneur"

This firm has close collaboration with both fish farmer and the most important national and international aquaculture related research institutes and universities. Innovations in this firm include developing breeding schemes to create fish that are the most resistant to diseases. The firm also has innovation projects taking into use DNA technology in order to use paternity tests on fish. Simultaneously the firm had started a breeding scheme for cod.

and they often are outsourced to others. Despite their highly technological orientation, these firms pursue an open strategy vis-à-vis technological developments (based on shared and generally accessible knowledge), and pursue joint projects with research institutes and universities.

The research-based firm often depends on access to scientific knowledge. "Research-based entrepreneurs" are often tightly connected to the publicly financed "knowledge infrastructure" in Norway and maintain close relationships with other research activities and institutions. Surprisingly, many of these firms do not try to shield new solutions internally within their boundaries, instead preferring an open and accessible knowledge base for the whole cluster. There are a number of reasons for this. No one company possesses the resources needed to privatize and control both new knowledge generation and the practical application of new knowledge. Instead, they depend on the collective knowledge base of the industrial cluster, and as such have every reason to keep the commons open. Of course, there are exceptions to this characterization. Some very competent research-based firms depend on open sources of knowledge while protecting their own knowledge and technology through secrecy and formal IPR protection.

"Research-based entrepreneurs" pursue innovations of a more radical nature (products or processes that are new or have been significantly improved, and/or also novelties in the market), since "new" knowledge is the driving force of the firms' investments/business activities (see Box 8.3). The projects are often long lasting, and interviews suggested that many research based entrepreneurs' utilized public or semi-public Regional Technology Offices for the purpose of obtaining co-financing of projects.

(4) "Science based process industry"

Increased demand for quality, reliability, and safety in farmed fish and shellfish has become an important source of pressure on innovation in recent years. Demands concerning food safety and traceability of products in the processing chain are growing stronger, and together with regulatory changes have influenced innovation-related activities. Important challenges for aquaculture firms include the establishment of high-quality and stable channels of access to the market, in order to be able to understand, "translate," and use market signals in their innovation-related activities. The fourth organizational model, "the science based process industry", includes firms that are able to exploit both horizontal and vertical integration and to control the development and application of knowledge, in contrast to the "research based entrepreneurs". This segment of the Norwegian aquaculture industry is only in its infancy, and no Norwegian firms yet can be placed in this category.

SUMMARY

Knowledge of fish farming has accumulated in different communities in Norway through time. The learning process was long and drew on both experience-based and scientific knowledge. Beginning in the 1970s, a combination of new knowledge and actors enabled solutions to specific problems that enabled commercial fish farming. Innovations and change have occurred in all parts of the value chain in aquaculture over the years, making the industry continuously more productive. The sector now employs approximately half the people that it did ten years ago while tripling sales during the same period. The value of exported aquacultured salmonids is approximately half the value of the total export of fish and fish products from Norway. Since 1990, the industry's structure has been transformed by increased horizontal integration and growing concentration in the ownership of licences. In 2001 the 10 largest companies accounted for 46 percent of total production, a significant increase from their 1989 share of 9.5 percent. There are also several examples of vertical integration in the supply chain of fish farming the last years. Norwegian fish farmers have become more internationalized and have bought up aquaculture firms in other countries.

Research activity in Norwegian public and semi-public institutions has been of great importance to the development of the aquaculture industry in Norway, and made possible the enormous growth and productivity gains experienced in the sector during the 1990s. Public sources provide more than half of the funding for aquaculture R&D conducted in Norway. Other important innovation drivers in aquaculture are the technology suppliers and the feed suppliers. Much of the innovation in aquaculture firms can be traced back to the supply industry.

By differentiating aquaculture firms according to their knowledge base and degree of formalization in management structure and procedures, we found that different categories of aquaculture firms have very different approaches to innovation; from "anti-innovation" strategies to strategies that seek to be in the forefront of innovation in the industry. Firms with very different approaches to innovation and very different linkages to the sectoral innovation system exists side-by-side within the industry and influence the overall functioning of the sectoral innovation system of aquaculture.

The discussion in this chapter reveals that knowledge generation is peripheral to the main activities of most Norwegian aquaculture firms, which depend instead on suppliers and the research community. Thus, most firms in this sector resemble Wicken's first path (Chapter 2), e.g. an innovation system characterized by localized learning and weak connections to the knowledge infrastructure. Such firms may face increasing competitive pressure in the future, however, especially if aquaculture firms in other countries are more effective in knowledge generation and exploitation than Norwegian firms. Measures to improve the Norwegian aquaculture industry's innovative ability might therefore be appropriate. Such measures may be directed at stimulating firms to expand their strategic and analytic knowledge base, in order to enable them to gain a better understanding of the uses of research as a tool for product- and process development to meet future market demands.

NOTES

The empirical work underlying the chapter was undertaken within the framework of two different projects focusing on innovation practices and the business system in Norwegian aquaculture. A project focusing on the innovation system in aquaculture was particularly important. This project was financed by the Ministry of Fisheries in 2002. Åge Mariussen, Heidi Wiig Aslesen, and Finn Ørstavik (all from the STEP group in Oslo) formed the core research team for the project together with Ulf Winther and Trude Olafsen from KPMG in Trondheim. The project was reported in Aslesen *et al.* (2002). Another important project was the OECD project focusing on knowledge intensive service activities on the aquaculture industry, reported in Aslesen (2004).

1. Salmonidae is a family of ray-finned fish, the only living family of the order Salmoniformes. It includes the well-known **salmons and trouts**; the Atlantic salmons and trouts of genus Salmo give the family and order their names.
2. Based on Olafsen *et al.* (2006), technology in this context is defined as technological solutions for land-based brood stock, fry, and smolt production, and sea-based food fish production. The transport of living organisms in well-boats and

tank lorries is also included. Examples of land-based technologies are the construction of buildings, energy installations, tank constructions, control systems, IT systems, feeding systems, etc.

3. By comparison, the export value of the Norwegian petroleum sector was 346 billion NOK in 2005 (crude oil, natural gas, suppliers, etc.) or 10 times more than the export value of fish and fish products (Olafsen *et al.* 2006).
4. WFE = Whole Fish Equivalent, i.e. whole fish that is starved and bled (Winther *et al.* 2005: 8).

REFERENCES

Aarset, B. (1998) "Norwegian salmon-farming industry in transition: dislocation of decision control," *Ocean and Coastal Management*, vol. 38, no. 3, pp. 187–206.

Aslesen, H. W. (2004) "Knowledge intensive service activities and innovation in the Norwegian aquaculture industry," STEP report 5/2004. Oslo: STEP.

—— Mariussen, Å., Olafsen, T., Winther, U., and Ørstavik, F. (2002) "Innovasjonssystemet i norsk havbruksnæring," STEP report 16/2002. Oslo: STEP.

Berge, D. M. (2000) "Samfunn, entreprenørskap og kunnskapsspredning i norsk fiskeoppdrett på 1970-tallet", in H. Gammelsæter (ed.) (2000), *Innovasjonspolitikk, kunnskapsflyt og regional utvikling*, Trondheim: Tapir Akademisk Forlag, pp. 159–78.

—— (2002) "Dansen rundt gullfisken. Næringspolitikk og statlig regulering i norsk fiskeoppdrett 1970–1997," PhD dissertation, University of Bergen and Møreforsking Molde.

Bjørndal, T., Pena, J., Tveterås, R., Tveterås, S. (2004) "Firm concentration and vertical integration in salmon farming: empirical evidence from Chile and Norway," Proceeding from EAFE XVIth annual conference, Rome, 5.-7. April 2004.

FAO (2002) "The State of World Fisheries and Aquaculture".

Jakobsen, S-E. and Aarset, B. (2007) "På en skyfri dag kan du se helt til Brussel. Om EUs innflytelse på norsk akvakulturpolitikk," in Aarset, B. and Rusten, G. (eds.) *Akvakultur på norsk*, Fagbokforlaget (forthcoming).

—— Berge, D. M., Aarset, B. (2003) "Regionale og distriktspolitiske effekter av statlig havbrukspolitikk," Working Paper 16/03, SNF Bergen.

Møller, D. (1996) "Forskningsprogrammet Frisk Fisk—et historisk tilbakeblikk," in *Frisk Fisk. Om liv og død i merdene*, Universitetsforlaget, Oslo.

Myrland, Ø. (2003) "EU avtalen—Gull og Gråstein"—Presentation Aqua Nor 2003, Trondheim 14. August.

Nielsen, M. (2003) "Beregningsgrundlag for uforarbejdet fisk i Danmark, Arbeidspapir til Fiskeriets Økonomi 2003," Fødevareøkonomisk Institut. Afdelingen for fiskeøkonomi og—forvaltning.

Olafsen, T., Sandberg, M. G., Senneset, G., Ellingsen, H., Almås, K., Winther, U., Svennevig, N. (2006) "Exploitation of Marine Living Resources—Global

Opportunities for Norwegian Expertise," report from a working group appointed by DKNVS and NTVA.

Reve, T. and Jakobsen, E. W. (2001) *Et verdiskapende Norge*, Universitetsforlaget, Oslo.

Sandberg, M. G., Olafsen, T., Sætermo, I. A., Vik, L. H., Nowak, M. (2005), "Betydningen av fiskeri- og havbruksnæringen for Norge—en ringvirkningsanalyse 2004," SINTEF REPORT.

Schwach, V. (2000) *Havet, fisken og vitenskapen. Fra fiskeriundersøkelser til havforskningsinstitutt 1860–2000*, Havforskningsinstituttet, Bergen.

Sundnes, S. L., Langfeldt, L., and Sarpebakken, B. (2005) "Marin FoU og havbruksforskning 2003," Skriftserie 3/2005, NIFU STEP.

Tveterås, R. (1999) "Production Risk and Productivity Growth: Some Findings for Norwegian Salmon Aquaculture," *Journal of Productivity Analysis*, Vol. 12(2), pp. 161–79.

—— (2004), "Markeder og produksjon i den globale sjømatnæringen," Unpublished paper.

—— and Kvaløy, O. (2004) "Vertical Coordination in the Salmon Supply Chain," Centre for Research in Economics and Business Administration, Working Paper No. 7/04.

Winther, U., Sandberg, M. G, Stokka, A., Setermo, I. A., Nowak, M., Vik, L. H., Hynne, H., Kvinge, T. (2005) "Employment in the EU based on Farmed Norwegian Salmon," SINTEF FAFO report.

Ørstavik, F. (2004) "Knowledge spillovers, Innovation and Cluster formation: The case of Norwegian Aquaculture," in Karlsson *et al.* (eds.) *Knowledge Spillovers and Knowledge Management*. London: Edward Elgar.

9

The Biotechnology Industry in Norway: A Marginal Sector or Future Core Activity?

Terje Grønning

Norway's policymakers frequently cite the huge potential of the nation's biotechnology industry. They single out marine biotechnology as an area of great potential, emphasizing the country's past experience in fisheries and its long coastline, which contains a large number of marine organisms whose commercial potential (particularly in the field of biotechnology) has not yet been fully explored (MTI 1998; RMV 2003; Eggset 2005). This chapter examines the extent, character, and future potential of the biotechnology industry in Norway. The biotechnology sector is not as closely connected to the historical patterns of Norwegian specialization as aquaculture, aluminum, or oil and gas, which represent the small scale decentralized and the large scale centralized paths within the Norwegian innovation system outlined in Chapter 2. It is reasonable to expect that the emergence of a biotechnology industry represents an example of the third "R&D intensive network based path," but it remains to be seen whether and in what ways this indeed is so.[1]

Any such inquiry must first investigate the characteristics of established biotechnology firms in Norway: What kinds of biotechnology R&D are being conducted, how extensive are these activities, and to what extent do they represent new activities, rather than developments rooted in past fields of specialization? Moreover, the overall character of the industry in Norway today has to be assessed, since many scholars argue that the potential of a biotechnology industry in any given country or region is associated with the emergence of a broader set of organizational actors (e.g. academic research, customers, and investors) in addition to the biotechnology firms themselves. Does the biotechnology industry in Norway today represent an industry with a strong sectoral innovation system, or does it instead consist of geographically and industrially diverse activities that lack a strong knowledge-based infrastructure?

With these broad research questions in mind, the next section discusses contrasting perspectives on the emergence of biotechnology industries and describes the methodology for selecting and analyzing data on the Norwegian biotechnology industry. Subsequent sections examine the emergence and current distribution of Norwegian biotechnology firms and trends in the environment of biotechnology related firms. The final section of the chapter assesses the dynamics and potential of biotechnology in Norway.

THEORY AND METHODOLOGY

Theory

In order to understand the current position and the future prospects for Norway's biotechnology industry, I first summarize the views of scholars who have contributed to the large literature on technological dynamics in biotechnology at the firm and at the national level. This summary provides some context for the challenges facing Norway's biotechnology sector in a national innovation system that contrasts sharply with those of such leading sites for biotechnology as the United States or the United Kingdom.

Research on the establishment and growth of biotechnology firms has long focused on the relationship between intrafirm competencies and the external environment (Pisano 2000; Owen-Smith *et al.* 2002). One side in this debate argues that appropriate intrafirm capabilities are the most important factors in the establishment and growth of firms. The opposing view asserts that environmental factors are crucial, reflecting (among other things) the science-based nature of biotechnology firms. As most proponents within the debate acknowledge, these two perspectives are more complementary than opposed, and more recently an alternative view has been proposed, based on a critique of the basic premises of each side in this debate.

Advocates of the importance of firms' internal capabilities argue that biotechnology firms can be established anywhere, as long as their internal capabilities are sufficient or can be supplemented through collaboration with partners that have the capabilities that the firm itself lacks (de la Mothe and Niosi 2000). Much of this literature (see e.g. Niosi 2003 for a review) accordingly highlights the role of alliances as a vehicle for firms to gain access to needed capabilities. Empirical studies of the role of intrafirm factors in firms' long-term development indicate that firms tend to expand in the same technical field or in fields close to the ones in which it is already proficient (Teece 2000).

The second view emphasizes the importance of the firm's external environment, because of the special science-based character of biotechnology firms. In spite of the globalized character of biotechnology knowledge, a successful firm requires a national or regional innovation system that has sufficient R&D investment, a labor market for scientists and technicians, and research facilities (Bartholomew 1997; Swann and Prevezer 1996). Public policy is important in supporting the growth of an R&D infrastructure, securing an adequate supply of science and technology graduates, and smoothing the relationships among firms, universities, and public research institutes. This view is encapsulated in Niosi's statement that "networks of DBFs (dedicated biotechnology firms), venture capital firms, large corporations and research institutions are key in the dynamics of biotechnology" (Niosi 2003: 738), although it is not entirely clear whether the "networks of DBFs" that he cites are restricted to regional, national or supra-national levels.

Still other scholars have expressed reservations concerning the validity of the sharp contrasts drawn by the opposing sides in this debate over the roles of intrafirm and external factors in the growth of biotechnology. These reservations are based in part on the observation that much of the analysis of the biotechnology industry has focused on the United States. The biotechnology industry in other nations, such as those in Europe, might evolve along significantly different trajectories in which intrafirm and external factors play very different roles (Senker 2000). Feldman (2003) also modifies the emphasis on external factors in arguing that successful "anchor firms can support the creation of regional agglomerations that attract new, specialized firms, reinforcing the path-dependent development of regional concentrations". Still another qualification to the debate highlights the fact that a large proportion of the literature on biotechnology focuses on biopharmaceuticals (McKelvey and Orsenigo 2006), ignoring other segments of the modern biotechnology industry. Although agglomeration and external relationships may be important to the growth of firms in biopharmaceuticals, the benefits of these environmental factors are less clear for other segments of the biotechnology industry that may be less dependent on science-rich locations, such as agricultural or industrial biotechnology. The debate over the importance of external infrastructure and agglomeration effects is clearly relevant to the case of Norway's biotechnology industry, which is geographically dispersed within Norway and operates within an institutional environment of public R&D infrastructure and industrial finance that have few analogues among industrial economies with strong biotechnology industries.

Methodology

This chapter's overview of Norwegian biotechnology includes an extended description of selected characteristics of Norway's existing biotechnology firms.[2] I also provide an overview of external factors, including a discussion of the past and current nature of public policy, universities, public research institutes, and venture capital firms. The discussion of the characteristics of Norwegian biotechnology firms highlights the fact that biotechnology in Norway thus far displays few of the agglomeration processes that appear to have been important in the industry's growth in other areas, such as the United States. Does the unusual pattern of development of Norway's biotechnology industry preclude the emergence of a viable network of biotechnology firms, venture capital firms, large corporations and research institutions? Does Norway's limited infrastructure of academic research organizations and the nation's remoteness from science based centers such as the United States, United Kingdom, and continental Europe mean that its biotechnology firms cannot achieve success while remaining independent? Does Norway's location on the European "periphery" mean that its biotechnology industry is doomed to commercial failure, or does Norway have special potential or strengths in new areas of biotechnology, such as marine biotechnology (biotechnology activities focused on organisms from the oceans).[3] Will Norwegian biotechnology pursue a development path that differs from those emphasized in studies of pharmaceutical biotechnology?

The chapter's discussion of Norwegian biotechnology firms uses data from a sample of 93 firms. Rather than a narrow definition of biotechnology activities that includes only the firms for which biotechnology is the principal activity, resulting in a low estimate of the number of Norwegian biotechnology firms (Critical I 2006), I use a broader definition that includes firms for which biotechnology is among their primary or secondary activities. The broader definition used for this study sacrifices precision for a more comprehensive picture of ongoing biotechnology-related activities in Norwegian industry. The resulting sample does not cover all Norwegian firms with biotechnology-related activities, but all of the included firms are active within biotechnology and span all of the important segments of Norway's biotechnology industry.[4] Firms are divided into seven basic product areas, ranging from pharmaceuticals to biochemical production and broad applications, including a category of firms that focus on both therapeutic and diagnostic applications as separate products or in combination (i.e. "theranostics," see Gilham 2002). The sample also includes a number of firms that focus on marine biotechnology, a segment of the biotechnology industry that has attracted considerable interest from policymakers. The sample

Table 9.1. Dedicated biotechnology industry and venture capital characteristics, selected countries, 2004–2005

	Employment	Revenues		Venture capital	
	Percentage of total civilian employment, 2004	Mill. (USD PPP) 2004	Per million GDP (USD PPP)	Venture capital investment as percentage of GDP, 2005	Health/ biotechnology venture capital investment as percentage of GDP, 2005
Norway	0.041	73	388	0.139	0.026
Denmark	0.671	4,773	27,354	0.401	0.231
Sweden	0.094	839	2,996	0.300	0.066
Finland	0.092	591	3,738	0.095	0.012
Austria	0.069	539	2,031	0.043	0.005
Belgium	0.087	688	2,108	0.038	0.013
France	0.037	2,403	1,326	0.082	0.007
Germany	0.041	3,122	1,318	0.057	0.007
Portugal	0.005	51	252	0.133	0.010
Italy	0.012	336	206	0.031	0.001
Netherlands	0.035	338	629	0.098	0.008
Spain	0.012	343	311	0.085	0.003
Switzerland	0.119	2,131	8,180	0.108	0.037
United Kingdom	0.074	5,059	2,713	0.292	0.018
United States	0.137	56,184	4,797	0.183	0.052

Notes: Percentage employment calculated based on employment numbers given in Critical I (2006) in relation to total employment (OECD 2006b). Revenues for Norway were particularly low in 2004 (down to 73 mill. USD from 128 in 2003, and down to 388 from 752 per mill. GDP in 2003). The "Employment" and "Revenues" data for Norway in Table 9.1 are based on a 42-firm sample that includes only biotechnology specialist firms, in contrast to the 93-firm sample used elsewhere in this chapter. PPP means conversion using purchasing power parity.

Sources: For revenues Science-Metrix (2006, based on Critical I 2006); for venture capital data OECD (2007).

of firms also includes firms in all of the four major geographic regions of Norway.

The "narrow" definition of biotechnology firms is nevertheless useful as a basis for comparing Norway's biotechnology industry with those of other European economies for which a similarly narrow definition of biotechnology firms has been used (Table 9.1). Norwegian firms specializing in biotechnology account for roughly the same share of total civilian employment within Norway as do the German, French, and Dutch biotechnology industries within their economies. Danish biotechnology represents a far larger share of employment within Denmark. Norway's biotechnology industry (using the "narrow" definition) accounts for a far smaller share of domestic employment than is true of the US, Swiss, Swedish, Finnish, and Belgian biotechnology industries.

Norway's specialized biotechnology firms also account for a relatively modest share of GDP as of 2004, exceeding that of the Italian, Spanish, and Portuguese biotechnology industries, but ranking well behind the biotechnology industries of the United States, Switzerland, and Denmark, among other industrial economies included in Table 9.1.

DISTRIBUTION OF BIOTECHNOLOGY FIRMS ACCORDING TO PRODUCT FIELDS AND GEOGRAPHY

The first segment of biotechnology firms in Norway, pharmaceuticals, includes a small number of firms but a large number of employees (Figure 9.1; Table 9.2). The Norwegian firms included in this product area are predominantly conventional pharmaceutical firms for whom biotechnology is a secondary focus. The second segment, therapeutics, includes a larger number of firms that collectively account for a small share of overall biotechnology-related employment. Examples include three smaller firms focusing on HIV

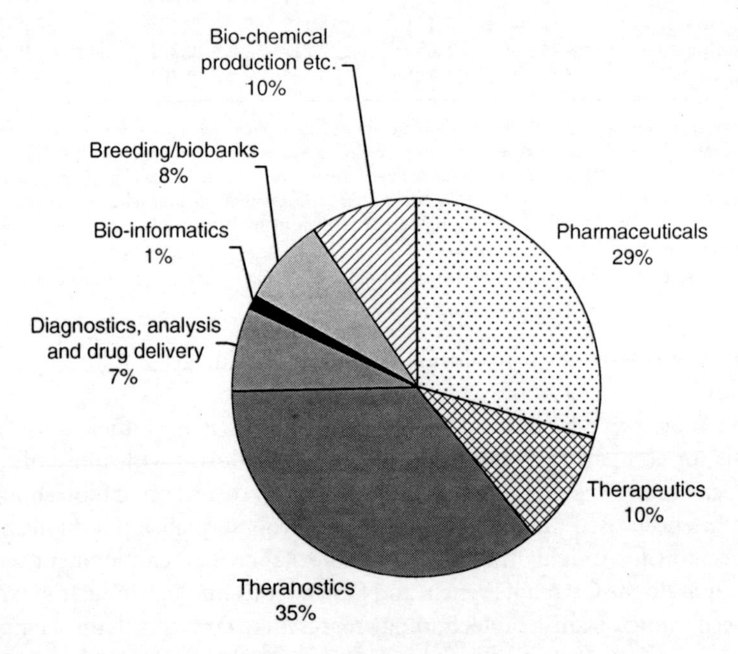

Figure 9.1. Approximate employment shares, Norwegian biotechnology segments, 2007.

Source: Table 9.2.

Table 9.2. Biotechnology firms in Norway, 2007

Product Area	Number of firms		Firms' headquarters in Norway				Total firms	Approximate number of employees
	Marine	Others	East	West	Central	North		
Pharmaceuticals	0	4	4	0	0	0	4	1,300
Therapeutics	7	16	14	4	3	2	23	440
Theranostics	0	4	4	0	0	0	4	1,635
Diagnostics, analysis and drug delivery	0	28	21	4	2	1	28	300
Bio-informatics	0	4	3	0	1	0	4	25
Breeding/biobanks	3	7	9	0	1	0	10	350
Bio-chemical production etc.	11	9	9	6	1	4	20	455
Total	21	72	64	14	8	7	93	4,505

Notes: This table includes all 93 firms used in the "broad" definition of biotechnology elsewhere in this chapter. For definition and explanation of segments see List of technical terms. For selection criteria, see Endnote no. 4. Firm headquarter refers to office in Norway for the firms with foreign ownership. "Bio chemical production etc." includes broad applications. Approximate employee numbers based on Proff.no/Purehelp.no as of December 2007 with the exception of 6 firms, where the source is Fastinfo.no or company home pages. Assignment of firms to industry segments has been based on author's assessment.

therapies with 2–7 employees each (A-viral, established in 1986; Bionor Immuno, established in 2000, and Lauras, established in 1998), and Clavis Pharma, a firm developing anticancer therapeutics, established in 1997 with approximately twenty employees as of 2007, that is based on research originally conducted at Norsk Hydro.[5] Marine biotechnology firms include Biotec Pharmacon (established in 1985), which develops several drug candidates based partly on marine biotechnology, and Pronova Biopharma, which developed the prescription drug Omacor, an Omega-3 based agent that was approved by the FDA in the US in 2004. Navamedic (established in 2001) emerged from a new method for producing glucosamine discovered by the company's founder and CEO while a student at the Norwegian Technical University, and the firm agreed in 2002 to co-develop a glucosamine drug together with one of the Norwegian pharmaceutical companies, Weifa. The product, which provides pain relief for osteoarthritis patients, was approved in 25 countries in the EU/EEA in December 2006. Two firms (Intervet Norbio, established in 1985 and Pharmaq, established in 2004) specialize in fish health products, such as vaccines for the aquaculture industry in Norway and abroad.

The third segment, "theranostics," includes firms active within both therapeutics and diagnostics, although it remains to be seen how successful these firms will be in developing integrated theranostics, i.e. "diagnostic tests

Box 9.1. Mergers, de-mergers, and acquisitions (highlights).

1984 Alpharma listed on NYSE.

1995 Norsk Hydro divests Swedish BioInvent (acquired in 1990).

1997 Merger between Amersham and Nycomed Imaging (subsequently acquired in 2004 by GE Healthcare).

1998 The firm Natural listed on Oslo Stock Exchange (OSE). Pronova Biocare acquired all the shares in Lipro, a firm established in 1877 as a producer of cod liver oil.

1999 Axis-Shield listed on London/OSE.

2002 FMC BioPolymer acquires a Norsk Hydro subsidiary that produces alginates and chitosans.

2004 Sarsia acquires majority of shares in Danish firm Medi-Mush.

2005 NorDiag, Biotec Pharmacon and DiaGenic listed on OSE. Dynal Biotech sold to Invitrogen. EPAX Sales and Production de-merged from Pronova Biocare.

2006 Clavis Pharma listed OSE.

2007 Algeta and Pronova Biopharma listed OSE. Nordiag acquires Genpoint. Aker BioMarine and Pronova signed a letter of intent to establish and develop pharmaceutical products from krill, and Natural merges with Aker Biomarine.

Source: Compiled by author based on company home pages and industry press.

directly linked to the application of a specific therapies" (Gilham 2002: 17). This segment dominates Norwegian biotechnology employment, since it includes the Norwegian division of GE Healthcare, which employs approximately 1,400 people,[6] as well as Invitrogen Dynal. The former traces its antecedents to Nygaard & Co (est. 1874) and its innovations in X-Ray and ultrasound related reagents (Amdam and Sogner 1994). Parts of this firm evolved into Nycomed Imaging, which was acquired by Amersham Health in 1997 and subsequently acquired by GE Healthcare. Invitrogen Dynal was first established as Dynal Biotech in 1986 on the basis of research conducted by the late Professor Ugelstad of the Norwegian Technical University in Trondheim. It was acquired by the US firm Invitrogen in 2007, and has approximately 150 Norwegian employees. Newer firms within this segment include smaller firms that have grown rapidly in terms of sales: PhotoCure (established in 1997, with roughly 40 employees as of 2007) targets research related to skin diseases and cancer related research, and develops and sells devices for these diseases based on photodynamic technologies. Affitech (also established in 1997, with 35 employees as of 2007) focuses on human antibody techniques. The majority of its development activities deal with therapeutics, but the firm's portfolio includes a "library" and high-throughput screening system for antibody discovery that can be classified as theranostics.

The fourth segment, diagnostics and drug delivery, includes a large number of small firms. The largest firm within the segment is Telelab, a laboratory with 55 employees that provides services to other firms. Other Norwegian firms in this category include Axis-Shield, which was founded in 1999 through the acquisition of Norwegian Axis by the British firm Shield and the subsequent acquisition of another small Norwegian company, Medinor (CGE&Y 2000: 14–15). Axis-Shield is now headquartered in Great Britain but is still partly financed by the Norwegian People's Pension Fund. The firm's Norwegian employment level, which stood at roughly 50 in 2003, was reduced after the firm reorganizations in the early 2000s (Critical I 2005: 9). Another firm with a relatively long history is Sero AS, established in 1963, which employed approximately thirty people in the early 2000s. It focuses on quality control of medical laboratories by using standardized biological solutions, and collaborated closely with Nycomed until 1998 (Amdam and Sogner 1994: 137). The firm DiaGenic est. in 1998 has 15 employees and focuses on early detection of diseases. In 2007 the firm announced the first validated prototypes for gene expression signature tests for the early diagnosis of breast cancer and Alzheimer's disease. The prototype enables detection of a disease by examining blood samples for specific gene expression signatures. Four Norwegian firms employing 2–25 persons each specialize in delivery technologies: CancerCure is developing a technique for delivering cancer drugs where they are needed by means of ultrasound mediated drug release; PCI Biotech, a subsidiary of PhotoCure (see above); Optinose, which is developing a nasal drug delivery system; and Inovio, which is developing delivery technology for skeletal muscles.

The fifth segment, bioinformatics, includes four specialized firms (Interagon, Lingvitae, Pubgene and Sencel Bioinformatics) that each employ 2–10 persons, all of which were established in the early 2000s. Interagon is a spin-off of one of the most renowned ICT firms in Norway, Fast Search and Transfer. The sixth segment, breeding and biobanks, consists of 10 firms whose average employment levels substantially exceed those of the firms in bioinformatics or diagnostics and delivery. Seven of these firms specialize in "Green" (i.e. agricultural) biotechnology, rooted in long-established Norwegian livestock breeding programs. One of these livestock breeding firms, Geno, employs approximately 250 workers and another, Norsvin, has roughly seventy employees. One smaller firm in this category (Genomar, established in 1996, and with five employees as of 2007) was founded by academics from the Veterinary College in Oslo with the goal of specializing in marine biotechnology. Genomar's business model initially focused on Atlantic salmon, but the firm now targets other farmed fish species, such as tilapia, and much of its business is conducted overseas. Two additional firms within this segment (Aqua Gen, established in 1985, and Genova Aqua, established in 1997) also focus on marine species.

The seventh and last segment of the Norwegian biotechnology industry includes firms specializing in biochemical production technologies, as well as miscellaneous products and processes with broad industrial and agricultural applications. This segment thus is a heterogeneous one, spanning industrial waste mechanisms, production of livestock feeds, and ingredients for other products.[7] Marine biotechnology firms account for a majority of this segment's firms and employees. Norwegian firms have entered nutraceuticals, i.e. foods that claim to have medicinal benefits, based in part on the possibilities within this area for expanded use of marine material, including the exploitation of byproducts from fisheries and aquaculture. Aker Biomarine focuses on food and feed industry, but seeks in the long term to develop pharmaceutical products as well. FMC Biopolymer, a subsidiary of the US-headquartered multinational firm, was established in 1999, and had roughly 180 employees as of 2006. This firm focuses on marine biotechnology and acquired two firms spun off by Norsk Hydro in 1999 and 2002.

The largest number of Norwegian biotechnology firms are in the third and fourth segments, consisting of theranostics, diagnostics and drug delivery, and accounting for slightly less than 2,000 workers (including 1,400 GE Healthcare employees) out of an approximate total employment in Norwegian biotechnology of 4,500 persons. Entry by smaller firms into these two segments increased sharply during 1996–1998 (Figure 9.2), partly as a result of the acquisition by Amersham of Nycomed Imaging in 1997 that led to the departure of a number of former Nycomed employees for Norway-based startups (Johannessen *et al.* 2005: 66; Grønning *et al.* 2006: 88). New-firm formation in marine biotechnology displays less year-to-year fluctuation, although there is some evidence of a slight increase in entry during 1996–2001 (Figure 9.2). The relatively large number of Norwegian marine biotechnology firms established before 1990 reflects the roots of several of these firms in older firms specializing in fisheries and aquaculture.

These growth patterns, particularly the surge in new-firm formation in the early 2000s, are reflected in the firm-size and—age distribution within Norway's overall biotechnology industry that is displayed in Figure 9.3. Not surprisingly, most of the remaining firms in the industry are young and small. But the nine largest Norwegian biotechnology firms (145–1,400 employees)

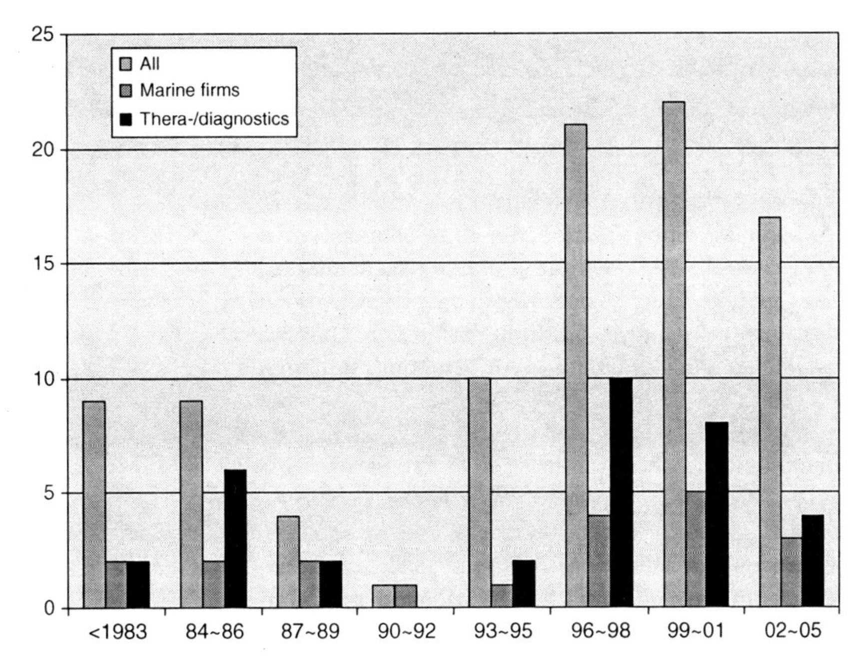

Figure 9.2. Year of establishment of Norwegian biotechnology firms.
Source: Table 9.2.

are also the oldest (in most cases these are firms founded in the nineteenth century that have developed biotechnology-based lines of business in the past 25 years), and include the four pharmaceutical firms, two theranostics firms, one marine technology firm developing therapeutic drugs, one firm producing ingredients based on marine technology, and one of the livestock breeding firms.

The data in Figure 9.4 show that marine biotechnology firms account for a markedly higher share of the biotechnology firms in Northern Norway, reflecting the important role of aquaculture and fisheries in the regional economy. Western Norway has a long tradition of fisheries, but also has a significant academic medical research complex at the University of Bergen and Haukeland Hospital. Central Norway has surprisingly few biotechnology firms, although several spin-offs from the Technical University of Trondheim (e.g. the current Invitrogen Dynal, located in greater Oslo) now are located in other regions.

In summary, the Norwegian biotechnology-related firms described in the previous paragraphs are a heterogeneous group in terms of age, areas

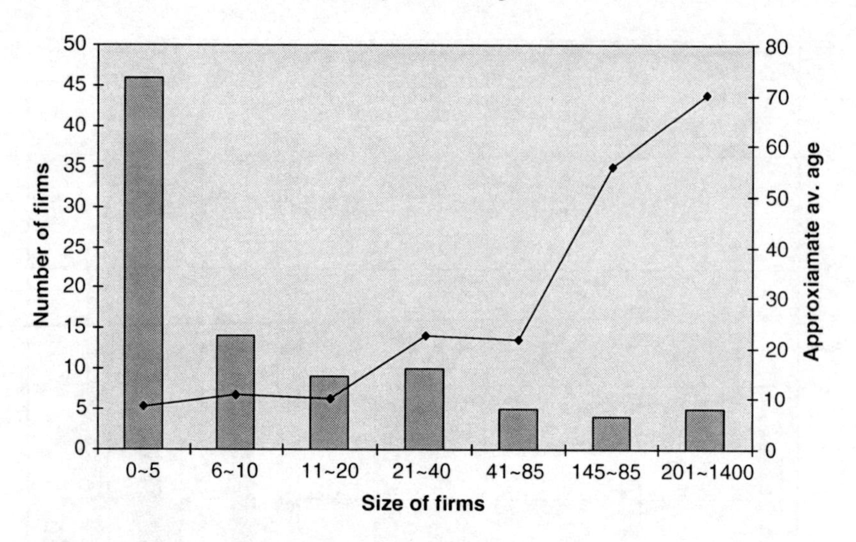

Figure 9.3. Size and age of Norwegian biotechnology firms.

Notes: "Age" refers to the year of establishment of a firm in Norway, rather than the founding year for the parent firm of foreign-owned Norwegian subsidiaries. The foundation-year data also include the year of establishment in Norway of nonbiotechnology firms that subsequently founded biotechnology subsidiaries (e.g. firms established in the 19th century within fish oil or pharmaceuticals). The nine largest firms in Figure 9.3 include two firms with the same Norwegian parent firm (Nycomed Pharma and GE Healthcare).

Source: Table 9.2.

of specialization, and geographic location. A recent survey of Norwegian biotechnology firms concluded that as of 2007, approximately one-tenth of the firms were profitable.[8] The low level of reported profitability reflects the fact that many Norwegian biotechnology firms focus on long-term development projects, especially in therapeutics and theranostics. Since a majority of these firms were established in the late 1990s and early 2000s, successful approval and marketing of products is unlikely before 2010 in most cases. A study of the medical segment of Norwegian biotechnology concluded that Norwegian firms have 30 products in the development pipeline, including one in late phase III clinical trials, as compared to 85 (including seven in phase III) in Sweden, 151 (including seven in phase III) in Denmark and 26 (including one in phase III) in Ireland (*E&Y 2007 Biotechnology Report*, as cited in Moran 2007).[9] Science-Metrix (2006: 18) assessed the volume and quality of Norwegian firms' patenting in biotechnology as of 2005, and ranked Norway in a tie for tenth place among 23 countries according to a multicriteria ranking system.[10]

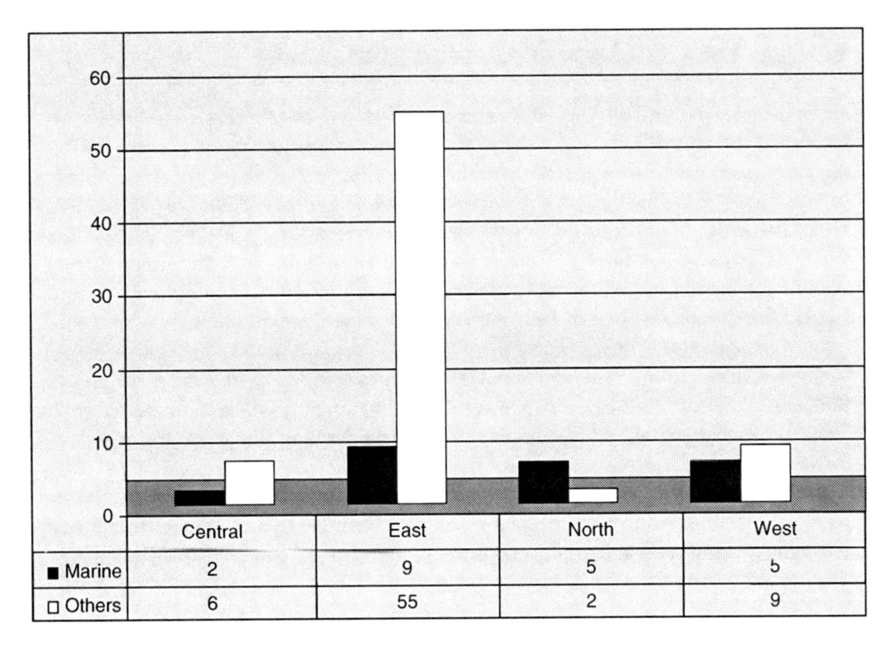

Figure 9.4. Geographical distribution of Norwegian biotechnology firms.
Source: Table 9.2.

TRENDS IN THE ENVIRONMENT OF BIOTECHNOLOGY RELATED FIRMS

This section describes selected aspects of the business environment firms in Norway, including the venture capital (VC) market, the institutional environment, and expenditures and types of public R&D.

Trends in venture capital markets

A 2003 OECD review concluded that there are relatively "low levels of risk capital in Norway as well as a scarcity of entrepreneurs" (Baygan 2003: 8), and attributed this alleged lack of entrepreneurial activity to Norway's "large public holdings, very small pension, insurance and other financial entities, and low levels of evenly distributed household wealth" (*ibid.* 8). Although the OECD study associated Norway's large state-controlled investments with the nation's low levels of VC funding, a number of public initiatives have attempted to increase public funding of venture capital in Norway (Box 9.2).

Box 9.2. The special case of public investment funds

State ownership of industry is relatively high in Norway, but there is no political consensus on the appropriate approach to public investments in start-up firms. Part of this discussion concerns what role a public equityholder should assume in nominally private firms. The Government privatized the major state-owned Norwegian VC firm (SND Invest) in 2003, consistent with the recommendations made in a 2002 White Paper on state ownership. SND Invest was acquired in 2003 by what is now known as Verdane Capital funds, formerly Four Seasons funds. Another initiative was Seed Capital Scheme established in 1997 with equal ownership shares held by private and public participants.

The remaining dominant state involvement in ownership as of 2007 is Argentum funds, established in 2001, which invests only in other funds and not directly in firms. Argentums is one of the largest investors in Verdane Capital funds as well as in Teknoinvest, a large Norwegian VC fund.

Another model of public–private venture is exemplified by the 2005 establishment of Sarsia Innovation, which focuses on high-technology investments distributed throughout Norway's geographic regions. 75 percent of Sarsia is owned by state investors and 25 percent by private investors.

Source: Compiled based on information material from Argentum, Sarsia Innovation, and Teknoinvest.

In 2005, venture capital (VC) and private equity investments in Norway was 0.14 percent of GDP (and had increased from 0.12 in 2003), compared to 0.40 percent in Denmark, 0.30 percent in Sweden and approximately 0.10 percent in Finland (Table 9.1). In Norway the total amount invested has been gradually increasing, from approximately 1.5 billion NOK in 2002 to 4.5 billion in 2006 (NAVC annual report).[11]

Norway's domestic VC sector is relatively young. The first professional venture capital firms in Norway emerged in the early 1980s, but it was not until 2001 that the Norwegian Venture Capital Association (NVCA) was established. As of 2005 the NVCA had 20 primary members and 21 associated members including the largest seed investor (Statoil Innovation) and the state-owned investment fund Argentum (see Box 9.2). The majority of Norwegian venture capital investment flows to resource extraction industries and ICT. Reiss (2006: 119) has noted that "towards the end of the 1990s there was strong confidence in biotechnology among VC investors in most European countries. However (...) Norway participated only belatedly." By 2005 combined health and biotechnology VC investments in Norway accounted for a larger share of GDP than is true in Belgium, the United Kingdom and Finland (Table 9.1), but one source suggests that dedicated biotechnology investments accounted for only five percent of all Norwegian VC investments in 2005 (approximately 155 million NOK), a far smaller share than the estimated 15 percent of Norwegian VC investment in energy-related projects (NVCA 2006). Sørheim *et al.* (2006: 3) claim that the biotechnology related VC investments in Norway in 2005 accounted for a smaller share than elsewhere in the Nordic region.

No systematic analysis has explained the apparent disinterest of Norwegian VC in biotech, but a number of issues have been cited by VC industry personnel. Trondsen (2007) argues that low investment activity in biotechnology by Norwegian VC stems from the scarcity of managers and entrepreneurs with experience in biotechnology, along with a lack of Norwegian VC firms that have specialized in the sector and a lack of early-stage financing from VC firms. In some cases, passive investors with limited experience in biotechnology development activities have become involved in early-stage financing, with limited success. Other Norwegian VC funds prefer to invest abroad in order to support projects with more experienced management teams and more robust business models (Trondsen 2007: 13).

General institutional environment

This section reviews the general institutional environment for Norway's biotechnology industry, including the role of public policy. Box 9.3 summarizes important Norwegian biotechnology-related policy initiatives since the 1970s. During the late 1970s and early 1980s, a number of public forums and study groups in biotechnology were established under official sponsorship in Norway in response to news of dramatic scientific advances in the field. These studies led to legislation on gene technology for living organisms (the Gene Technology Act of 1993), and on medical applications of biotechnology (Biotechnology Act of 1994). A revision of the Biotechnology Act that allows limited use of pre-implantation diagnostics and research on surplus fertilized embryos was passed by the Parliament in 2007 and relaxed somewhat the regulatory environment for biotechnology research. At present, Norway also maintains a liberal policy towards conventional clinical tests of drug candidates, and the number of clinical tests being performed in the country each year is comparatively high (Dobos *et al.* 2004: 37).

An area of continuing controversy, however, concerns the commercial handling of genetic information obtained from the Norwegian population. The 2003 Act on biobanks restricts access by scientists to human material stored in Norwegian public biobanks. The Act has been criticized by public research groups and business, who argue that the possibilities for scientific and

Box 9.3. Chronology of events, policies, and institutional changes

1977–78 SINTEF establishes Norway's first biotechnology group with support from research council.

1980–84 Series of committees on biotechnology led by Laland (1980), Kleppe (1982) and Prydz (1984).

1985 Biotechnology became one of five "main target areas" in the 1985 Government White Paper on research.

1987 Norway participates in Nordic cooperation program on biotechnology.

1991 The Norwegian Biotechnology Advisory Board established.

1993 Gene technology Act passed; White Paper on humans and biotechnology.

1994 Biotechnology Act passed; European Laboratory for Marine Biotechnology established in Bergen.

1996 RCN perspective analysis and action plan for biotechnology for 1995–2005.

1997 RCN presents Strategy for biotechnology.

1998 RCN memorandum, "Research for the Future," proposes a national strategy for Norway's biotechnology industry that includes research directed towards medicine and health sector, food production, and marine biotech.

1998 RCN funds biotechnology research programs planned to last through 2011, including large scale FUGE—Functional genomics program (2002–2011).

1999 Government White paper on Research advocates expanded support for marine biotechnology research aimed at expanding use of marine resources in food products, pharmaceuticals and cosmetics.

2002 SkatteFUNN tax scheme for SME R&D tax credits; University Act amendment passed, assigning ownership of faculty inventions to universities.

2003 Act on biobanks and amendments to Biotechnology Act passed; Government Council on Marine Value Creation established.

2005 Government White paper on Research supporting an expanded role for biotechnology in "securing that Norway exploits its naturally given prerequisites among e.g. marine resources, and strengthening good and promising milieu within biology and biomedicine" (p. 31).

2006 Eurobarometer study shows increased optimism towards biotechnology amongst Norwegians.

2007 Amendments to Biotechnology Act.

Source: Compiled based on Kallerud (2004) and on Norwegian government and Research Council of Norway strategy plans, white papers, information material, and news releases.

commercial exploitation of Norwegian data equal or surpass those currently being pursued in Iceland, because of their greater volume and the potential for linking data from the Norwegian biobank with data from other registries (Olsen 2006: 29).

Another institutional issue that may affect the attractiveness of Norwegian sites for biotechnology firms is university–industry research cooperation and collaboration. The 2002 University Act amendments gave universities and colleges the primary responsibility for facilitating the use and development of their research, and the University of Oslo subsequently established a technology transfer office. These policy changes obviously are not specific to biotechnology, but the legislation has brought Norway into line with other countries in Europe when it comes to potential academia—business collaboration. Thus far there is little evidence of increased commercialization of academic research in biotechnology resulting from these reforms. Indeed, the most visible academic spin-offs within Norwegian biotechnology, including Biotech Pharmacon (a Tromsø University spinoff), Invotrogen Dynal (a spinoff from the Technical University in Trondheim), and Photocure (a spinoff from Det Norske Radiumhospitalet, a Norwegian cancer-treatment hospital), all occurred before the 2002 reforms.

Public funding for biotechnology R&D

The number of Norwegian biotechnology research programs has grown significantly since 1998, but with the exception of the FUGE program (see Box 9.3) and MabCent (see below), public funding for these programs has remained relatively modest. Analyses disagree on the relative levels of private and public funding of biotechnology R&D in Norway. One OECD assessment of Norwegian biotechnology R&D concluded that, "over 70 percent of biotechnology R&D expenditure in 2003 came from public sources" (OECD 2006a: 112). The OECD study argued that the overall amount of public funding for biotechnology R&D in Norway in the early 2000s was modest by comparison with funding in other high-income economies, and lower than public biotechnology R&D funding in Finland, New Zealand and Denmark (*ibid.* 19). It is possible that increased Norwegian public spending on biotechnology R&D has shifted these rankings, but more current comparative data are not available.

Biotechnology R&D accounted for 6 percent of total Norwegian public R&D spending in 2003, rising to approximately seven percent in 2005 (Sundnes and Sarpebakken 2007).[12] Within Norway's universities and public research organizations (PROs) in 2005, the largest share of the approximately

826 million NOK in biotechnology related expenditures went to medical biotechnology (449 million NOK), followed by agricultural, fisheries-related, and veterinary R&D (227 million NOK). Engineering and mathematics/ natural sciences received approximately 118 million NOK each (Gunnes and Sandven 2007: 14). An indicator of the increased attention to marine biotechnology in recent years is the designation of Tromsø's marine biotechnology research centre MabCent as a "national centre of excellence" in 2007.

The presence in Eastern Norway of large academic medical research centers such as the University of Oslo and several hospitals resulted this region receiving the largest share of public biotechnology research funding (68 percent) in 2005, followed by Central Norway with 14 percent, West Norway (including the Southwest and South) with 10 percent, and North Norway with 8 percent of the expenditures (Gunnes and Sandven 2007: 16, 21). More recent growth in public funding for marine biotechnology R&D is likely to increase North Norway's share of overall public funding for biotechnology R&D.

Biotechnology R&D within Norway's public sector is dominated by university-based R&D. Total employment of biotechnology researchers in the Norwegian public sector R&D infrastructure (including universities and public research institutes) amounted to 1,440 in 2003, and grew to almost 1,800 in 2005 (Sundnes and Sarpebakken 2005: 3; 2007: 30).[13] Universities and university hospitals[14] accounted for approximately 80 percent of R&D performed within Norway's public R&D infrastructure, while research institutes performed 20 percent of public-sector R&D in both 2003 and 2005 (*ibid.* 2005: 31; 2007: 30).

Although the research institutes account for a relatively modest share of overall biotechnology R&D within Norway, some recent policy initiatives, most notably in marine biotechnology, are likely to expand their role. Akvaforsk (Institute of Aquaculture Research), which was established in 1960, founded the Akvaforsk Genetics Center (AFGC) in 1999 to commercialize breeding programs on a worldwide basis. Akvaforsk itself also participates in the "Akvaforsk alliance," founded in 2003 and staffed by 80 researchers. The consortium includes among its members the Research Institute for Food Research (i.e. Matforsk) and the University for Environment and Bio-sciences in Ås outside Oslo (Akvaforsk-alliansen 2003). The consortium was established to support "A profitable aquaculture sector with competitiveness and value added when it comes to breeding and value added" (*ibid.* 11), and its research efforts cover marine biotechnology, marine biology and resource management. Another recent public-sector initiative in marine biotechnology was the government establishment in 2005 of an umbrella organization, NOFIMA AS (Norwegian Fisheries and Food Research Inc.), although it is too early to assess the direction or performance of this new initiative.

The usual indicators for research performance suggest that Norwegian academic institutions are below average in productivity and reputation for biotechnology in general, but fare rather well within selected fields.[15] Assessments gave high rankings to Norwegian academic research performance in the broad field of "aquatic sciences" (Hinze 2002), but this field includes a wide range of topics apart from marine biotechnology.[16] The Norwegian R&D system seems to face particularly significant challenges in linking pre-existing fields of knowledge, e.g. marine biology, with emergent fields such as marine biotechnology, as an evaluation of genome sequencing of salmon and cod noted in its 2003 report: "The strong impression of the evaluation committee was that ongoing research in the field of marine molecular biology in Norway is limited. This implies that there are only few research groups that could really take advantage of a full genome sequence of the cod or the salmon and make interesting biological research of such a resource and/or use the resource for useful practical applications" (Andersson 2004: 11).[17]

DISCUSSION AND CONCLUSION

Although Norwegian biotechnology seems to represent an example of the "R&D intensive network based path" of development discussed in Chapter 2, this characterization requires some qualifications. In important respects, Norway's emergent biotechnology sector builds on long-established trajectories of economic and technological development that are rooted in the "small-scale decentralized" and "large-scale centralized" paths. Nonetheless, it remains to be seen whether this foundation is sufficient to sustain a competitive biotechnology industry. Norway may be defining a genuinely new path of development that will produce a viable biotechnology sector, but the sector's long-term viability remains uncertain.

Parts of Norway's emerging biotechnology industry build on past capabilities developed within the large scale centralized path outlined in Chapter 2. Notable examples of these links include the initiatives to diversify into biotechnology that were undertaken by Norsk Hydro before this company terminated its diversification efforts. These initiatives live on, however, in the independent firms FMC Biopolymer, Clavis Pharma and Pronova Biopharma. One recent example of a "large-scale developmental path" company entering the biotechnology sector is the Aker consortium's Aker Biomarine, which acquired the biotechnology company Natural in 2007. Several modern Norwegian biotechnology firms emerged from chemically and engineering based enterprises that long predate the modern biotechnology revolution. Nygaard & Co., established in the nineteenth century, was the ancestor of Nycomed Pharma and GE Healthcare, both of which are active in research reagents, theranostics, and pharmaceuticals. The acquisition of former Nycomed Imaging by GE Healthcare also produced an exodus of scientific and engineering talent from the acquired Norwegian firm that contributed to the foundation of newer startups. The examples of startups PhotoCure and PCI Biotech also point towards a potential Norwegian capability that consists of combining academic knowledge about diseases and their diagnosis or treatment with electronics and reagents.

Other segments of Norway's biotechnology industry are rooted in agriculture and fisheries, industries that typify the small scale decentralized path of development. Several Norwegian aquaculture firms and processors of fish-based products such as fish oils were the predecessors of marine biotechnology firms. Although Norwegian marine biotechnology draws on more than the basic technology for processing fish oils, at least one firm (Pronova Biopharma) has exploited techniques related to fish oil extraction in the development of prescription drugs, and a second firm (Aker Biomarine) aspires to follow suit. Other firms (e.g. Biotec Pharmacon, Navamedic, and FMC Biopolymer) have applied knowledge from other marine organisms in production of drugs and other product, and firms like Intervet Norbio and Pharmaq cater to the aquaculture industry both in Norway and abroad by offering health-related fish products. All of the seven Norwegian "Green biotechnology" firms within the breeding-related segment originated in their exploitation of basic breeding techniques dating to the early 1900s, and now are applying more advanced biobank techniques to their long-established core businesses.

These links between modern biotechnology and older trajectories of economic and technological development in Norway provide some support for arguments concerning the path-dependent development of intrafirm capabilities and the importance of these capabilities for industrial change. Support for the *agglomeration thesis*, which emphasizes the importance of factors external to the firm, is less obvious in Norway. The nation's biotechnology firms are spread throughout the country, which seems inconsistent with predictions that the importance of academic researchers, venture capital, and a highly trained workforce should lead to a majority, if not all, of Norwegian biotechnology firms located in one of the university cities, such as Oslo, Bergen, Tromsø, or Trondheim. Nevertheless, a majority of Norwegian health-related biotechnology activity is located in the Oslo area, and this regional grouping may benefit from proximity to resources (e.g. research organizations, labor) in the nearby Gothenburg region in Sweden as well.

An interpretation of Norwegian biotechnology as dominated by *nonmedical biotechnology*, a segment whose dynamics and development may differ from those of pharmaceutical and medical biotechnology, receives only limited support The population of firms and (even more so) employment in Norwegian biotechnology are primarily medical, although Norway's biotechnology sector does include a number of relatively large nonmedical segments, most notably within livestock breeding and the non-medical parts of marine biotechnology. Nevertheless, Norwegian biotechnology cannot be characterized as specialized in these segments, rather than theranostics and diagnostics. The recent interest in and strengthening of basic research in marine biotechnology, such as the establishment of the Akvaforsk research consortium and the recently designated national center of excellence MabCent might stimulate developments in this field, such as products developed on the basis of marine bioprospecting (i.e. the search for commercially exploitable molecules, genes, and mechanisms). Such products and firms nevertheless are rare today, with the exception of firms developing ingredients from older and better-known marine organisms.

The *anchor firm hypothesis* may be relevant to the theranostics and diagnostics segments. Although no direct proof exists of a strong relationship between established firms on the one hand and new start-ups on the other hand, it is likely that the foundation of new Norwegian firms in these segments was related to the existence of anchor firms such as Nygaard & Co. in earlier times and Nycomed Pharma and GE Healthcare in more recent times.

At present, it is not clear that Norway has achieved a "critical mass" for the realization of the economic and developmental benefits of agglomeration at the national or regional levels in biotechnology, whether this "critical mass" is measured in terms of the number of firms, academic medical research centers, funding from public and private sources, or trained personnel. But if the external networks that Swann and Prevezer (1996) and Niosi (2003) describe as essential to the development of a biotechnology sector are lacking in Norway, alternative developmental trajectories may nonetheless be feasible. One unusual element of the Norwegian biotechnology industry is the significance of marine biotechnology, even if the status of marine biotechnology might be more modest than is implied in some policy documents (RMV 2003). Norway currently has significant research efforts and companies in this field, and may well be able to exploit its long coastline in future research efforts focused on marine organisms. Nevertheless, it seems premature to classify the marine biotechnology sector as a success, inasmuch as its technological and economic benefits have yet to be realized.

Although medical biotechnology is present in Norway, the auxiliary segments of medical biotechnology (theranostics, diagnostics, and drug delivery) account for a greater share of employment within overall biotechnology in Norway. These segments may not have the same requirements as medical biotechnology for geographical proximity to venture capital firms and science centers. Thus, perhaps in conjunction with advances in marine biotechnology and increased differentiation of industrial applications, might Norwegian biotechnology prove less dependent on agglomeration effects for its development? If so, it might be possible to foster a viable biotechnology industry in a location relatively remote from the science-based centers of the world with a small number of firms and a relatively small R&D base.

Any answer to this question requires further research on the Norwegian experience in biotechnology. The number of new firms began to grow from the mid 1990s, and the emergence of the biotechnology industry in Norway is therefore in its early stages. Nevertheless, the future evolution of the Norwegian biotechnology industry is likely to build on current and past strengths, especially in advanced diagnosis and drug delivery techniques and in phototherapeutic solutions. The relatively large number of Norwegian biotechnology firms in the nutraceuticals field also builds on national and regional capabilities in fisheries and aquaculture. The current and likely future path of development for biotechnology in Norway thus exhibits significant elements of path-dependency.

Norway's biotechnology industry, however, is rooted in all three of the paths of development of the Norwegian innovation systems that were discussed in Chapter 2, rather than being identified solely with the third, science-intensive path. Parts of the contemporary biotechnology industry in Norway have elements in common with high technology engineering, and in the case of marine biotechnology, there are some links with the nation's long-established trajectory of resource-based industry. Moreover, the recent entry of a large number of new firms into Norway's biotechnology industry has occurred in parallel with the development of a Norwegian venture capital industry.

LIST OF TECHNICAL TERMS

alginate ingredient derived from seaweeds which may be used within e.g. research, drug delivery and tissue engineering

antibody protein used by the immune system to identify and neutralize foreign objects like bacteria and viruses

biobank storage of genetic material and information for public health sector research or commercial purposes

biochemical production utilizing the active substances in material for the purpose of producing altered material

bioinformatics developing or using high throughput technology for processing great amounts of e.g. genetic information

bioprospecting the search for commercially exploitable molecules, genes, and mechanisms

biotechnology as defined by the OECD the application of science and technology to living organisms, as well as parts, products and models thereof, to alter living or non-living materials for the production of knowledge, goods, and services

breeding selecting on the basis of genetic information for the purpose of generating improved generations

chitin and chitosan ingredient derived from crustacean shells which may be used in e.g. tissue engineering functions as protectors or "scaffolds" for cells (wound-healing), in health foods, and as soil conditioners in agriculture

diagnostics indicate existence or potential for disease

drug delivery technique for delivering drugs in a precise and efficient manner

enzymes protein molecules which can catalyze chemical reactions within cells without being altered itself

glucosamine a derivative of chitin used in pain reliever drugs and the like

lipids class of hydrocarbon-containing organic compounds (molecules that contain carbon)

nutraceuticals combination of the words "nutrition" and "pharmaceutical" and refers to foods claimed to have a medicinal effect

omega-3 group of fatty acids

theranostics diagnostic tests directly linked to the application of specific therapies

therapeutics drugs aimed at providing treatment, prevention or pain relief of disease

VC/venture capital capital which is provided by specialized financial firms acting as intermediaries between primary sources of finance (such as pension funds or banks) and firms

NOTES

1. I would like to thank Eva Dobos for conducting interviews and research assistance in a related study (Dobos *et al.* 2004; Grønning *et al.* 2006) that is partly referenced within the present study; Yukie Hara for help in cataloguing material; Thor Amlie for advice on the nature and extent of the biotechnology industry in Norway; and Solveig G. Ericson, Jan Fagerberg, and David C. Mowery for comments on previous versions of the manuscript. All remaining errors are, however, my responsibility.
2. Resource constraints precluded any systematic study of the number and types of alliances.
3. Marine biotechnology is unlike other areas of biotechnology in that it is defined in terms of its source material, rather than the market it serves (Biobridge Ltd. 2005). ESF (2001) divides into the five different marine biotechnology research and business sub-areas developing novel drugs; producing diagnostic devices for monitoring health; discovering new types of composite materials, biopolymers and enzymes for industrial use; ensuring safety of aquaculture and fisheries; and providing new techniques for management of marine environments; (ESF 2001: 6) It should be noted that in the case of Norwegian policy documents and the industry press (see e.g. MTI 1998 and RMV 2003: 14), it is indeed highlighted that the country might have special potential within this area rather than claiming that it is already in a leading position.
4. There is some controversy over which firms should be counted as biotechnology firms, and in the case of Norway, estimates of the population of biotechnology firms vary from *c.*85 firms (Research Council of Norway according to Berntsen and Dalen 2004), 110 firms (Marvik 2005), and 125 firms (FUGE 2007). For the sake of consistency and validation the 93-firm sample used for this study was arrived at by selecting the 42 firms (Hodgson 2006) defined as dedicated biotechnology in the sense of being "companies whose primary commercial activity depends on the application of biological organisms, systems or processes, or on the provision of specialist services to facilitate the understanding thereof" (Critical I 2006: 41), as well as biotechnology related firms listed in CGE&Y (2000), OECD (2001), Biobridge Ltd. (2005), and Nor-Bio (2007a). Eliminating overlaps and nine firms that have been acquired, discontinued or have shown minimal activity (and are thus dormant firms) since these overviews were made, the subtotal of firms based on these sources is 85 firms. In addition, seven firms within agricultural biotechnology (breeding related firms) are included based on Hernes *et al.* (2004), and one environmental biotechnology firm is included based on its inclusion in Norgren *et al.* (2007). The 93 firms sample thus captures developments in Norwegian biotechnology through roughly 2004–2005, and is not particularly comprehensive in its coverage of firms established after 2005, e.g. the biobank firm Hunt Biosciences est. in 2007, and firms registered as recipients from the Norwegian research program FUGE such as AlgiPharma AS, PatoGene Analyse AS, and

Salmobreed AS (Andersen 2007; for the latter firm see also Science-Metrix 2003: 19).

5. Norsk Hydro during the 1980s and early 1990s attempted to establish itself within pharmaceuticals (Andersen and Yttri 1997: 292–7) as well as within biomarine ingredients (Biomarint Forum 2003: 4), but these efforts were later abandoned. Several independent firms (including Clavis Pharma as well as Pronova) were formed out of Norsk Hydro's divestiture of various initiatives.
6. Technically the Norwegian divisions of this firm are divided into GE Healthcare and GE Vingmed Ultrasound, although these are treated together in here.
7. Ingredients are divided into basic ingredients for food additives, additives to human food or animal feed, and bioactive ingredients, and consist of oils, proteins, minerals, gelatine, chitin, chitosan, glucosamine, and enzymes.
8. The FUGE (2007) survey includes 125 firms. It is plausible that the FUGE sample includes all of the 93 firms in this chapter's sample of biotechnology firms, but the study did not disclose the identities of the 125 firms that it included.
9. Another survey provides an even more optimistic assessment, reporting that 31 candidates are in the clinical trials phases I-III (NorBio 2007b), whereas the E&Y 2007 Biotechnology Report (as cited in Moran 2007) captures all 30 product candidates including 10 in the so called preclinical phase.
10. In this ranking Denmark came out first, followed by USA, whereas Sweden was sixth and Finland fourteenth (Science-Metrix 2006: 18)
11. It was 1.9 billion in 2003, 2.2 billion in 2004, and 3.1 billion in 2005 (NAVC annual)
12. Sundnes and Sarpebakken (2005, 2007) state that the 2003 and 2005 estimates of business R&D investment are not directly comparable (since statistics until 2003 are separately reported for marine R&D and biotechnology R&D, whereas starting in 2005 marine biotechnology as a sub-field is to be classified as biotechnology). Roughly 75 percent of Norwegian biotechnology-related R&D was financed by government, and this share grew to 80 percent in 2005 (*c*.960 million NOK). Gunnes and Sandven (2007: 7), on the other hand, estimate that total biotechnology R&D expenditures were approximately 1940 million NOK as of 2005, and report that more than 50 percent of total Norwegian biotechnology R&D was financed by industry (approximately 926 million NOK public and 1,014 million NOK private expenditures). This huge discrepancy may reflect the use by respondents in the 2007 study of very broad definitions of biotechnology-related R&D. The 2007 study also uses a broader definition of biotechnology related firms than this paper employs, since the study uses a results from a sample of 115 firms that are scaled up to an estimated total Norwegian population of 167 biotechnology firms (*ibid.* 37).
13. These figures exclude technical and administrative personnel.
14. University hospitals include Rikshospitalet, Ullevål Universitetssykehus, Radiumhospitalet med Institutt for Kreftforskning, Aker Universitetssykehus, Akershus Universitetssykehus and Diakonhjemmet sykehus.

15. Comparative assessments of publications and citations found that Norwegian academic institutions were below average within the overall field of biotechnology (Science-Metrix 2006: 13), and an 11-country comparison of publications in biopharmaceuticals per million population 11999/2000 ranked Norway in sixth place (Reiss 2006: 107) In a study of 2006 citation frequency of articles published 2002–2006 Norway was above the world average in botany, clinical medicine, biology, biochemistry as well as zoology and veterinary sciences (RCN 2007, Tables A9.2 and A9.3). Moreover, in a study of publishing within cancer molecular epidemiology Norway ranked tenth out of 35 countries in the number of papers and their impact (based on citations) (Ugolini *et al.* 2007)
16. Biobridge, while acknowledging that "Norway has a very strong presence in marine bio[techno]logy" (Biobridge 2005: 169, their brackets), ranks both USA, the UK, France, and Germany ahead of Norway within marine biotechnology (*ibid.* 43).
17. One major objective of the large scale research program FUGE which started in 2002 is to address the fostering of functional genomics competencies.

REFERENCES

Akvaforsk-alliansen (2003) *Akvaforsk-alliansen: Forskning, utvikling og utdanning innen akvakultur*, Ås: Akvaforsk-alliansen.

Amdam, R. P., and K. Sogner (1994) *Rik på kontraster: Nyegaard & Co.—En Norsk farmasøytisk industribedrift 1874–1985*, Oslo: AdNotam Gyldendal.

Andersen, Elisabeth Kirkeng (2007) "Støtter norske bioteknologibedrifter," *Forskning.no*, 28 March, 2007.

Andersen, K. G. and G. Yttri (1997) *Et forsøk verdt: Forskning og utvikling i Norsk Hydro gjennom 90 år*, Oslo: Universitetsforlaget.

Andersson, L. (ed.) (2004) *Report: Evaluation of the Idea to use Cod as a Species for Genome Sequencing and Functional Genomic Studies in Norway*, Oslo: RCN.

Bartholomew, Susan (1997) "National Systems of Biotechnology Innovation: Complex Interdependence in the Global System," *Journal of International Business Studies*, Vol. 28, pp. 241–66.

Baygan, G. (2003) *Venture Capital Review Norway*, Paris: OECD.

Berntsen, W. and E. Dalen (2004) *Foresightanalyse bioteknologi: Rapport utarbeidet for Norges Forskningsråd*, Oslo: MMI Unero.

Biobridge Ltd. (2005) *A Study into the Prospects of Marine Technology Development in the United Kingdom, Volume 1—Strategy*, London: FMP Marine Biotechnology Group.

Biomarint Forum (2003) *Hvordan styrke innovasjonskjeden i den biomarine næringen*, Oslo: Biomarint Forum.

CGE&Y (2000) *Næringsrettet bioteknologi i Norge: Statusrapport*, Oslo: RCN.

Critical I (2005) *Biotechnology in Europe: 2005 Comparative Study*, Critical I.
—— (2006) *Biotechnology in Europe: 2006 Comparative Study*, Critical I.
de la Mothe, J. and J. Niosi (2000) "Issues for Future Research: Measurement and Policy," in de la Mothe, John, and Jorge Niosi (eds.), *The Economic and Social Dynamics of Biotechnology*, Boston: Kluwer Academic Publishers, pp. 229–33.
Dobos, E., T. Grønning, M. Knell, D. S. Olsen, and B. K. Veistein (2004) *Norway Vol. 1: Biopharmaceuticals*, Report submitted to the OECD in the series 'Case Studies in Innovation' (DSTI/STP/TIP(2002)) Paris: OECD.
Eggset, G. (2005) "Norskekysten som konkurransefortrinn," in Johne, Berit and Erik F. Øverland (eds.) *Leve av, leve med, leve for? Vår bioteknologiske fremtid*, Oslo: Cappelen Akademiske Forlag.
ESF [European Science Foundation] (2001) *Marine Biotechnology: A European Strategy for Marine Biotechnology*, ESF Marine Board Position Paper 4, Strasbourg: ESF.
Feldman, M. (2003) "The Locational Dynamics of the U.S: Biotech Industry: Knowledge Externalities and the Anchor Hypothesis," *Industry and Innovation*, September 2003.
FUGE (2007) *Nytt fra FUGE- Funksjonell genomforskning*, No. 3, 2007.
Gilham, I. (2002) "Theranostics: An Emerging Tool in Drug Discovery and Commercialization," *Drug Discovery World*, Fall 2002: 17–23.
Grønning, T. and D. Sutherland Olsen, with Eva Dobos, Mark Knell, and Bjørg K. Veistein (2006) "Norway," in Enzing, C. M. (ed.), *Innovation in Pharmaceutical Biotechnology: Comparing National Innovation Systems at the Sectoral Level.* Paris: OECD.
Gunnes, H. and T. Sandven (2007) "Tematiske prioriteringer og teknologiområder i det norske forsknings- og innovasjonssystemet," NIFUSTEP Report 22/2007.
Hernes, T., P. I. Olsen, and A. Espelien (2004) *Jordbruksvaresektorens muligheter i sjømatbasert næringsutvikling*, Forskningsrapport 3/2004 Handelshøyskolen BI Senter for samvirkeforskning, Sandvika: Handelshøyskolen BI.
Hinze, S. (2002) *Bibliometric Analysis of Norwegian Research Activities*, Oslo: Report No. 2 in evaluation of the Research Council of Norway.
Hodgson, J. (2006) "Europe must do Bigger: Critical I's 2006 Biotechnology Comparative Study," presented to the *Norwegian Bioindustry Forum*, 13th June 2006.
Johannessen, T. A., Ø. Enger, and N. Vogt (2005) "Klynger og Nycomed-effekten," in Johne, Berit and Erik F. Øverland (eds.), *Leve av, leve med, leve for? Vår bioteknologiske fremtid*, Oslo: Cappelen Akademiske Forlag.
Kallerud, E. (2004) "Science, Technology and Governance in Norway Case Study no. 1: Biotechnology in Norway," *STAGE (Science, Technology and Governance in Europe) Discussion Paper 15.*
McKelvey, M. and L. Orsenigo (eds.) (2006) *The Economics of Biotechnology.* Cheltenham: Edward Elgar.
Marvik, O. J. (2005) *Norwegian Life Science Industry: Overview and Status*, Oslo: O. J. Marvik, 4bio AS, on behalf of Innovation Norway.

Moran, N. (2007) "Setting the scene," in Richard Hayhurst and Ole Jørgen Marvik (eds.) *Life Sciences in Norway: Naturally Inspired*, Oslo: Innovation Norway and RCN.

MTI (1998) *Nasjonal strategi for næringsrettet bioteknologi*, Oslo: MTI.

Niosi, J. (2003) "Alliances are not enough Explaining Rapid Growth in Biotechnology Firms," *Research Policy*, 32, 737–50.

NorBio (2007a) *Biotekbedrifter 2003–2005*, Unpublished Overview of Financial Results of Norwegian Biotechnology Firms.

——(2007b) *Analysis of the clinical development pipeline of product candidates based on Norwegian R&D*, Oslo: Norbio, October 2007.

Norgren, L., R. Nilsson, E. Aperez, H. Pohl, A. Andström, and P. Sandgren (2007) *Needs-Driven R&D Programmes in Sectorial Innovation Systems*, Stockholm: Vinnova Analysis (VA) 2007: 15.

NVCA (annual) *Analyse av aktivitet i der norske venture- og private equity markedet*, Oslo: NVCA.

OECD (2001) *Biotechnology Statistics in OECD Member Countries: Compendium of Existing National Statistics*, OECD Science, Technology and Industry Working Papers 2001/6, Paris: OECD.

——(2006a) *OECD Biotechnology Statistics 2006*, Paris: OECD.

——(2006b) *OECD Employment Outlook 2006*, Paris: OECD.

——(2007) *Science, Technology and Industry: Scoreboard 2007*, Paris: OECD.

Olsen, C.R. (2006) "Skreddersydd medisin," *Horisont* 3/2006, 24–33.

Owen-Smith, J. M. Riccaboni, F. Pammolli, and W. W. Powell (2002) "A Comparison of US and European University–Industry Relations in the Life Sciences," *Management Science*, 48(1): 24–43.

Pisano, G. (2000) "In Search of Dynamic Capabilities," in G. Dosi, R. R. Nelson and S. G. Winter (eds.), *The Nature and Dynamics of Organizational Capabilities*, Oxford: Oxford University Press.

RCN (2007) *Det norske forsknings- og innovasjonssystemet 2007*, Oslo: RCN.

Reiss, T. (2006) "Comparison of Performance in National Biopharmaceutical Innovation Systems," in Enzing, C. M. (ed.), *Innovation in Pharmaceutical Biotechnology: Comparing National Innovation Systems at the Sectoral Level*, Paris: OECD.

RMV (2003) *Forslag til strategi for kommersialisering av marin bioteknologi*, Oslo: Fiskeridepartementet.

Science-Metrix (2003) *Genomics in Norway: Overview of Research in Genomics in Norway Prepared for Genome Canada*, Montréal: Science-Metrix

——(2006) *Canadian Biotechnology Innovation Scoreboard 2006*. Montréal: Science-Metrix.

Senker, J. (2000) "Biotechnology: Scientific Progress and Social Progress," in J. de la Mothe, and Jorge Niosi (eds.), *The Economic and Social Dynamics of Biotechnology*, Boston: Kluwer Academic Publishers, pp. 53–68.

Sundnes, S. L. and B. Sarpebakken (2005) *Bioteknologisk FoU 2003: Ressursinnsats i universitets—og høgskolesektoren og instituttsektoren*, NIFU Report 7/2005. Oslo: NIFU.

—— (2007) *Bioteknologisk FoU 2005: Ressursinnsats i universitets—og høgskolesektoren og instituttsektoren*, NIFUSTEP Report 8/2007. Oslo: NIFUSTEP.

Swann, P. and M. Prevezer (1996) "A Comparison of the Dynamics of Industrial Clustering in Computing and Biotechnology," *Research Policy*, Vol. 25, 1139–57.

Sørheim, R., L. Ø. Widding, and K. Havn (2006) *Nordic Private Equity: An Industry: Analysis*, Oslo: Nordic Innovation Centre.

Teece, D. J. (2000) *Managing Intellectual Capital*, Oxford, Oxford University Press.

Trondsen, A. (2007) "Muligheter for økt kommersialisering innen bioteknologi fra en institusjonell investors ståsted, "Presentation to symposium *Bioteknologi i sentrum*, Oslo, 24–25 September, 2007.

Ugolini, D. R. Puntoni, F. P. Perera, P. A. Schulte, and S. Bonassi (2007) "A Bibliometric Study of Scientific Production in Cancer Molecular Epidemiology," *Carcinogesis*, vol. 28, no. 8, pp. 1774–9.

10

Slow Growth and Revolutionary Change: The Norwegian IT Industry Enters the Global Age, 1970–2005

Knut Sogner

Although Norway is one of the world's richest nations and consequently a large user of IT equipment and IT solutions, it remains a relatively small producer of IT equipment.[1] Of the four rich Nordic nations, Norway has the smallest production of IT goods, both in terms of absolute value and as percentage of total industrial output.[2] Norwegian imports of IT goods at the start of this century constituted about three times the value of Norwegian IT exports. In terms of the quantitative significance of its "national IT competence," Norway is the worst performer among the Nordic countries. This chapter seeks to shed light on the development of the Norwegian IT industry and why Norway has become such a comparatively poor IT performer.

It is not because of a late start. Norwegian firms were early entrants into what (much later) became the IT industry.[3] As early as 1882 a Norwegian maker of telecommunications equipment, Elektrisk Bureau, was founded in response to the early and intensive use of telephones in Norway. Elektrisk Bureau also soon began to export its products and invested in foreign production operations by the turn of the last century.[4] Prior to the Second World War, several new electronic companies were started, mostly making radios, and gradually a fully fledged industry formed and developed, influenced to a substantial degree by Government policies promoting the production of electronic goods. Both Tandbergs Radiofabrikk and Norsk Data were at times quite successful in foreign markets, and Tandberg in particular was well known. But the success disappeared as time went on, and Norway is the worst Nordic performer in IT.

Perhaps the most dramatic difference in 2007 between Norway and its Nordic neighbors Sweden and Finland is the absence of any large Norwegian multinational company in the IT sector. Sweden's Ericsson and Finland's

Nokia are not matched by any comparable Norwegian enterprise, and even fellow Nordic IT laggard Denmark can boast the presence of world-famous consumer electronics company Bang and Olufsen.

Why has Norway not been more successful in IT? Could it be because of lack of competence? Probably not, but more will be said about that later. The cost-increasing and crowding out effects of the large oil and gas sector in the Norwegian economy since the mid 1970s unquestionably have made it difficult for Norwegian IT companies to export successfully, and this factor is discussed extensively below.[5] But Norway is not the only Nordic exception to industrial success in IT. Oil-free Denmark is a rich, small nation with a large IT market that manufactures a relatively small output of IT products. Clearly Denmark and Norway have used IT for economic advantage, but the benefits of IT for these economies are more likely to be found in general economic growth figures than in IT industry statistics.

This discussion focuses on the IT *industry*, i.e. the industrial and research-based part of the IT sector, arguably the best way to gain insight into the Norwegian national IT competence and its competitiveness. The following chapter is divided into five sections. First, the current situation is discussed, leading to an examination of the historic development of the Norwegian IT industry, with emphasis on understanding the situation at the beginning of the 1970s. Then the development during the troubled 1970s and 1980s is discussed in two parts, before the ensuing changes of the IT industry are analyzed. The concluding section considers the long-term implications of the analysis.

LARGE USER, SMALL PRODUCER (1970–2005)

The four Nordic countries are very similar in terms of population, economic structure, and history. Sweden, with around nine million inhabitants, is roughly twice as large as each of the other three nations, and Finland—with its later start as an independent nation and its different language—is the odd nation out. But by and large the four nations resemble one another in many respects, and among these one *has* to be the worst IT performer in terms of industrial performance. That position belongs to Norway.

The gross turnover in the ICT sector of these four nations is remarkably similar, as is underlined by Table 10.1, which shows per capita turnover and employment in each Nordic nation's IT industry during 2000–2003. While Norway has the smallest turnover in absolute terms, its small population means that on a per capita basis, Norwegian IT sector turnover was nearly

Table 10.1. Turnover and number of employees in the ICT sector (normalized by total population)*

Country\Year	Per capita turnover (EUR)				Per capita number of employees			
	2000	2001	2002	2003	2000	2001	2002	2003
Norway	5056.1	5536.7	5217.5	4354.7	0.0189	0.0214	0.0187	0.0158
Finland	4791.7	5137.1	5041.5	5247.8	0.0211	0.0225	0.0207	0.0209
Sweden	7112.3	6388.5	5823.3	5338.4	0.0234	0.0258	0.0224	0.0200
Denmark	4987.3	5107.0	5060.0	4765.9	0.0195	0.0196	0.0181	0.0173

* Included in the ICT sector are manufacturing, services, telecommunications and consultancy services.

Source: OECD and Nordic Information Society Statistics 2005, Copenhagen: Nordic Council of Ministers.

80 percent as large as that of Sweden in 2003. Sweden, a global player in telecommunications equipment with a thriving Internet sector up until 2000, shows the highest per capita turnover. Although it was industrialized later than the other three nations, the table shows that Finland's IT industry by 2003 was as large on a per capita basis as that of Sweden.

Sweden and Finland have more people employed in the IT sector than Norway and Denmark. Although the absolute number of people employed in Norway is significant, totalling around 80,000 in 2001–3, this is by far the smallest number of the four nations. Finland and Sweden, each of which is home to a large IT multinational, have the largest number of employees in either absolute or per capita terms. Per capita employment in IT declined in all 4 Nordic economies during 2000–2003, partly because of the "dot com crisis" of 2000–2001, but the decline in per capita employment was smallest in Finland. The employment figures suggest that the collapse of Internet-related businesses that was associated with the broader stock-market crisis in 2000 affected Sweden most severely among the Nordic economies. Sweden's entrepreneurial effort in the Internet business was huge in the late 1990s, and the sector's collapse accordingly had its greatest effects in Sweden. Sweden's national telecommunications flagship, Ericsson, also restructured in the period after 2000 and reduced its domestic employment considerably.

By and large Table 10.1 underlines the importance of ICT in the four Nordic countries. The *World Economic Forum*, which organizes the annual Davos meetings, compiles an annual ranking of global ICT capabilities in cooperation with the French business school Insead. The four Nordic countries score very well in this "Networked Readiness Index", which is based on a broader set of indicators than production of IT-related goods and services.[6] Denmark was on top of the 2007 rankings, with Sweden in second place, Finland ranking fourth and Norway tenth. Their strong performance in this ranking reflects the high levels of general use of ICT in all four countries, along with a large ICT service-sector and significant production of software and programming tools.

Extensive use of ICT—broadly defined—in Norway began early in the twentieth century and had great impact in a number of businesses, not least insurance and banking.[7] The Bull name in Honeywell-Bull came from the Norwegian Fredrik Rosing Bull who began making tabulator-machines for the Norwegian insurance company Storebrand in the late 1910s, before moving to France. More recently retailing and other services have made extensive use of ICT, although the economic impact of widespread ICT use within Norway remains unclear.[8] The very large portion of Norway's ICT sector comprising service providers, software houses, and ICT departments in large ICT-using firms is beyond the scope of this chapter. Nevertheless, turnover among Norwegian ICT service providers is significant – 45,7 billion Norwegian kroner in 2004.[9]

Norway now has a domestic software sector of some importance, though it is difficult to measure its size. Most of the people doing research in Kongsberg Gruppen, the largest Norwegian ICT enterprise, are in some fashion engaged in programming.[10] Many of the people in "the software sector" elsewhere are providing services that are only remotely connected to production of standardized IT goods of the type produced by Kongsberg Gruppen. There are very few Norwegian makers of standardized software that are internationally competitive, but well-known Norwegian firms include the producer of the Internet-browser Opera, the data-search maker Fast and the developer of programming language Trolltech.

The Trolltech programming language is based on Simula, an internationally successful Norwegian programming language developed during the 1960s. Simula was a language introducing so-called modularity and has been used extensively for regulatory purposes.[11] The two men behind the language were professors at the University of Oslo, and one of them had his background at the National Defence Research Establishment. Simula never became a commercial success, but was a technical breakthrough for a way of programming. The history of Simula underscores the decades-long history of Norwegian involvement in software development, and highlights the question of why Norway has failed to achieve greater export success in ICT.

The remarkable differences among the Nordic countries are most apparent in a comparison of their indigenous production of ICT hardware (Table 10.2), which includes "Audio and Video equipment," "Computers etc.," "Components," "Telecommunications," and "Other", omitting consultancy and software. Table 10.2 reveals two clear leaders in ICT hardware production, Finland and Sweden, and two laggards, Denmark and Norway. These sharp

Table 10.2. ICT Goods

Country\Year	ICT goods*				
	1973	2000	2001	2002	2003
Norway	2.2	3.4	3.6	3.9	3.1
Finland	1.1	20.9	19.6	19.8	17.7
Sweden	4.0	14.1	11.9	9.2	7.4
Denmark	3.2	4.4	4.3	4.5	4.3

* As percentage of total national industrial sales.

Sources: Elektronikkindustri, Oslo: Norges Offentlige utredninger, 1976: 30 and personal communication from Ole-Petter Kordahl, Statistics Norway as a supplement to: Nordic Information Society Statistics 2005, Copenhagen: Nordic Council of Ministers.

contrasts reflect the global success of Finland's Nokia and Sweden's Ericsson. When sales of these two firms are removed from the data in Table 10.2, Finland and Sweden do not differ greatly from Norway and Denmark. Yet the success of Ericsson and Nokia must to some degree reflect their national roots and raises the question of why there are no large Norwegian IT manufacturers.

Table 10.2 contains data on ICT production as a share of industrial output, and covers a longer period than Table 10.1. The Table vividly depicts the rise of Finland to Nordic leadership in ICT production. The 1973 data for the "electronics industry," which during this period essentially is the IT industry, show that Finland's ICT industry in 1973 accounted for a share of industrial output that was one-half the share of Norway's ICT industry. By 2000, however, ICT accounts for more than 20 percent of Finnish industrial output, by far the largest such share within the Nordic region. Yet Sweden, too, the most successful ICT producer in 1973, grew rapidly in the ensuing years, reflecting its large telecommunications equipment sector. Indeed, the data for 1973 are drawn from a 1976 Norwegian Public Report of the future of the electronics industry that remarked on the importance of Ericsson for Sweden.[12]

Much of the 1970s policy discussion on the future of the Norwegian IT industry revolved around how to create successful exporters. Efforts to support the growth of Norwegian IT exports were hardly an unqualified success. Yet the data in Table 10.2 also reveal that Norway's "ICT goods" production has increased as a share of the nation's industrial output since 1973, and that Norwegian performance in this respect does not differ significantly from that of Denmark. Nevertheless, the 1973–2000 period covers the industrial transformation from "electronics" to IT, i.e. from analog to digital technology, and the sector's growth within the economies of such rich and productive economies as Denmark and Norway [neither country can be defined as

R&D-intensive] reasonably could have been expected to be higher. Finland and Sweden were global innovators in telecommunications in the period, while Norway and Denmark remained fairly small producers of IT equipment.

This chapter's discussion of the Norwegian experience in ICT since the early 1970s reveals—as seen in Figure 10.1—a steady path of development over long periods of time that is punctuated by sudden and short-term shocks. That development will be expanded upon below, but here a few words should be said about the data reported in Figure 10.1, which are based on two slightly different classification systems. The 1972–1992 data are based on the ISIC-classification, while the 1992–2005 data utilize the NACE-definitions.[13] These data for the first period in the Figure may overestimate the size of the IT industry, as a comparison of the 1992 data from the two sources indicates. But by and large the numbers from the two classifications fit relatively well, and they show that Norway's ICT industry since 1972 more than doubled its output. Employment in the industry in 2005 was lower than in 1972, but (as was noted earlier) some of this apparent shrinkage reflects differences in classification of IT.

The Norwegian ICT industry's growth came, however, without large and visible companies. Part of the very different evolution of the ICT industries of Norway, Finland, and Sweden reflects Norway's lack of a large IT company, and in the following the Norwegian development will be analysed from a company perspective. Why did the Norwegian entrants into ICT during the 1970s and 1980s not succeed? Is there a common reason behind these failures? After all, the development of Norway's ICT industry, combining slow growth and revolutionary change, has yielded significant economic benefits. Yet the data indicate that in 1973, Norway's ICT industry was larger than that of Finland, and Norwegian prospects in the industry appeared very promising.

AN INDUSTRY SET TO GROW? (1900–1973)

The best explanation for Norway's failure to spawn a success like Nokia or Ericsson may be the presence of Nokia and Ericsson. There simply was not room for a Norwegian version of this Nordic path to international success. But Nokia's rise to dominance is relatively recent, and Norway has had broad and deep technological competence in mobile telecommunications since the early years of the twentieth century.

Norway was an early starter. The aforementioned Elektrisk Bureau was fairly successful until the 1920s, when the Swedish firm L. M. Ericsson bought

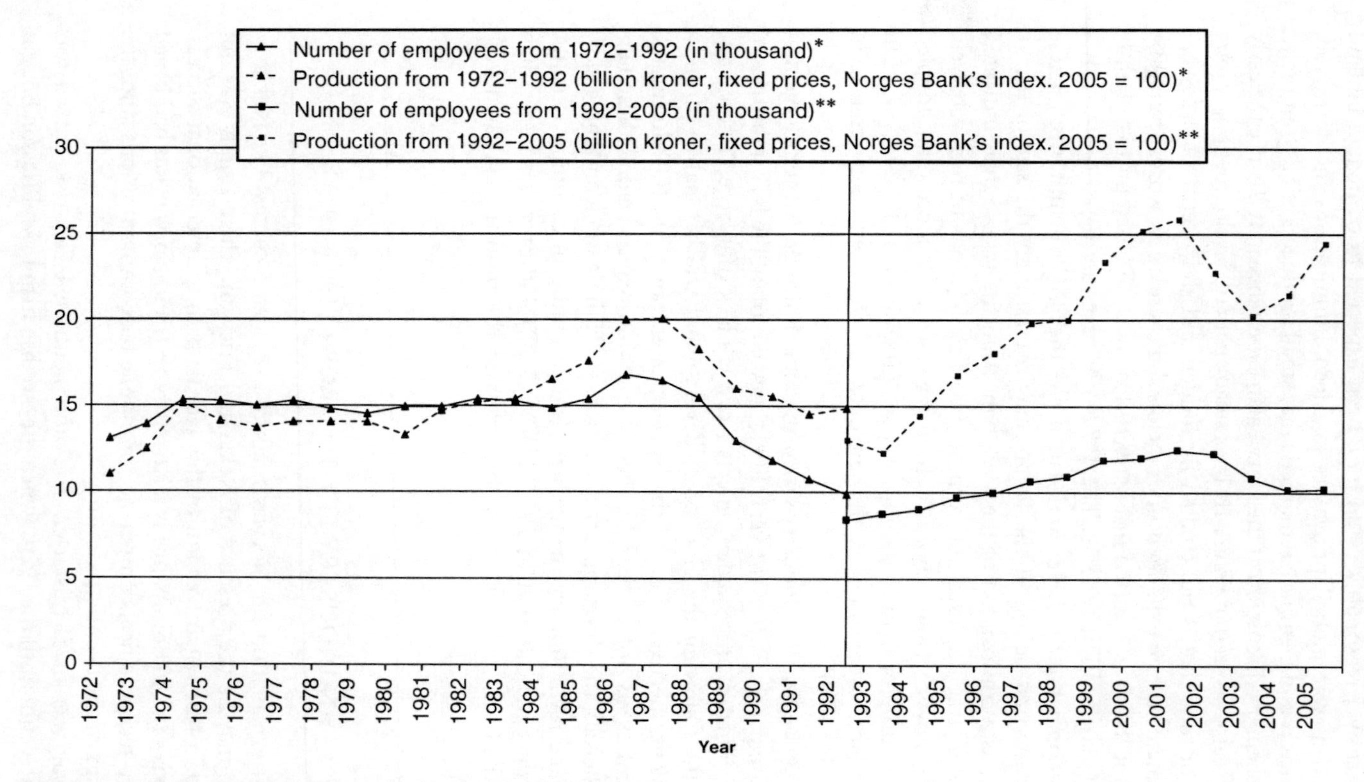

Figure 10.1. Norwegian ICT-goods industries by year, 1972–2005.

**Source:* Statistics, Norway. From Nils Petter Skirstad. Based on the ISIC-classification (1973–1992).

***Source:* Statistics, Norway. From Nils Petter Skirstad. Based on the NACE-classification (1992–2005).

a controlling share in its crisis-ridden Norwegian competitor. In the early 1930s, in the reorganization of the business empire of Swedish entrepreneur Ivar Kreuger following his death, another foreign-owned Norwegian telecommunications company, STK (Standard Telefon og Kabel) was created by ITT from the merger of an import company and a Norwegian cable-maker.[14] L. M. Ericsson-controlled Elektrisk Bureau and ITT-controlled STK were from the early 1930s through 1990 the providers of telecommunications to the Norwegian telecom operator, Televerket. Both of them had Norwegian minority shareholders, both of them had large production in Norway, both of them had some Norwegian research activity, and in the early 1970s they were two of the five largest Norwegian electronics companies in terms of turnover and employment, with well over 1,000 employees each. They were also controlled by an oligopoly comprising the international operators L. M. Ericsson and ITT. Norway's telecommunications industry, like its aluminum industry for much of the twentieth century, thus was controlled from abroad.

Norway's strongest national asset may have been its broad technological competence in radio communications. In addition to strong domestic demand for the radio receivers that were the foundation of the modern consumer electronics industry in the 1920s, Norway's large merchant- and fishing-fleets had formed a large market for communication equipment since the 1910s. In the interwar years this demand was met by an array of research-intensive small operations—Elektrisk Bureau, STK, the Norwegian Marconi-company, the Norwegian Telefunken-company and some wholly Norwegian companies—making radio equipment.[15] The Second World War and ensuing developments were not helpful to the fortunes of this industrial segment, but it nonetheless remained important until the 1970s when satellite-communication technology was adopted. Norway also had a strong national and Government-funded commitment to space research for meteorological and telecommunication purposes during the post-1945 period,[16] helping Nera—a wholly Norwegian continuation of Norwegian Telefunken—to become the main Norwegian operator in satellite-telephony, building on its long involvement in the maritime radio industry.

Equally important was the change in Nera's fortunes just after the Second World War and its growing role in national industrial policy. Norwegian engineers had been involved in large radar projects during the Second World War as part of Allied R&D teams.[17] Their expertise in electronics and related technologies was strengthened with the 1946 creation of NDRE (National Defense Research Establishment), which soon established collaborations with Nera. The knowledge from military radar projects was transferred to military and civilian "radio-link" technology, wireless radio-transmission of telecommunications and an alternative to cable that was highly desirable in a mountain-rich country like Norway.[18] Nera grew into a medium sized company with several hundred employees, depending on Norwegian and NATO markets for civilian and military products, as well as the occasional projects for developing nations.

Nera and Simrad (discussed in greater detail below) were perhaps the only important Norwegian companies up until the 1970s that were heavily influenced by Government R&D funding. A host of new government organizations were established after 1945, creating a new Norwegian public research infrastructure (see Chapters 3 and 4 for further discussion). The new organizations included the aforementioned NDRE, as well as two national research councils and two important research establishments for the technical research council (SI, literally "The Central Institute") and the Norwegian Technical University (called SINTEF) respectively.

Nera was closely linked to the post-1945 defense-related R&D efforts of Norway and NATO allies on radar, but Simrad, another Norwegian electronics firm founded after 1945, was influenced by defense-related R&D projects in telecommunications and hydro-acoustics. The founder of Simrad, Willy Simonsen, came from a prominent Norwegian business family and had been a technical leader in British walkie-talkie production operations during the war. Simrad initially focused on selling radiotelephones to the large Norwegian fishing-fleet. Simonsen was himself involved in technical issues of concern to the company in the beginning, not least when Simrad modified naval echosounders and sonar-equipment licenced from NDRE to produce small and efficient fish-finding equipment that was sold throughout the world.

Willy Simonsen was a transitory figure in the nascent Norwegian electronics industry, bridging the gap between the prewar entrepreneurial community of founders of electronics companies based on competence obtained through education abroad (in Simonsen's case, Dresden) or in Norway, and the postwar industrial effort that to a much larger degree involved public research institutions. The prewar entrepreneurial efforts of Simonsen's predecessors had created a Norwegian radio industry that produced receivers for the home use. Until the postwar crisis of the Norwegian radio industry in the early 1950s, there were tens of Norwegian producers of radios. Some of these had been founded by tinkerers who had little education, and other founders, like Vebjørn Tandberg, were well-educated engineers with contacts in the small number of prewar Norwegian research institutions, but who were independent entrepreneurs. Vebjørn Tandberg's Tandbergs Radiofabrikk, producer of "Tandberg" radios, was hugely successful, and merged with long-time competitor Radionette in 1972 to create the last of these Norwegian radio firms.[19] In 1972 Tandberg was among the three largest electronics firms in Norway, with a healthy export of tape recorders and stereo equipment and a growing output of color televisions and a research program in computers. Tandberg was a small version of "Philips"—the diversified Dutch consumer-electronics giant—and a source of national pride.

The last of the prominent Norwegian companies from the early postwar period that must be mentioned is Kongsberg Våpenfabrikk, the state-owned weapons-manufacturer that since the late 1950s had become a producer of advanced rockets and command and control systems as well as other products for automation applications.[20] Kongsberg Våpenfabrikk worked in tandem with some of the public research institutions mentioned earlier, but the main provider of new technologies for its development and production was the NDRE, which had become an independent and influential promoter of new electronics technology in Norway.[21] Kongsberg Våpenfabrikk was involved in a host of new technological activities reaching beyond IT, but its capabilities in digital technologies led the organization to specialize in signal-processing and systems technologies.[22] Kongsberg Våpenfabrikk also began to produce minicomputers in 1970, relying on computer technology developed by the NDRE.

NDRE and other Norwegian electronics-related public research institutions played an increasingly important role in the Norwegian IT industry. NDRE was particularly known for its role in the modernization of Kongsberg Våpenfabrikk, its relationship with Simrad, and the fact that several of its engineers in the late 1960s left and established the minicomputer maker Norsk Data. Norsk Data was a very early entrant into minicomputers, and had by the early 1970s established itself technologically and internationally, even though it was at the time a very small company. Elektrisk Bureau, STK and Tandberg were quite large and until the mid-1960s, pursued R&D programs that were managed and funded by these firms. Elektrisk Bureau and STK also had international partners. In hindsight, the influence of the public research institutions almost certainly was outweighed by the R&D efforts of the larger industrial enterprises until the end of the 1960s. In addition to Norsk Data, the most important new Norwegian electronics company founded during the 1960s was AME (A/S Mikro-Elektronikk and originally called Akers Electronics), a maker of integrated circuits established in 1965 as a spin-off of SI (the technical research council's research establishment).[23]

The influence of the Norway's public electronics research institutions grew in importance during the 1960s and 1970s, and when the national telecommunications provider, Televerket (Telenor since 1994) started a research institute in 1967, both Elektrisk Bureau and STK felt compelled to start their own R&D laboratories to collaborate with Televerket.[24] Televerket's R&D Institute wanted, among other things, to help Nera to become a satellite communications company, and to help Elektrisk Bureau and STK to become more independent of their mother companies and possibly, gain a technical leadership role within their respective parent multinational corporations.[25]

Overall, the development of the Norwegian IT industry until 1973 may be described as an amalgamation of private enterprise, entrepreneurship, and public research policy, with public research policy playing an increasingly important role.[26] Up until around 1960, when an important national debate over Norway's economic performance and future growth began, Norwegian IT policy rested on *research* policy. The industrial policy of Norway—which since 1945 had been quite interventionist—devoted little attention to the electronics industry. However, due to a number of developments, not least the discovery by economists of the "residual" and the growing acknowledgement of the impact of technology on economic growth, electronics moved to a central position in debates over industrial policy.[27] The modernization of state owned Kongsberg Våpenfabrikk, and the start of AME, as well as expanded Governmental funding for development projects undertaken by Norsk Data, Simrad and other companies were part of this policy change.

The Ministry of Industry had little direct role in these policy changes, as the new industrial policies of the 1960s were designed and financed by four government-appointed organizations for regional development (Distriktenes Utbyggingsfond), mergers (Tiltaksfondet), export collaboration (Omstillingsfondet) and development activities (Utviklingsfondet). The overall strategy constituted a kind of "indicative planning" system in which government funding indicated priorities and thereby influenced development activities, export collaboration, mergers and regional issues when firms made contact for these purposes. During the 1970s, the Ministry of Industry became increasingly involved in these technology oriented industrial policies, which relied on collaboration among companies, the Government and various public organizations. In 1973 the Government initiated a Norwegian Public Report of the electronics industry, a testament to the priorities of the time, and a sign of challenging developments.[28]

REVOLUTIONARY CHANGE, PART I

There was considerable basis for optimism about the prospects for what might be called "a national plan" for Norwegian electronics in the 1970s: Norway had initiated a concerted effort in new digital technology, through the creation of new companies, research organizations and modernization of older companies like Kongsberg Våpenfabrikk. Several companies, in addition to Tandbergs Radiofabrikk and Simrad, Kongsberg Våpenfabrikk must be mentioned along with Nera, enjoyed some export success. Yet by 1990, almost everything was in ruins. None of the main Norwegian companies existed in the same form as they had in the early 1970s, and in no instance was their transformation remotely positive.

What happened was the combined result of several factors, but perhaps the most significant single difference between Norway and other industrial nations, especially other Nordic nations, was the complex and dramatic effects of the domestic oil and gas industry on the Norwegian economy. One crucial consequence of the Norwegian oil and gas boom that began in the 1970s was to increase the cost structure of the whole economy, the IT exporters included. But this is only a partial reason for the IT industry's decline, and other factors also contributed. The Norwegian IT industry's development during the 1970s contrasted with its experience during the 1980s. The companies most severely (and negatively) affected by the 1970s were Tandbergs Radiofabrikk, Simrad and Nera as well as a host of smaller companies, while Elektrisk Bureau, STK, Norsk Data and Kongsberg Våpenfabrikk were severely hit by the developments of the 1980s.

The initial economic effects of the exploitation of offshore oil and gas in Norway during the 1970s were indirect.[29] Policymakers' expectations of large oil revenues underpinned the strongly counter-cyclical Keynesian economic policy applied during the downturn following the OPEC oil price increases of 1973. The situation for a new oil producer like Norway was contradictory of course, for the very price shocks that negatively affected the world economy positively affected the new oil sector of the Norwegian economy. Nonetheless, inflationary pressures developed in the Norwegian economy during 1975–78 as a result of growing state budget deficits, subsidies for export industries, and the rapid rise of a domestic oil production and processing industry.

The effects of this new environment on IT companies like Tandberg were complex: On the one hand, Governmental financial support was available for nearly every constructive effort to restructure, but on the other hand such "gifts" were associated with a broader economic transformation that increased the firm's operating costs to an increasingly painful level. Tandberg started to lose money and was taken over by the Norwegian state by 1978. Yet almost simultaneously with the government takeover of the firm, the Minister of Industry realized that Tandberg could not be saved and allowed the company to go bankrupt in December 1978.

Tandberg's problems were, of course, not solely related to its Norwegian location, but reflected competitive and technological changes as well. But similar technological and strategic challenges faced most European consumer-electronics industry firms during this period, and many of these enterprises (e.g. the Danish firm Bang and Olufsen) survived. The avenue followed by

Bang and Olufsen in its successful restructuring, which emphasized exterior design, increased reliance on purchased technology and a major reorganization and layoffs—was not taken by Tandberg, for a number of reasons.[30] First, Tandberg's commitment to a particular design approach was deeply rooted. Although Bang and Olufsen pursued a minimalist modern design approach as part of its restructuring, the techno-Scandinavian flair (metal and polished wood) connected to Tandberg's products retained considerable appeal to the firm's managers and its customers. There is no evidence that Tandberg's products had any design disadvantages, and Tandberg's sales slightly exceeded those of Bang and Olufsen until the early 1970s. Prior to Bang and Olufsen's large reorganization in 1974/75, Tandberg had much higher turnover per employee than Bang and Olufsen, consistent with the claims by Tandberg's last "pre-crisis" CEO, Andreas Skogvold, after the bankruptcy, that Tandberg by early 1977 was better at the technical aspects of automation of colour television production than the Japanese.[31]

Skogvold was an expert on automation and had visited several Japanese plants. But Skogvold also argued that the Japanese worked harder and were more disciplined than the Tandberg employees, highlighting a central element of Tandberg'a competitive problems. Tandberg was a model enterprise, famous for its harmonious industrial relations that had been conscientiously nurtured since the 1940s by founder and CEO (until 1974), Vebjørn Tandberg. Layoffs were out of the question, and new products were seen as replacements for products with saturated markets and tools for growth in sales and maintenance of employment levels. In the 1970s Tandberg planned to expand its production of color televisions (Tandberg was the second foreign firm to produce color televisions in Great Britain in the 1970s, after Sony).[32] In the 1980s, when Tandberg predicted that the color television market would be saturated, computer products, streamers (based on Tandberg's tape recorder technology) and monitors (based on the firm's expertise in television receivers) were to support corporate sales growth.

Governmental subsidies, part of the growth-oriented policy that built on its study of the electronics industry that had been released in 1976, supported Tandberg's efforts to develop new products. The Ministry of Industry even proposed a strategy for electronics in 1976 that would concentrate the national R&D efforts in three "cornerstone enterprises" in the electronics industry: Tandberg in consumer electronics, state owned Kongsberg Våpenfabrikk in defence electronics and a partly nationalized Elektrisk Bureau in telecommunications.[33] When Tandberg managers faced difficult decisions about the company's future in early 1976, the availability of public financial support led to a decision to reject proposals for restructuring and layoffs within the firm. CEO Andreas Skogvold's plan to cut costs and reorganize was overruled by the board of directors, a decision supported by Vebjørn Tandberg (member of the board), the Government and the labor unions, all of whom agreed that Tandberg should persist with its growth plans that were financed in part from public sources. From then on everything went to pieces.

Could Tandberg have survived? Bang and Olufsen did. Some of Tandberg's businesses have survived to the present day, mainly in the computer products produced by two publicly traded Norwegian companies, Tandberg Data and Tandberg Storage. Several other former parts of the firm continue to produce high-end stereo components, but the products (radios, stereo equipment, and color television receivers) aimed at large quality mass markets where Tandberg was famous in Norway and abroad no longer are produced. Tandberg's survival, like that of Bang and Olufsen, almost certainly would have required painful restructuring, including substantial layoffs and a radical shift in the organizational culture. Even in 1976, however, the firm's position within the consumer-electronics market, its R&D capabilities and its technical expertise in production automation provided considerable strengths on which to build a smaller enterprise. But the political support and financing for continued reliance on past strategies in a new global market meant that the firm never pursued such a radical (and risky) strategy and Norway's single most important electronics company went bankrupt.

Two of Tandberg's problems—the economic pressures in the Norwegian economy resulting from oil and gas exploitation, and shifts in the technologies underpinning electronics products for consumer and industrial applications—affected most of Norway's IT companies. Perhaps Kongsberg Våpenfabrikk, which had had a digital IT strategy since 1960, was the least vulnerable to the technological shifts, inasmuch as it did not have any old electronics products in its portfolio, and the government's 1976 "cornerstone plan" made much more sense for this defence-centered company than it did for Tandberg. State financial assistance enabled Kongsberg Våpenfabrikk to purchase a couple of troubled Norwegian electronics companies, including Norcontrol, a firm specializing in electronic systems for navigation and operation of oceangoing ships that had been founded in 1965, in 1978.[34]

Elektrisk Bureau, the large maker of telecom equipment, also faced challenges from rapid change in the technologies underlying its products and was able to shed parts of its workforce as production became more automated.[35] The 1976 cornerstone plan had great promise for Elektrisk Bureau, which faced growing demand from the large programs for modernizing the national and Nordic telecom infrastructure that were underway at the time. This confidence in the prospects for Norwegian telecommunications-equipment firms was widely shared, and approximately one-quarter of Elektrisk Bureau's shares were bought by Norwegian investors from L. M. Ericsson, reducing the

Swedish company's share by 50 percent.[36] Elektrisk Bureau also took over Nera, a company experiencing a multitude of problems in the 1970s. For Elektrisk Bureau and STK, the situation in the 1970s was characterized by a close relationship with the authorities, not least the Ministry of Transport, that provided some expectation of long-term market stability for their operations and supported restructuring. The cornerstone-plan seemingly enhanced this situation, forging a stronger linkage between the national industrial policy overseen by Norway's Ministry of Industry and the traditionally close relationship between the telecom sector and the Ministry of Transport.

Simrad, like Tandberg, was challenged by Japanese competition.[37] Simrad, a producer of small hydro-acoustic equipment for all kinds of vessels, was a global innovator in a small niche. Traditionally its main competitors had been the British firm Kelvin Hughes and Atlas from Germany—with their naval roots—but the 1960s saw the arrival in international markets of Japanese competitors such as Furuno. Simrad met this competition in much the same way as Tandberg had sought to, developing more advanced products and acquiring new product lines. Nevertheless, by 1980, Simrad faced a situation similar to that confronted by Tandberg in 1976—falling sales and falling prices.

Unlike Tandberg, Simrad responded with rationalization, reorganization, outsourcing of production, and a division of the company into several smaller units. The unit providing deep-water hydro-acoustic positioning equipment for the growing oil business became a huge success and was listed on the Oslo Stock Exchange in 1982. The unit making fishing equipment went into a deep crisis that was only solved when young engineers from the NDRE joined it and radically redesigned Simrad's sonar-equipment to exploit the digital technologies that they—unlike Simrad's older engineers—had mastered. Building on that new technical platform, as well as the earnings from the company serving the oil industry that purchased the fishing-equipment business in 1983, Simrad enjoyed considerable success throughout the 1980s. Its revival arguably would have been impossible without the growth of a domestic offshore oil exploration and production sector in Norway during the period.

There were several differences between Simrad and Tandberg, not least in terms of the political climate that each firm faced during its period of crisis. Simrad's troubles emerged after Tandberg's problems and bankruptcy had sent shockwaves through the Norwegian political system. The decision by Norway's government in late 1978 to let Tandberg go bankrupt, after the state had formally became the owner of the firm, reflected a radical shift in policy in Norwegian industrial policy under the Labor government. 1978 was also the last year of strongly counter-cyclical macroeconomic policy measures, which reflected some of the same disillusionment with their effects and more general concern over the loss of competitiveness within Norwegian industry during the late 1970s. Although the firm's politically conservative owner almost certainly would have rejected it, by 1980 government financial assistance for Simrad was no longer politically feasible.

In the IT industry, as in other sectors, the industrial policies of the Norwegian governments of the 1960s had become very interventionist by the mid-1970s, a posture that was largely reversed by 1978. The several state agencies for assisting companies were still very much in place, but they were not coordinated anymore. The two most important Norwegian IT companies formed in the 1960s and with an aim for being helpful to the whole IT industry—AME (integrated circuits) and Norsk Data (minicomputers)—functioned more or less independently and did not play important roles in relationships with the bigger Norwegian IT companies.

Two new developments at the end of the 1970s exemplified the shift in government industrial policy away from the posture that it had assumed for almost all of the post-1945 period. The Norwegian Government began to promote the equity markets as *alternatives* to industrial policy and state controlled funding, reflecting free-market liberal attitudes that were gaining popularity in other western countries at the time, not least Great Britain and USA. The second change followed in the same ideological path, namely a gradual shift to a new public-procurement policy that no longer favored domestic "national champions." Although the Norwegian public sector had never been a rigidly nationalistic purchaser, reflecting strong liberal values,[38] this policy was liberalized still further during the 1980s, with significant consequences for Norsk Data in particular.

REVOLUTIONARY CHANGE, PART II

The 1970s had ended with strongly negative developments in Norway's electronics industry, but the first half of the 1980s started with a boom. The 1980s were the decade of the entrepreneurial IT firm in Norway, often listed on the stock exchange, and Simrad and Norsk Data were preeminent actors in this new era. But the decade was disastrous for the larger Norwegian ICT corporations, and by the start of the 1990s, Elektrisk Bureau, Norsk Data, and Konsgsberg Våpenfabrikk—along with several new start-ups—were gone, and STK was a shadow of its former self, fated to fail later in the decade.

In retrospect the failures of the 1980s were a result of the shift to a less interventionist government procurement policy, in addition to the effects of continued technological innovation. The first casualty was Elektrisk Bureau. After fifty years as co-provider for the Norwegian telecom monopoly as one part of a politically influential duopoly (STK being the other "national champion" supplier), the company failed to win the contract for the first fully digital switch for the Norwegian telecom infrastructure, which was to be provided by one company through an international tender.[39] Elektrisk Bureau, partly nationalized in 1976–77 through the Government's cornerstone-plan, was expected to win the contract based on the AXE-system of its parent firm Ericsson. But the contract went to STK, which promoted ITT's still-incomplete "System 12". Elektrisk Bureau was forced to merge with the Norwegian producer of electrical equipment Elektro Union. When the great merger between Swedish ASEA and Swiss Brown Boveri happened in 1987, L. M. Ericsson's 25 percent ownership share in Elektrisk Bureau and the Wallenberg-family's strong position in both ASEA and Ericsson meant that Elektrisk Bureau became part of ABB Norway, and its telecom business became a minor part of ABB Norway.

STK did not fare much better.[40] Winning the 1983 contract turned out to be something of a Pyrrhic victory. Getting "System 12" into operation was a huge undertaking that strained relations between STK and majority-owner ITT. Even though "System 12" was a product of ITT, STK had to bear a large share of its development costs. When in 1986 "System 12" went into operation as a fully digital switch, the benefactor was ITT, or more precisely ITT's owners who subsequently were able to get a good price for the firm's telecom operations (including STK) in a sale to French Alcatel. STK was renamed Alcatel Norway and what had until then been a semi-independent company with an assured Norwegian market declined further. The next big contract for Norwegian switches in 1990 went not to STK/Alcatel but to Ericsson Norway, the telecom equipment remainder of Elektrisk Bureau that had been sold to Ericsson. From then on Alcatel Norway was reduced to little more than a sales organization for the large French multinational.

The absorption of Norway's former "national champions" in telecommunications equipment into two multinationals in the same year—1987—was hardly coincidental, but reflected the changing posture of Norwegian government policy in procurement and support for these "champions." According to Sverre Christensen's recent doctoral thesis about STK, Norway was by the 1980s one of the countries most open to international competition and modernization of its telecom network.[41] For STK, the Norwegian contract for System 12 should have enabled the firm to leapfrog its competition and dominate future procurement competitions. Instead ITT and Alcatel reaped the benefits. The Norwegian Government's more liberal policy in telecom equipment procurement may have changed private investors' attitudes towards participating in this industry, and hastened the sale of these supplier firms. Although both STK/Alcatel and Elektrisk Bureau/Ericsson Norway were foreign-owned, they had had a significant Norwegian management and technical component within them for decades. Following their sale, this Norwegian role within both firms declined significantly.

Several new companies have been created in the aftermath of the dismantling of Elektrisk Bureau and STK, the most significant of which was Nera, listed on the stock exchange in 1994 after seventeen years in the Elektrisk Bureau/ABB fold. More than anything, however, the attitude of the Norwegian Government helps to explain why there was no Norwegian contender in the commercialization of the next great innovation in telecommunications, GSM.

Mobile telephony—with its use of radiowaves—was "natural" for Norway, that is to say the Norwegian topography with high mountains and long coastline influenced the development of Norwegian telecommunications competence.[42] In addition to maritime radios, satellite-communication, and radio links, the Norwegian Defence forces had been particularly concerned with radio communication.[43] Televerket's R&D Institute was also very active, as were both Elektrisk Bureau and STK, Elektrisk Bureau being an expert on radio communication while STK made military switches.[44] Up until the early 1990s Norway also had a maker of mobile telephones, Simonsen Elektro, the company that Simrad-founder Willy Simonsen started when he left his original company in 1967. Simonsen Elektro was the Norwegian counterpart to Ericsson and Nokia in the mobile telephone system developed by the Nordic telecom providers called NMT.

During the latter part of the 1980s, when the international community had chosen the wireless technology originally developed by a *Norwegian* scientist as its GSM-system, Simonsen Elektro, Elektrisk Bureau/ABB and a Norwegian research organization tried to develop a coordinated system of products to compete internationally. Problems at Simonsen Elektro and the negative effects of the ABB merger on Elektrisk Bureau led to the termination of the project, however, and Norway failed to introduce any products based on the GSM architecture.

The situation in 1990 for Norwegian mobile telephony resembles that following Tandberg's bankruptcy in 1978.[45] Mass market opportunities for mobile telephony and color televisions both were not exploited by Norwegian firms. The two situations were very different, but in some respects they both reflected a lack of managerial strength. Tandberg's management was weakened by its long decline, and the new owners of Tandberg after the bankruptcy saw the television challenge as too big—which it probably was for people without industry experience.[46] In mobile telephony there was no Norwegian company that could take the lead. Norway boasted an abundance of technical competence in companies and research institutions to support a Norwegian entry into mobile telephony around 1990. But entry demanded entrepreneurship and financial commitment of a kind that was harder to find.

Simonsen Elektro was a natural candidate for leadership in mobile telephony because of its market position (Simonsen made highly regarded phones), but increased international competition in the NMT-business, along with the Norwegian economic downturn at the end of the 1980s, discouraged the firm's entry. CEO and part-owner Simonsen, who had the experience and competence needed for such an effort, was by 1990 nearly 80 years old and ended up selling the company. The comparison between Norway and Finland at the inception of GSM technology is instructive. Both countries possessed similar competences in relevant fields, but Nokia was financially more powerful, had a broader competence base than Simonsen, and was more favorably positioned commercially within NMT than Simonsen.[47] The other company that could have played such a role in Norway, Elektrisk Bureau/ABB Norway, chose in 1990 to sell its Telecom business to Ericsson, creating Ericsson Norway. It is possible that ABB's sale of Elektrisk Bureau was motivated by a desire to eliminate a potential Norwegian mobile-telephone competitor for Ericsson, but in 1990 Ericsson had not yet formulated its strategy for GSM.[48]

The fall in oil prices in the second half of the 1980s created economic problems in Norway. It was a period of consolidation, and no company consolidated more than ABB Norway, the successor to Elektrisk Bureau: It reorganized its electricity business, sold its telecom business and launched a large-scale entry into petroleum-related activities.[49] As the ABB example suggests, Norway's oil and gas industry has since the 1970s been a magnet for new business formation and entry, and as a result may have "crowded out" private investment in business opportunities in other fields. ABB Norway was very successful during the 1990s, suggesting that its oil-oriented strategy was commercially sound, at least in the short run. STK/Alcatel did try to pick up the pieces, e.g. by continuing its participation in state-supported research programs, after Simonsen and other firms had given up, but the Norwegian subsidiary's efforts received no support from its French parent.[50]

By the 1990s, Norwegian investors had become reluctant to mount the costly effort needed to penetrate mass markets for electronics-related products, a reluctance that was compounded by the presence of attractive domestic investment opportunities in the oil and gas sector. Why enter a risky business based in a high-cost nation when there were plenty of investment opportunities in the sector (oil and gas) that had made Norway such a high-cost location, even if oil prices had fallen since 1986? Indeed, why mount a risky entry into telecom where no national-government purchases could be taken for granted?

The fall in oil-prices hit the remaining big IT companies of Norway hard.[51] According to Figure 10.1 the only real downturn in Norwegian employment after 1972 occurred after 1988, highlighting the severity of the economic contraction. Although the 50 percent fall in IT industry employment from 1987 to 1992 also reflected the broader international downturn during this period, most of the downturn was related to problems in Norwegian IT. Elektrisk Bureau and STK changed fairly gradually, but other Norwegian IT companies restructured more abruptly, with significant layoffs and bankruptcies.

The first of the companies to face liquidation in 1987 was Kongsberg Våpenfabrikk, the state owned weapons-manufacturer that had been in operation since 1814. The official investigation following its liquidation revealed a company that for many years had exceeded its budgets and had met its financial needs with funding from the Government, justified by a continuous process of investing in new technological areas.[52] But the new political climate exemplified by the 1978 Tandberg bankruptcy finally caught up with Kongsberg Våpenfabrikk when oil prices fell in late 1986, ending the economic boom of the 1980s. Government spending was dramatically cut both to stabilize the economy and to reflect the lower level of oil-related revenues. Kongsberg, which (remarkably) had been refinanced several times during the 1980s, was rejected for the first time when it asked for more money in 1987.

Several new companies were created after the demise of Kongsberg Våpenfabrikk; the largest were Kongsberg Gruppen (defence electronics, state owned), Kongsberg Offshore Systems/FMC Kongsberg (for underwater oil-production systems), Kongsberg Automotive (for car parts) and a few others. A great success during the 1980s until the oil price fell, Kongsberg Albatross, the world leader in dynamic positioning equipment, was sold for a very low price to collaborator Simrad. Albatross had been viewed as a great success in the modernization of Kongsberg Våpenfabrikk in its extensive use of IT.

The next big IT industry casualty during this period was Norsk Data, the most successful Norwegian IT company of the 1980s.[53] Norsk Data's initial success depended on being an innovator in the mini computer market, but its strategy during the 1980s had left it very vulnerable when the cuts in Norwegian Government spending hit the Norwegian public sector, an important market for Norsk Data. Founded as a seller of minicomputers to a wide variety of users in many countries, Norsk Data tried to become a systems provider, selling programs and packages that had been developed in collaborations with Norwegian government agencies and partly financed through R&D contracts. The coming of the PC (personal computer), open systems and the firm's dependence on the Norwegian market, all contributed to Norsk Data's fall. When the Norwegian public sector stopped purchasing from Norsk Data, the company failed.

Norsk Data was not alone in failing, of course, and to some degree the downfall of Norsk Data and Tandberg reflected new developments in highly competitive international markets that affected electronics firms around the world. In contrast to Tandberg, however, Norsk Data had no strategy for surviving such international and technological competition. Norsk Data and its emphasis in developing closed-system applications for its computers resembled market leader Digital Equipment Corporation, another minicomputer pioneer that failed.[54] Norsk Data was hardly alone in failing as a minicomputer specialist, but like Kongsberg Våpenfabrikk, it failed to meet market expectations. Norsk Data and Kongsberg Vapenfabrik both also failed to recognize how the new political realities of Norwegian technology policy, which had weakened the positions of "national champions," had increased their vulnerability to market competition.

The 1980s changed the Norwegian system of innovation for IT. The numerous contacts among Government loan institutions, the Norwegian research council, and the Norwegian IT sector remained, and produced some positive outcomes. But Government procurement practices had shifted by the end of the 1980s, as "national preferences" were abandoned, limiting the influence of research policy and favorable loans. Even more dramatically, all the large actors of the sector, including many that had been meticulously built up with direct and indirect government support since 1945—were gone. Tandberg and Norsk Data went bankrupt. Elektrisk Bureau and STK were dramatically reorganized, not least because there was little to be gained by having a R&D operation for telecommunications in Norway in the face of the government's new procurement policies. Televerket's R&D Institute changed its role to serving only Televerket, rather than the national industry. Kongsberg Våpenfabrikk did not technically go bankrupt, but ended its existence as a legal entity. Overall, the digital revolution and the global competition that began to intensify in the 1970s had contributed to a revolutionary change for the Norwegian IT industry. In addition, government policy shifted from political steering of this sector towards a much more market oriented situation. Despite their drastic consequences, these changes in competition and policy ultimately led to the revival in the early twenty-first century of a very different Norwegian IT industry.

SLOW GROWTH, SMALL COMPANIES (1991–2006)

For the Norwegian economy, the 1987–1993 period was very problematic. The fall in the oil prices depressed the overall Norwegian economy, and when the international downturn following the Gulf war took effect, a recession was transformed into a full-blown property and banking crisis. On another level the political leadership of Norway realized during the early 1990s that most of the high income from oil and gas had to be kept outside the Norwegian economy, so the post-1990 period has been characterized by a stable economic climate, not a bad domestic environment within which to revive an internationally competitive IT industry.

To some degree the "oil problem" was replaced by the "oil solution" within Norwegian IT, as some of the more successful IT companies found a profitable market in the Norwegian oil industry. Simrad was a pioneer here, and in addition to its products for dynamic positioning equipment, the firm introduced systems for seabed mapping through the use of hydro-acoustics.[55] Many of the new IT products followed in the wake of the so called "Deals for technology", a Government-induced system after 1978 that gave oil companies goodwill in allocating new concession for oilfields if they purchased or supported the development of Norwegian technology (see Chapters 4 and 7).[56] Some of the deals made between Norwegian companies and oil companies did not concern oil-related activities, and Norsk Data benefited from several of these. But some new products were developed under the sponsorship of the international oil companies under the terms of the "goodwill program," and the full effects of this influx of fresh research money are still being felt within Norway's IT industry.

The "goodwill policy" reflected the changes in economic policy that began in the late 1970s. Gone were the days when the Government directly supported companies like Tandberg, a policy that proved dangerous for the Government when Tandberg failed. Gone were the days (there were never many) when large Government contracts directly supported companies; this too could be dangerous for the Government. But the application of political pressure to support the construction of new networks between foreign oil companies and Norwegian technology-providers in exchange for future benefits for foreign oil companies, was for some years after 1978 a viable alternative. Norway's industrial policy most certainly did not convert to anything like pure and principled liberalism in the late 1970s, and the limited nature of these shifts in policy may have contributed to misperceptions of the nature and extent of change in the political climate by Kongsberg Våpenfabrikk's leadership, among others.

The companies that benefited the most in various ways from the oil economy may have been the successors of Kongsberg Våpenfabrikk. The producer of underwater production systems (and a firm whose products and technologies spanned much more than IT), Kongsberg Offshore Systems/FMC, experienced a commercial breakthrough when underwater production systems replaced large platforms in Norway's offshore oilfields in the late 1980s and early 1990s.[57] Kongsberg Gruppen, the defence-products company and the heir to Kongsberg Våpenfabrikk's core businesses, adapted to the changing geopolitical climate after 1990 with a series of acquisitions of companies in maritime electronics, including Norcontrol, previously under the Kongsberg umbrella. The largest and most important purchase was Kongsberg's unfriendly takeover of Simrad in 1996. Simrad became the cornerstone in Kongsberg Maritime, which along with Kongsberg Defence and Aerospace was one of the two most significant parts of Kongsberg Gruppen. Kongsberg Maritime had a range of offshore customers, but oil-related activities were its most significant market.[58]

It is difficult to quantify the positive role of oil for the Norwegian IT industry of the 1990s. The oil industry proved to be a significant market for Kongsberg Maritime. IT was an integral part of larger technical systems and therefore important for the development of new and complex products in the oil-industry products of both Kongsberg Offshore System and ABB. For the telecommunications sector, the development of an infrastructure for the Norwegian petroleum sector has been a large project covering several technological phases and firms.[59] The single most compelling illustration of oil's helpful role may be the crucial importance of oil-industry contracts for the largest IT company of Norway, Kongsberg Gruppen. This is also the one company most clearly building its future on the technical past of two of the most important companies of the postwar era, Kongsberg Våpenfabrikk and Simrad.

The rebuilding of a strong Kongsberg group created the only major new industrial enterprise in Norwegian IT during the 1990s. Otherwise the 1990s and the ensuing years of the new decade were characterized by organic growth of old and new enterprises. Most of the largest enterprises of this period were remnants of former large enterprises. The revival of the Tandberg-companies was the most remarkable example of this tendency. Apart from the computer company, Tandberg Data, which recently was split in two, the company that was the continuation of Tandberg Radiofabrikk had by the early 1980s become dormant. Its revival, which eventually led to the creation of two companies, was based on new technology from Televerket's R&D Institute in the area of moving pictures. One of the two Tandberg successor companies makes equipment for live-picture telephony (still called Tandberg), and the other makes equipment for satellite transmission of live pictures (television, and called Tandberg Television, now part of Ericsson). There are several spinoffs from Elektrisk Bureau, STK, and Norsk Data, and the commercially unsuccessful maker of integrated circuits, AME, has spawned a number of different companies.

The biggest commercial success of all, albeit one that was in services rather than manufacturing, was Telenor, the partly privatized old Televerket that was able to grow significantly abroad, not least through becoming a mobile-telephone operator in several countries. Telenor has been a clever strategist, but all three observers of Telenor and the Norwegian telecom industry have thought that the Norwegian telecom equipment industry's collapse laid the foundations for Telenor's success.[60] While STK and Elektrisk Bureau went down, Telenor went up. Telenor's success relied on a number of factors such as the input from Televerket's R & D Institute, its own work to modernize the Norwegian infrastructure, which reached a high point with the specifications for the digital switch contract in 1983, and a continued input of highly qualified and very experienced people who were laid off after the demise of Elektrisk Bureau and STK.

Industrial reorganization also had other ramifications. In the wake of its dramatic reorganization in the early 1980s, Simrad has adopted a strategy that Tandbergs Radiofabrikk rejected, outsourcing a large share of its production activities. Other IT companies have followed Simrad in outsourcing production. Several new Norwegian companies have been formed as specialists in contract manufacturing, and Kitron has become a fairly large specialist in contract production of IT goods. The appearance of specialized Norwegian production companies underlined the transformation of traditional, manufacturing-oriented electronics firms into firms that relied on their expertise in technology creation, design and commercialization, relying on strong in-house engineering capabilities. Much of this restructuring required a generational shift within the senior management of the companies involved. If the case of Simrad is representative, the R&D departments of these firms now are filled with "young people" skilled in digital technology.[61] A specialized company like Simrad still has many positions for "old" engineers, however, not least in its large global sales force.

To some extent, of course, the IT industry built on past achievements and new market-opportunities during the 1990s. But most Norwegian IT companies had to reinvent their market positions and restructure their R&D, product development, and manufacturing activities. With the exception of Kongsberg, these tasks required new ideas and talent rather than continued reliance on accumulated corporate knowledge and culture. The survival and continuing development of a small scale Norwegian IT industry reflects the continuous input of newly educated engineers able to master new technological and commercial challenges. Indeed, the influx of highly educated young people willing and able to adapt to rapidly changing circumstances, more often than not in a national rather than an international context, has been a historic hallmark and strength of the Norwegian IT industry throughout the past century.

Norway's IT industry experienced a quiet revival during the 1990s and early 2000s. The stock-market crisis of 2000 was remarkably good to the Norwegian IT industry, perhaps because of the lack of a "bubble frenzy" in Norway comparable to those in other industrial economies, such as the USA. Indeed, one positive factor in the industry's growth during this period arguably is the more realistic attitude within the industry, an attitude that had ample opportunity to develop during the 1970s and 1980s. There is nothing bold and flashy about the Norwegian IT industry of 2007, and the various hypes of history's yesterdays—"information" or "knowledge" economy, "high tech industry," "new economy"—seem left behind. That is partly because there is so little that is flashy left, and the Norwegian IT industry of today is finding its niches, many of which are related to large, growing, and profitable domestic markets.

CONCLUSION

From a comparative perspective, Norway is the Nordic IT industrial laggard, albeit a laggard that is not too far behind Denmark. Set against the ambitions that senior industrial managers and politicians articulated during 1970s and 1980s, the Norwegian IT industry is a failure. Measured against all the company failures and its many crises during the 1970s and 1980s, the Norwegian IT industry has left former employees and shareholders in problems. But the Norwegian industry has survived and has resumed growth. It was able to make painful changes to adapt to technical change, global challenges and new opportunities. It has throughout the twentieth century, and not least during Norway's oil boom, been an important source of technical competence in the broader transformation of the Norwegian economy. Comparing Norway to other nations gives interesting perspectives, the most important lesson to be learned from the history of the Norwegian IT industry is the importance of its integration with the rest of Norway's economy. The failure of the Norwegian IT industry to develop products for huge foreign markets paradoxically may have benefited other sectors of the Norwegian economy, whose firms were able to draw on Norwegian IT-industry partners to make specialized products needed for particular domestic purposes.

The development of Norway's IT industry has been characterized by both continuity and change. The most dramatic change has come in the fortunes of the companies. Failures of some firms have meant the extinction of whole sectors of the Norwegian electronics industry, as consumer electronics, computers, mobile telephony, have vanished. But smaller organizations, making equipment for the maritime sector and/or developing products for the oil industry have arisen from the industrial wreckage of larger enterprises and have contributed to larger economic activities within Norway—running boats, the process industry, the oil business. The influx of oil in the Norwegian economy has been particularly important in this recent history, contributing a complex mix of higher factor costs (1970s and 1980s in particular) and volatile markets (rapid changes in public sector spending in 1980s) on the one hand, while on the other hand providing large new markets for specialized products, providing financial grants, and supporting industry and research-institute R&D. The Norwegian oil sector has also sucked managerial and engineering talent away from IT production, and has provided a more attractive market than IT for many sources of private capital seeking investment opportunities in Norway.

Seen against this background, the Norwegian IT industry represents something of a success story, as the industry has contributed to Norway's resource-based economic growth since the 1980s with advanced technology. Overall Norwegian economic growth—and Norway is one of the richest nations in the world per capita—has been strong during this period, and the IT industry has contributed significantly to this success.

NOTES

1. My interpretation of the development of the Norwegian IT industry has benefited from the advice of many people over the years, most recently from doctoral candidates Sverre A. Christensen, Stein Bjørnstad, and Gard Paulsen. Comments from Jan Fagerberg, Sjur Kasa, David Mowery, and Olav Wicken at the IPP workshop at Leangkollen were helpful.
2. See table 7.6 and figure 7.17 in Nordic Council of Ministers, Nordic Information Society Statistics 2005, Copenhagen 2005.
3. See Knut Sogner, "Næringspolitikkens betydning for fremveksten av norsk elektronikk–og IT-industri," in Olav Spilling (ed.), *Kunnskap, næringsutvikling og innovasjonspolitikk*, Bergen: Fagbokforlaget 2006.
4. Harald Rinde, *Et telesystem blir til: Norsk telekommunikasjonshistorie*, bind I, Oslo: Gyldendal Fakta 2005.
5. This is the main argument in: Knut Sogner, *En liten brikke i et stort spill. Den norske IT-industrien fra krise til vekst 1975–2000*, Bergen: Fagbokforlaget.
6. The Global Information Technology Report 2006–7, World Economic Forum's homepage.
7. Gunnar Nerheim and Helge W. Nordvik, *Historien om IBM i Norge 1935–1985*, Oslo, etc.: Universitetsforlaget 1986; Sverre Knutsen, Even Lange, and Helge

Nordvik 1998, *Mellom næringsliv og politikk. Kreditkassen i vekst og kriser 1918–98*, Oslo: Universitetsforlaget 1998.

8. Keith Smith, "Hva slags økonomiske virkninger har IKT skapt?," in Helge Godø (ed.), *IKT etter dotcom-boblen*, Oslo: Gyldendal akademisk 2003.
9. Statistics Norway, "Databehandlingsvirksomhet... 2004", sent by Ole-Petter Kordahl.
10. Interview with Tom Gerhardsen November 30, 2000.
11. See Jan Rune Holmevik, Educating the machine. A study in the History of Computing and the Construction of the SIMULA Programming Language, report number 22, University of Trondheim: STS1994. Gard Paulsen is currently writing the doctoral thesis "Innovating code and coded innovations: The dynamics of the software industry 1965–2005 at the Norwegian School of Management BI.
12. *Elektronikkindustri*, Oslo: Norges Offentlige utredninger, 1976: 30.
13. See Keith Smith (2003) for more about the difficulties with ICT statistics.
14. Sverre A. Christensen, *Switching relations. The rise and fall of the Norwegian telecom industry*, Oslo: Handelshøyskolen BI 2006.
15. Knut Sogner, *God på bunnen. Simrad-virksomheten 1947–1997*, Oslo: Novus forlag 1997.
16. John Peter Collett (ed.), *Making Sense of Space. The History of Norwegian Space Activities*, Oslo: Scandinavian University Press 1995.
17. Tor Arne Eilertsen, "Fra FOTU til FFI. Grunnleggingen av norsk forsvarsteknologisk forskning 1942–46," unpublished hovedoppgave in history, University of Bergen 1987.
18. Per Fremstad, "40 år med radiolinjer 1955–1993," unpublished booklet, Bergen 1993.
19. Knut Sogner, "Veksten mot idealfabrikken, rapport nr. 5," *Teknologihistorieprosjektet, NAVF-NTNF*, Oslo 1990.
20. Olav Wicken has written extensively on this subject, see for example: Olav Wicken, "Norske våpen til Natos forsvar," *Forsvarsstudier*, nr. 1, 1987 and Olav Wicken, "Stille propell i storpolitisk storm," *Forsvarsstudier* nr. 1, 1988.
21. Olav Njølstad and Olav Wicken, *Kunnskap som våpen. Forsvarets forskningsinstitutt 1946–1975*, Oslo 1997.
22. Sogner (2002).
23. Håkon With Andersen, *Fra det britiske til det amerikanske produksjonsideal. Forandringen av teknologi og arbeid ved Aker mek.Verksted og i norsk skipsbyggingsindustri 1935–1970*, Trondheim: Tapir 1989; Knut Sogner, *Fra plan til marked. Staten og elektronikkindustrien i 1970-årene*, Oslo: TMVs skriftserie/Pensumtjeneste 1994.
24. John Peter Collett og Bjørn O.H. Lossius, *Visjon-Forskning-Virkelighet. Televerkets Forskningsinstitutt 25 år*, Kjeller 1993.
25. Christensen (2006).
26. Sogner (1994).

27. Kjersti Jensen, "Moderniseringsmiljøet som pådriver i norsk industriutvikling på 50 og 60-tallet," unpublished hovedoppgave in history, Universitetet i Oslo 1989.
28. *Elektronikkindustri*, Oslo: Norges Offentlige utredninger, 1976: 30.
29. Sogner (1990).
30. Sogner (1990).
31. Letter from Andreas Skogvold to Knut Sogner, July 1989.
32. Erik Arnold, *Competition and Technological Change in the Television Industry. An empirical Evaluation of Theories of the Firm*, London: MacMillan1985. For the Japanese development, see Gene Gregory, *Japanese Electronics Technology*, Chichester: John Wiley and Sons 1986.
33. Sogner (1994) and Sogner (2002).
34. Signy Overby, "Fra forskning til industri. Utviklingen av skipsautomatiserings-bedriften Norcontrol," uupublished hovedoppgave in history, Universitetet i Oslo 1988; Norges Offentlige Utredninger 1989: 2, Kongsberg Våpenfabrikk.
35. Sogner (1994) and Christensen (2006).
36. The large-scale share purchases resulted from negotiations between the Norwegian Government and the Swedish holding company owned by the Wallenberg family that controlled L. M. Ericsson.
37. Sogner (1997).
38. John Peter Collett and Bjørn Lossius, *Visjon—forskning—virkelighet. Televerkets Forskningsinstitutt 25 år*, Skedsmo: Televerkets forskningsinstitutt 1993.
39. Christensen (2006).
40. Christensen (2006).
41. Christensen (2006).
42. Sogner (2002).
43. Sogner (1994).
44. Christensen (2006).
45. The following discussion is based on Sogner (2002). who draws upon several other works.
46. They had taken over Tandberg after receiving a huge help-package from the Government.
47. I am basing this comparison on my own and Christensen's (2006) work in addition to Ari Hyytinen, Laura Paija, Petri Rouvinen and Pekka Ylä-Antilla, "Finland's Emergence as a Global Information and Communications Technology Player: Lessons from the Finnish Wireless Cluster," in John Zysman and Abraham Newman, *How Revolutionary was the Digital Revolution? National Responses, Market Transitions and Global Technology*, Palo Alto, CA: Stanford Business Books 2006.
48. Maureen McKelvey and Francois Texier, "Surviving technological discontinuities through evolutionary systems of innovation: Ericsson and mobile telecommunication," in Charles Edquist and Maureen McKelvey, *Systems of Innovation: Growth, Competitiveness and Employment*, Vol. II, Cheltenham, UK 2000.

49. Sogner (2002).
50. Christensen (2006).
51. Sogner (2002).
52. Norges Offentlige Utredninger: "Kongsberg Våpenfabrikk," NOU 1989: 2.
53. Sogner (2002).
54. For DEC, see Alfred D. Chandler Jr., *Inventing the Electronic Century. The Epic Story of the Consumer Electronics and Computer Industries*, New York: The Free Press 2001.
55. Sogner (2002).
56. Kjell Grønhaug, Torger Reve and Tor Fredriksen, "Teknologiavtalene: samarbeidsaktiviteter og samarbeidsvirkninger", rapport 1/86, *Senter for anvendt forskning*, NHH. See also Chapter 4 in this volume.
57. Stein Bjørnstad, "Forklaringsmodeller for 'århundrets største ingeniørbragder' ", in Olav Spilling Olav Spilling (ed.), Kunnskap, næringsutvikling og innovasjonspolitikk, Bergen: Fagbokforlaget 2007.
58. In addition to Sogner 2002, Stein Bjørnstad is writing a doctoral thesis about the oil-related activities of the Kongsberg companies that goes detailed into the history of dynamic positioning and under-water production systems.
59. Gard Paulsen, "Innovasjon over Nordsjøen: Telekommunikasjoner på norsk sokkel," Forskningsrapport 3/2005, Handelshøyskolen BI.
60. Sogner (2002), Christensen (2006), Lars Thue, *Nye forbindelser (1970–2005)*, book 3 in Norsk telekommunikasjonshistorie, Oslo: Gyldendal fakta 2005
61. Sogner (2002).

REFERENCES

Andersen, H. W. (1989) *Fra det britiske til det amerikanske produksjonsideal. Forandringen av teknologi og arbeid ved Aker mek.Verksted og i norsk skipsbyggingsindustri 1935–1970*, Trondheim: Tapir.

Arnold, E. (1985) *Competition and Technological Change in the Television Industry. An empirical Evaluation of Theories of the Firm*, London: MacMillan.

Bjørnstad, S. (2007) "Forklaringsmodeller for 'århundrets største ingeniørbragder'," in Olav Spilling Olav Spilling (red.), *Kunnskap, næringsutvikling og innovasjonspolitikk*, Bergen: Fagbokforlaget.

Chandler, A. D., Jr. (2001) *Inventing the Electronic Century. The Epic Story of the Consumer Electronics and Computer Industries*, New York: Free Press.

Christensen, S. A. (2006) Switching relations. The rise and fall of the Norwegian telecom industry, Oslo: Handelshøyskolen BI.

Collett, J. P. (ed.) (1995) *Making Sense of Space. The History of Norwegian Space Activities*, Oslo: Scandinavian University Press.

—— and Bjørn, O. H. L. (1993) *Visjon-Forskning-Virkelighet. Televerkets Forskningsinstitutt 25 år*, Kjeller.

—— and Lossius, B. (1993) *Visjon–forskning–virkelighet. Televerkets Forskningsinstitutt* 25 år, Skedsmo: Televerkets forskningsinstitutt.

Eilertsen, T. A. (1987) "Fra FOTU til FFI. Grunnleggingen av norsk forsvarsteknologisk forskning 1942–46," unpublished hovedoppgave in history, University of Bergen.

Elektronikkindustri (1976) Oslo: Norges Offentlige utredninger: 30

Fremstad, P. (1993) "40 år med radiolinjer 1955–1993," unpublished booklet, Bergen 1993.

Gerhardsen, T. (2000) Interview, November 30, 2000

Global Information Technology Report (2006–7) World Economic Forum's homepage.

Gregory, G. (1986) *Japanese Electronics Technology*, Chichester: John Wiley and Sons.

Grønhaug, K., Reve, T. and Fredriksen, T. "Teknologiavtalene: samarbeidsaktiviteter og samarbeidsvirkninger", rapport 1/86, *Senter for anvendt forskning*, NHH. See also Chapter 4 in this volume.

Holmevik, J. R. (1994) "Educating the Machine. A Study in the History of Computing and the Construction of the SIMULA Programming Language," report number 22, University of Trondheim: STS1994.

Hyytinen, A., L. Paija, P. Rouvinen, and P. Ylä-Antilla (2006) "Finland's Emergence as a Global Information and Communications Technology Player: Lessons from the Finnish Wireless Cluster," in John Zysman and Abraham Newman, *How Revolutionary was the Digital Revolution? National Responses, Market Transitions and Global Technology*, Palo Alto, CA: Stanford Business Books.

Jensen, K. (1989) "Moderniseringsmiljøet som pådriver i norsk industriutvikling på 50 og 60-tallet," unpublished hovedoppgave in history, Universitetet i Oslo.

Knutsen, S., E. Lange and H. Nordvik (1998) *Mellom næringsliv og politikk. Kreditkassen i vekst og kriser 1918–1998*, Oslo: Universitetsforlaget 1998

Kordahl, O. P. (2004) Statistics Norway, "Databehandlingsvirksomhet . . . 2004".

McKelvey, M. and F. Texier (2000) "Surviving Technological Discontinuities through Evolutionary Systems of Innovation: Ericsson and Mobile Telecommunication," in Charles Edquist and Maureen McKelvey, *Systems of Innovation: Growth, Competitiveness and Employment*, Vol. II, Cheltenham, UK.

Nerheim, G. and Nordvik, H. W. (1986) *Historien om IBM i Norge 1935–1985*, Oslo etc.: Universitetsforlaget.

Njølstad, O. and Wicken, O. (1997) *Kunnskap som våpen. Forsvarets forskningsinstitutt 1946–1975*, Oslo.

Nordic Council of Ministers (2005) Nordic Information Society Statistics 2005, Copenhagen.

Overby, S. (1989) "Fra forskning til industri. Utviklingen av skipsautomatiseringsbedriften Norcontrol," uupublished hovedoppgave in history, Universitetet i Oslo 1988; Norges Offentlige Utredninger: 2, Kongsberg Våpenfabrikk.

Paulsen, G. (2005) "Innovasjon over Nordsjøen: Telekommunikasjoner på norsk sokkel," Forskningsrapport 3, Handelshøyskolen BI.

——(2008) "Innovating Code and Coded Innovations: The Dynamics of the Software Industry 1965–2005", PhD thesis, Norwegian School of Management BI.

Rinde, H. (2005) *Et telesystem blir til: Norsk telekommunikasjonshistorie*, bind I, Oslo: Gyldendal Fakta.

Skogvold, A. (1989) Letter to Knut Sogner, July.

Smith, K. (2003) "Hva slags økonomiske virkninger har IKT skapt?,"in Helge Godø (ed.), *IKT etter dotcom-boblen*, Oslo: Gyldendal akademisk.

Sogner, K. (1990) "Veksten mot idealfabrikken, rapport nr. 5," *Teknologihistorieprosjektet, NAVF-NTNF*, Oslo.

——(1994) *Fra plan til marked. Staten og elektronikkindustrien i 1970-årene*, Oslo: TMVs skriftserie/Pensumtjeneste.

——(1997) *God på bunnen. Simrad-virksomheten 1947–1997*, Oslo: Novus forlag.

——(2006) "Næringspolitikkens betydning for fremveksten av norsk elektronikk- og IT-industri," in Olav Spilling (red.), *Kunnskap, næringsutvikling og innovasjonspolitikk*, Bergen: Fagbokforlaget.

——(2002) *En liten brikke i et stort spill. Den norske IT-industrien fra krise til vekst 1975–2000*, Bergen: Fagbokforlaget.

Thue, L. (2005) *Nye forbindelser (1970–2005)*, book 3 in Norsk telekommunikasjonshistorie, Oslo: Gyldendal fakta.

Utredninger, N. O. (1989) Kongsberg Våpenfabrikk, NOU.

Wicken, O. (1987) "Norske våpen til Natos forsvar," *Forsvarsstudier*, nr. 1.

——(1988) "Stille propell i storpolitisk storm," *Forsvarsstudier* nr. 1.

Part III

Innovation Policy and Institutions in Contemporary Norway

11

University–Industry Relations in Norway

Magnus Gulbrandsen and Lars Nerdrum

INCREASED INTEREST IN UNIVERSITY–INDUSTRY RELATIONS

In this chapter we analyze the changing *relationship between industry*[1] *and universities* in Norway. Using different empirical sources, we look at the three main tasks of universities, i.e. *research*, *teaching*, and the "*third mission*" of knowledge transfer in a broad sense. The expansion of fields of research like ICT, biotechnology, and nanotechnology provide arenas for increased, closer and potentially novel forms of university–industry relations. Biotechnology is an example of a new industry that owes its existence to university science and where academic patenting and spinoff companies founded by university faculty are central means of collaboration.

In Norway, policy discussions among policymakers and industry managers of "improving university–industry relations" are not recent, and were prominent topics at the beginning of the twentieth century. Indeed, in Norway and elsewhere within the industrialized economies, the three decades following the Second World War may be seen as an exception to a trend of relatively close university–industry relations (Martin 2003). Since the early 1980s, many industrialized economies introduced new policies to encourage closer links between university and industrial research, inaugurating what Guston and Kenniston (1994) have called "a new social contract for science."

The "old social contract" stated that "government promises to fund the basic science that peer reviewers find most worthy of support, and scientists promise that the research will be performed well and honestly and will provide a steady stream of discoveries that can be translated into new products, medicines, or weapons" (Guston and Kenniston 1994: 2). This was based on a linear model of innovation, and arguably applied to no more than a modest share of total government-funded R&D (Martin and Etzkowitz 2000). Reduced faith in the linear model, pressure on public budgets during the 1970s and 1980s, and increased faith in liberalization and market control have changed this social contract. The new social contract implies expectations that universities, in exchange for public funding, should be responsive to the "needs" of "users" in industry, public sector, and elsewhere (Guston and Kenniston 1994). In addition, governments frequently demand accountability for the funds that higher education institutions receive, often basing funding allocations on assessments of "results" or "quality" (see Nowotny *et al.* 2001).

In many countries, the new contract was embodied in part in new legislation, with the US Bayh–Dole Act from 1980 as a prime example (see Mowery *et al.* 2004). Many other countries have emulated this legislative change, encouraging their higher education institutions to commercialize research results and often changing intellectual property rights regimes. In 2003, Norway followed the example of many other countries, including Denmark, Germany, Belgium and the Netherlands, removing the "teacher exemption clause" that prevented academic institutions from gaining title to intellectual property resulting from faculty research. These changes in Norway were associated with a broader trend to provide greater autonomy to higher education institutions (HEIs) in exchange for these organizations assuming responsibility for more tasks. Norwegian White Papers and other documents emphasize that these changes are not an expression of unhappiness about university–industry relations or commercialization, but rather are intended to create a coherent support structure for commercialization of all public R&D. As in other countries, these shifts in Norwegian policy toward HEIs have been accompanied by reductions in core (noncompetitive) institutional funding, increased accountability, standardization of degrees and duration of study programmes through the Bologna process and more.

A number of Norwegian policy mechanisms now support closer university–industry relations. These have grown organically over many years, often spinning out of special programmes and initiatives. Some of the most important mechanisms in late 2007 are:

- *Programmes in the* ***Research Council of Norway***, including "user-controlled projects," "regional innovation" collaboration support, a programme for commercialization of R&D results, centres of excellence in basic research (firms are partners in some of the centres) and centres for "research based innovation" (public–private partnerships).
- *Programmes in the innovation agency Innovasjon Norge* including centres of expertise and other "cluster" support mechanisms, innovation project support and various initiatives oriented at small and medium-sized enterprises (SMEs).

- *Infrastructure support in the industrial development agency SIVA*, which supports 18 incubators, 18 regional “knowledge parks,” 9 investment companies, 8 science parks and more. The 8 science parks are all located in the vicinity of a major university, while most of the knowledge parks have a link to a State College.
- *Tax deductions* for private R&D. The SkatteFUNN scheme allows a tax refund of up to 20 percent of total costs, limited to 4 million NOK. The deductible amount is doubled if the firm collaborates with an “approved R&D institution”—including all Norwegian and some foreign universities, colleges, and research institutes.

There are other sources of support for university–industry collaboration that are related to regional support, EU funding and regular research council programmes. In spite of these programmes, however, the incentives for HEIs to engage in collaboration with Norwegian industry are not strong. Core funding for HEIs is largely based on the number of students, and the tiny “performance indicator”-based part of basic funding favours scientific publishing. A new “third mission” component involving stronger links with public and industrial research issues has recently (autumn 2007) been suggested, and could include financial incentives for university faculty to pursue popular science publications, newspaper articles and patents, but it has not yet been implemented and has been opposed by the largest universities. Scientific publication and other academic credentials are still the only basis for promotion decisions for staff in most HEIs.

State colleges have little R&D funding from industry and low R&D funding in general. Some of these institutions are seeking to advance to university status, with the strong support of regional industry. The oil industry in Stavanger helped finance the transformation of the city's state college into a university, and firms in other regions have taken similar initiatives.

Norway's higher education system is relatively young. It consists in 2007 of 7 universities, 6 scientific colleges, 24 state colleges and 26 private institutions. Most of the private institutions are very small, but this group includes one large business school (see Table 11.1). The latest government White Papers on research (e.g. UFD 2004–5) promote a “binary system” in which universities are responsible for basic research and state colleges for regional industrial relations. In practice, however, the state colleges have less contact with industry than the universities. They are practically oriented, reflecting the origins of many of them in mergers of teacher training colleges, nursing schools and engineering schools.

Table 11.1. Key characteristics of the Norwegian higher education institutions

	Total income 2005 (MNOK)	Share external funding 2003	Total staff 2003 (FTE)	Of which: Professors	PhD students 2003 (FTE)	Gained doctorates (2005)	Total students 2005	R&D expenditures 2003 (MNOK)
U. of Oslo	4,534	35%	5,405	815	1,868	319	30,289	2,149
U. of Bergen	2,616	39%	2,940	478	918	157	15,838	1,255
U. of Tromsø	1,545	34%	1,828	237	551	60	5,763	717
U. of Stavanger	729	47%	810	64	102	6	7,066	n.a.
NTNU	3,613	40%	4,321	602	1,808	218	19,736	1,631
UMB	756	46%	874	115	293	49	2,784	317
Total universities	13,791	38%	16,178	2,311	5,540	809	81,476	6,069
State Coll. (25)	7,673	n.a.	8,766	253	85	10	83,410	896
Scientific Coll. (6)	1,015	n.a.	1,177	165	284	39	4,772	n.a.
Private Coll. (26)	n.a.	n.a.	1,542	94	125	17	24,469	n.a.

Notes: For universities, total staff and PhD students are full-time equivalents (FTE). For the others they are head-counts. NTNU's full name is the Norwegian University for Science and Technology, a merger between the technical university NTH and the University of Trondheim from the mid 1990s. UMB is the Norwegian University for the Life Sciences, formerly the University College of Agriculture. UMB and the University of Stavanger received university status in 2005. Agder State College received university status in 2008 but is included in the State Coll. row in the table.

Sources: DBH (www.dbh.nsd.uio.no/), RCN S&T Indicators, RCN and R&D statistics at NIFU STEP.

UNIVERSITY–INDUSTRY RELATIONS: RESEARCH

University–industry research interactions assume many different forms, from unpaid consultancy to expensive projects lasting many years. Funding indicators, output indicators and innovation indicators for Norwegian university–industry collaboration are examined in this section.

Industrial funding of research in HEIs

Industry funding of university research is probably the most widely used indicator of university–industry interaction, not least because these data are available in statistical time series that often support international comparisons. Recent studies find that industrial funding may even be a good predictor of related university activities such as patenting and the creation of spinoff companies (van Looy *et al.* 2004; Gulbrandsen and Smeby 2005). Table 11.2 contains data on industrial funding of higher education R&D for a number of OECD countries.

Industrial funding of research in HEIs in Norway is slightly below the OECD average, but there are fairly large differences between countries. Overall, however, most nations display increases in the share of university R&D spending that is funded by industry. Indeed, this share more than doubled during the 1981–91 period, although it has grown much more slowly or has declined since 1991 in most countries, including the United States, the United Kingdom, and Norway. Germany, by contrast with these three countries, displays a steady increase in the industrial funding share for the entire 1981–2004 period. Variation in the industry-funded shares among OECD member nations reflect numerous factors such as the size and organization of the HEI sector, the scale and mechanisms for core institutional funding, and differences in the structure of national higher education systems.

The data in Table 11.2 prompt two questions. First, why did the industry-funding share increase sharply during the 1980s? Second, why has there been no increase or even a decline in this share since the mid 1990s, despite the intense focus of public policy in many OECD member nations on university–industry collaboration during the period?

The increase in industry funding of academic R&D during the 1980s in Norway reflects increased industrial interest in university research in some disciplines and the effects of early policy initiatives to improve cross-sector relations. In Norway during the 1980s, the technological challenges of the North Sea oil and gas production, including deep sea and horizontal drilling technologies, created the grounds for common projects, as did the emergence

Table 11.2. Industrial funding of higher education R&D (BEHERD), selected OECD countries, 1981–2004

Country	1981	1983	1985	1987	1989	1991	1993	1995	1997	1999	2001	2003	2004
Australia	1.4	1.6	2.1	2.3	2.2	2.5	3.5	4.7	5.3	4.9	5.1	—	—
Belgium	—	9.4	8.7	8.7	12.6	15.4	12.1	13.1	11.2	10.5	12.7	11.6	—
Canada	4.1	3.9	4.3	5.0	4.9	7.0	8.6	8.1	9.8	9.1	9.4	8.4	8.4
Denmark	0.7	0.9	1.0	1.3	1.5	1.6	1.8	1.9	3.4	2.1	3.0	2.7	3.0
Finland	2.1	2.6	—	3.8	4.8	3.6	4.6	5.7	5.2	4.7	6.7	5.8	5.8
France	1.3	1.3	1.9	3.6	4.6	4.2	3.3	3.3	3.1	3.4	3.1	2.7	—
Germany	1.8	5.2	5.4	6.4	7.1	7.0	8.4	8.2	9.7	11.3	12.2	12.6	12.8
Ireland	7.1	7.2	6.9	7.1	9.2	8.6	7.1	6.9	6.5	5.9	4.4	3.0	2.6
Iceland	1.2	1.9	0.6	24.3	6.8	5.0	4.3	5.4	9.2	4.0	10.9	9.5	—
Japan	1.5	1.8	2.4	2.8	3.3	3.7	3.8	3.6	2.4	2.3	2.3	2.9	2.8
Netherlands	0.3	0.6	1.0	1.1	1.1	1.2	1.5	4.0	4.3	5.1	7.1	6.8	—
Norway	2.9	3.5	5.0	4.5	3.9	4.7	5.7	5.3	5.3	5.1	5.8	5.0	—
Spain	0.0	—	1.1	2.7	9.2	10.0	5.9	8.3	6.5	7.7	8.7	6.4	7.5
Sweden	2.3	3.7	5.5	5.9	7.9	5.2	5.1	4.6	4.8	3.9	5.5	5.5	—
UK	2.8	3.1	5.2	5.7	7.7	7.8	7.6	6.3	7.1	7.3	6.2	5.6	—
US	4.4	5.2	6.1	6.5	6.8	6.8	6.8	6.8	7.3	7.4	6.5	5.3	5.0
Total OECD	*2.9*	*3.5*	*4.2*	*4.8*	*5.7*	*6.0*	*6.1*	*6.2*	*6.4*	*6.5*	*6.4*	*6.1*	—
EU 15	*2.0*	*3.0*	*3.7*	*4.3*	*5.9*	*5.8*	*5.8*	*5.9*	*6.1*	*6.5*	*6.8*	*6.6*	—
EU 25	—	—	—	—	—	—	—	*6.0*	*6.1*	*6.6*	*6.7*	*6.5*	—

Source: OECD—Main science and technology indicators.

of large Norwegian firms in computers and electronics. The scientific and technological research council, NTNF, established "user-controlled" research programmes in the early 1980s, for which company needs and specifications were the basis for the design of publicly subsidized collaborative R&D projects. Many of the most technologically advanced Norwegian companies took part in this first wave of user-controlled research (cf. Hervik and Waagø 1997). These firms had a history of collaboration with HEIs and the capability to articulate their technological needs and to absorb and apply the results of collaborative R&D projects.

New policy mechanisms introduced during the 1990s in Norway focused on SMEs and low-tech industries and at colleges with weak research traditions. Some programmes have sought to build up new relations from scratch, often involving firms with little or no experience in HEI interaction and within "less favoured regions"; others have been focused on transfer of knowledge through graduates, which does not show in the R&D statistics. One explanation for the decline in industry's share of academic R&D funding in Norway emphasizes the strong growth in public funding for higher education since the late 1990s that reflects increased enrolments (Smeby and Sundnes 2005). Similar growth may explain the declines in the industry-funded share in some other OECD countries, although this issue remains to be fully explored. Yet another speculative explanation for the trends in Table 11.2 (see Larsen 2007) argues that the interaction between academic and industrial R&D may have reached a level during the 1980s and first half of the 1990s that was "optimal" for both parties. Further intensification of interaction might lead university researchers away from basic research and teaching and/or lead firms too far away from market-based innovation activities. Once again, however, this speculative hypothesis has not been investigated systematically for Norway or other OECD economies.

Data on industrial funding and interactions for individual faculty shed further light on trends in Norway. The share of faculty members at Norwegian universities with industrial funding and/or co-operation has increased from 7 percent in 1982 to 21 percent in 2001 (see Gulbrandsen and Smeby 2005). This increase spans natural science, medicine and social science, but not the technological disciplines (including agriculture) where the share of faculty with industrial funding went down from 78 to 71 percent from 1992 to 2001 (1982 data not available here) nor the humanities where the share of faculty with industrial funding has long been low. Individual faculty in all academic fields receiving industry funding are also much more likely to receive external funding from other sources, such as EU, the Research Council of Norway and non-profit foundations. These faculty members are also more likely to be engaged in "commercial" activities such as developing

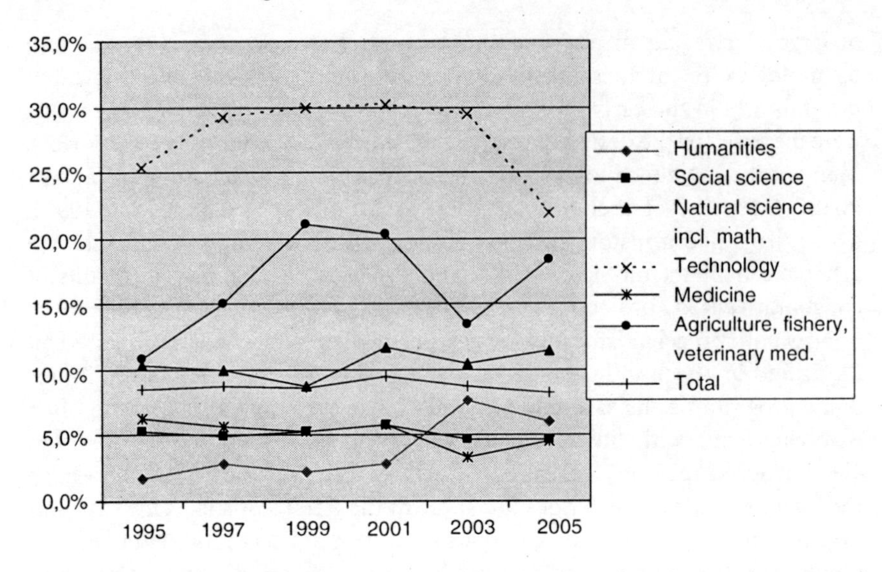

Figure 11.1. Academic fields: share of industrial funding of total operating costs of R&D in the HEI sector in Norway.

Source: R&D statistics, NIFU STEP.

new products, obtaining patents, founding spinoff companies and pursuing consultancy activities (Gulbrandsen and Smeby 2005). Finally, national surveys conducted in 1982, 1992, and 2001 indicate that professors in all academic fields who receive industrial funding publish significantly more than their peers without such funding. This relationship between industrial research support and publishing in Norway is consistent with the evidence from other countries (*ibid.*).

Figure 11.1 shows the share of industrial funding in the total R&D operating costs in different academic fields.[2] Technology and natural science fields are the main recipients of industrial funding, both receiving more than 100 MNOK in 2005,[3] although the share of costs supported by industry is much higher in the "technology" fields than in the natural sciences and mathematics where the share of core funding and funding from public sources like research councils is higher. Academic departments in agriculture, fisheries, and veterinary medicine also display relatively high shares of industrial funding that exceed 20 percent in several years during the mid 1990s. The low share of industrial funding in medicine may reflect Norway's lack of advanced domestic or foreign-owned pharmaceutical companies, but the volume of industrial funding in medicine is still substantial (around 75 MNOK in 2005).

The share of operating costs accounted by external funding from all sources (including funding from the research council, non-profit foundations,

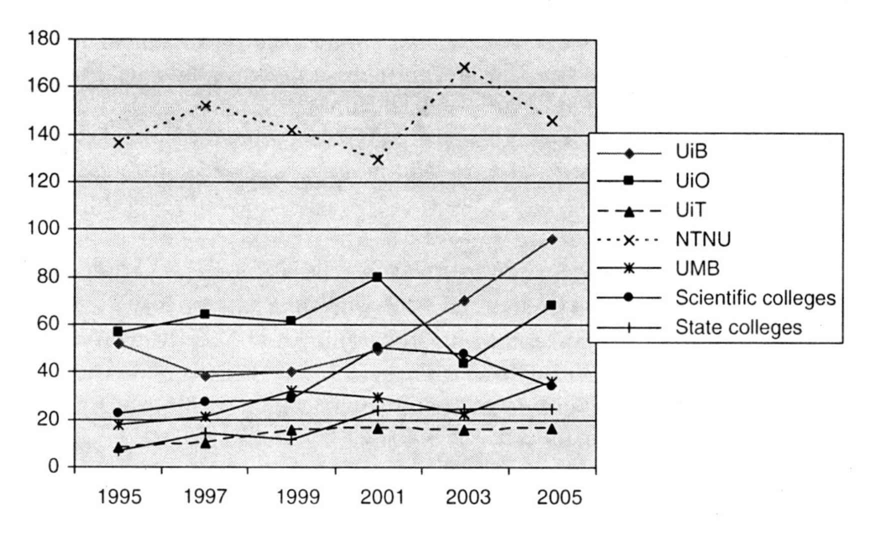

Figure 11.2. Level of industrial funding in the HEI sector in Norway, 1995–2005. Million NOK, fixed (2005) prices.

Source: R&D statistics, NIFU STEP.

EU, etc.) is much higher in the technology, agriculture and natural sciences fields than in medicine, social sciences and the humanities, reaching roughly 40 percent in the three first areas and 20 percent in the last three. There has been an increase in industrial funding from 40 MNOK in 1995 to close to 60 MNOK in 2005 within the social sciences, but this has coincided with growth in core funding and external funding from other sources, slightly reducing industry's share.

Figure 11.2 shows the volume of industrial research funding for each of the leading Norwegian academic institutions during 1995–2005.[4] The dominance of NTNU is apparent here where industry funding averaged roughly 150 million NOK annually during the period, nearly 17 percent of R&D operating costs in 2005. The University of Tromsø received the smallest level of industrial funding among the major academic institutions during this period, an amount exceeded by that for all 26 state colleges counted together. The University of Bergen experienced considerable growth in industrial funding during the 1995–2005 decade, and by 2005 received 30 million NOK more than the University of Oslo which is almost twice as big in budget or enrolment. NTNU, with total enrolment of 20,000 in 2005, received nearly 90 MNOK in industry funding for salaries alone in 2005, largely for support staff and PhD students. The University of Oslo with 1.5 times as many students, received less than one-third as much industry funding for salaries in the same year. Interestingly, industry funding accounted for a relatively high share of operating costs at

UMB (formerly the University College of Agriculture), although the smaller size of this institution means that the absolute amount of industry funding was much smaller than that flowing to NTNU.

Projects with budgets of less than 100,000 NOK (about 12,500 EUR) constituted around half of the total industry funding of HEI R&D in most years during 1995–2005. Surprisingly, however, Norwegian universities house more large projects than do the research institutes. Only two percent of the projects undertaken in Norway's research institutes in 2000 had budgets above 1 million NOK, but in the universities and colleges, this share was seven percent. Nearly one-fourth of the projects in the universities and colleges had budgets of 250,000–999,999 NOK, far more than in the institutes. Industry-funded research projects in HEIs thus may differ systematically from those undertaken in Norwegian research institutes. An investigation of Danish firms' support for external R&D projects found that firms used institutes for problem definition and the explication of technological needs, and relied on universities and colleges for problem solving (Valentin and Jensen 2003). Although no evidence on the Norwegian case has been developed, a similar pattern of use by industry of research institutes and universities could account for the observed differences in project budgets, on the assumption that "problem-solving" is on average more costly than "problem definition." But more research on this topic is needed.

These funding data suggest that Norway's university system does not differ greatly from those of other developed OECD countries in the sources of funding and especially, in the role of industry funding. There was a sharp increase in industry's share of HEI research funding in the 1980s but a stabilization and slight decline after the mid 1990s, reflecting more rapid growth in funding for academic operating costs from non-industrial sources. In Norway as in other OECD economies, the aggregate figures hide large differences between institutions and disciplines. Industrial funding is higher in the technological disciplines and at NTNU in Trondheim, although industry also accounts for a substantial share of funding for operating costs in the social sciences and at other universities.

Co-publication patterns

Co-publication, that is, patterns of co-authorship of published scientific papers is an important measure of research collaboration. It may not provide an accurate measure of university–industry interaction if such interactions are more proprietary, more applied, or of shorter duration, characteristics that may not promote scientific publication. We have used data from ISI, which

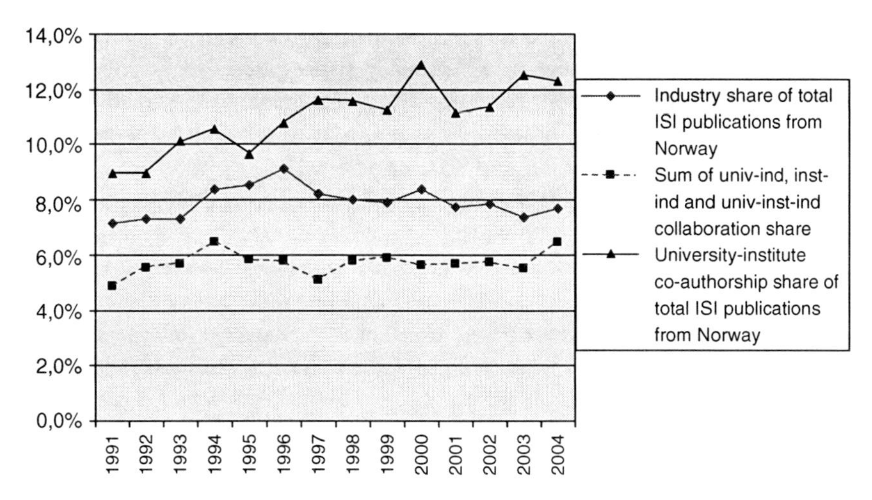

Figure 11.3. Co-publication patterns between sectors in the Norwegian research system.

Source: Based on data from Frölich and Klitkou (2006).

mainly covers leading English-language journals, to analyze Norwegian co-publication based on papers that include co-authors from different sectors of the research system (i.e., universities, industry, or the institutes).

In Norway, "all-academic" co-author teams account for a little over 80 percent of all Norwegian ISI publications, which is slightly higher than in the United States. Norwegian industry's share of ISI publications is around 8 percent, comparable to that in Canada, the Netherlands and the UK (see Calvert and Patel 2003), and the "all-industry" share is fairly stable during 1991–2004 despite an 80 percent increase in annual ISI publications from Norway during this period. Industry's share as well as co-publication patterns are shown in Figure 11.3. A particularly strong increase in co-publishing between universities and institutes can be seen. The share of university–industry and institute–industry co-authorships within total publishing more than doubled in absolute numbers since 1991, and these collaborative articles' share of total publishing has increased from 4.9 to 6.5 percent in the period. This could reflect policy initiatives like joint/user-controlled research programmes but also needs of firms to collaborate with public sector research.

The share of articles from Norwegian industry[5] that have co-authors from a public research organization (including universities and research institutes) fluctuates between 62 and 84 percent during 1991–2004 with higher scores towards the end of the period. Forty-five percent of Norwegian industry papers in 2004 were co-authored only with researchers from universities/

colleges, 28 percent only with researchers at institutes and 11 percent with both sectors—i.e. 55 percent of all industry papers were co-authored with HEIs. Data from the late 1990s from the UK and Japan show 46 and 45 percent university–industry co-publication, i.e. somewhat lower than in Norway (cf. Calvert and Patel 2003; Pechter and Kakinuma 1999).

Conversely, the share of all Norwegian HEI publications during 1991–2004 that have at least one industry co-author is relatively stable at around 4 percent, a share that again is similar to the 4.5 percent share observed in UK data in 1996–2000 (Calvert and Patel 2003) and Canadian data, where the share was 3.5 percent in 1998 (see Godin 1998). US data do not always separate the private for-profit sector from other sectors, but the share of university–industry co-authored papers within all university papers in 1999 was 7.3 percent, a considerable increase from the 1989 share of 4.9 percent. The similarities in university–industry co-publication shares within university publications in the United Kingdom, Canada, and Norway may reflect the similarly low levels of GDP in each economy accounted for by industry-funded R&D spending (in 2003, 1.24 percent, 1.0 percent, and 1.0 percent respectively, according to the OECD's *Main Science and Technology Indicators*).

Innovation survey data

Finally, data from the Community Innovation Survey (CIS) allow for additional examination from an industry perspective of university–industry research relationships in Norway and elsewhere in Europe. The CIS asks firms whether they co-operate on innovation with various actors and whether universities, institutes and other organizations constitute "highly important sources of information for innovation."[6] Data from the fourth survey (CIS4) are presented in Table 11.3.

The three highest scores in each column are marked in bold typeface. Finland scores very high in all of the measures of collaboration, but Norwegian firms report fairly high levels of innovation co-operation with HEIs and institutes. The share of firms from Norway rating the research institutes as "highly important information sources" for innovation is higher than that for firms in any of the other nations for which data are reported in Table 11.3, and the share rating HEIs as "important in innovation cooperation" is the third highest in the table. These data suggest that Norway has a highly collaborative innovation system.

Other questions in the CIS provide additional data on Norwegian firms' use of other information sources for innovation. Most of the information sources like clients, customers, competitors, consultants etc. receive an average

Table 11.3. Share of firms with innovation activities which report innovation cooperation with public sector research 2002–2004, and share of firms with innovation activities which report public sector research as "highly important source of information for innovation" 2002–2004. All industries incl. services

Country	Innovation cooperation		Highly important information source	
	HEIs %	Institutes %	HEIs %	Institutes %
Belgium	13.2	9.2	3.8	2.3
Denmark	13.7	6.9	3.3	0.5
Germany	8.5	4.1	3.4	1.4
France	10.1	7.3	2.3	2.0
Italy	4.7	1.5	2.0	1.0
Netherlands	12.4	9.4	2.6	2.0
Austria	10.0	5.2	n.a.	n.a.
Finland	33.2	26.4	4.9	2.4
Sweden	17.4	6.4	n.a.	n.a.
United Kingdom	10.0	7.6	n.a.	n.a.
Norway	14.8	16.3	3.1	3.2

Source: Eurostat (CIS4).

or slightly above average score from Norwegian firms. But "scientific journals and trade/technical publications" receive a very low ranking from Norwegian firms, well below that from firms in other high-income European economies. This low ranking is consistent with the stereotypical view of Norwegian industry as dominated by small units with limited absorptive capacity, although the reasons for the score have not been analysed.

Table 11.4 disaggregates by industry (NACE classifications) the responses of Norwegian firms concerning "innovation partners" and "important information sources" that were displayed for different European nations in Table 11.3. Innovation cooperation between firms and public sector research is particularly common in high technology industries like chemicals, communications equipment and instruments and in key national industries like oil and gas, basic metals and pulp and paper. Higher education institutions tend to be more important partners and sources of information than research institutes in high technology industries, and vice versa, but there are also exceptions. Institutes are slightly more important than HEIs in manufacturing industries and slightly less so in construction and services. Specialization patterns, e.g. the small number of Norwegian research institutes in the biomedical field, may account for some of the differences. Collaboration between firms and institutes may complement rather than substitute for university–industry interaction, since the two institutional sectors' importance as partners and information sources is similar in most industries.

Table 11.4. Innovation collaboration/information source distributed on industries. Total figures for each main category of industries (bold) and selected sub-categories. Data for Norway

Industry	Co-operation		Information source	
	HEIs %	Institutes %	HEIs %	Institutes %
Mining and quarrying (all)	**26.3**	**26.3**	**12.3**	**10.5**
Extraction of oil and natural gas	32.5	30.0	12.5	15.0
Manufacturing (all)	**17.2**	**19.6**	**2.8**	**4.6**
Food and beverages	17.2	25.8	1.6	3.7
Textiles	9.3	9.3	0	0
Wearing apparel, dressing, fur	23.1	30.8	0	n.a.*
Wood and wood products	10.4	14.1	n.a.*	4.4
Pulp, paper, and paper products	23.8	38.1	9.5	9.5
Publishing, printing, media	2.5	5.9	3.4	2.5
Chemicals/chemical products	37.7	27.9	11.5	6.6
Non-metallic mineral products	27.6	32.8	1.7	5.2
Basic metals	30.8	33.3	5.1	5.1
Machinery and equipment	18.6	20.6	2.5	6.4
Radio, television, com. equipm.	39.5	34.2	5.3	2.6
Medical, precision, and opticalinstrum.	27.1	23.7	8.5	5.1
Furniture	8.6	9.4	0.9	1.7
Electricity, gas and water supply (all)	**22.0**	**22.0**	**3.3**	**1.1**
Construction (all)	**6.3**	**6.6**	**5.0**	**2.6**
Services (all)	**11.8**	**12.9**	**3.2**	**2.0**
Wholesale and retail trade	6.7	11.5	0.9	0.7
Hotels and restaurants	0	0	0	0
Transport, storage, and communication	5.4	6.3	2.5	2.9
Financial intermediation	7.3	4.9	1.2	2.4
Real estate, renting and business activities (includes R&D consultancy)	20.0	18.0	6.1	2.8
All industries	**14.8**	**16.3**	**3.1**	**3.2**

Note: *means confidential information (not made publicly available).
Source: Eurostat (CIS4).

UNIVERSITY–INDUSTRY RELATIONS: TEACHING

The distinction between research and teaching often is blurry, especially for graduate education in research-oriented universities. Other research on university–industry interaction in Norway found that "access to graduates" is an important motive for industrial support of academic R&D (Gulbrandsen and Larsen 2000).

Few investigations of university–industry relations have focused specifically on this relationship in the area of teaching. Economists have of course been aware of the productive power of knowledge at least since Adam Smith, and various perspectives, e.g. human capital theory, have analyzed the private and social returns from schooling (e.g. Becker 1964). This research does not seem to have influenced theories of innovation and technical change, but synthesizing these perspectives may provide a fruitful area for future research. Below we use data from labour market studies to further examine university–industry relations in Norway.

Rosenberg and Nelson (1994) claim that a key contribution of US universities to industrial innovation during the late nineteenth and early twentieth centuries was their development of industry-specific training programmes that responded to employer needs and technological developments. There are no broad investigations of this area in Norway, and there almost certainly is considerable variation among HEIs.

At NTNU, for example, all departments have had industry/external representatives on their board for decades; and such boards have been created more recently at other Norwegian universities and colleges too. NTNU has assumed national leadership in higher education in technology related fields within Norway—85 percent of Norway's chartered engineers (Master degree holders) were NTNU graduates in 2005. This mission influences NTNU's teaching in other fields as well. The university also has developed curricula in the social sciences and humanities that support training that is relevant for graduates seeking employment in firms as part of its effort to become the "private sector university" of Norway (Sotarauta *et al.* 2006).

In interviews, industry representatives emphasize the importance of informal contacts for creating changes in curricula and study approaches, for example by making data (geological surveys, case material, simulation models, etc.) available to students (Gulbrandsen and Larsen 2000). Industry interviewees also express confidence that the quality of Norwegian university graduates is good. Indeed, large Norwegian companies typically express greater concern about shortages in certain fields than about the quality of university graduates (*ibid.*).

University graduates' transition to work

Most studies in the innovation systems tradition assume that private firms are the most important actor in the creation and commercialization of innovations. The share of graduates finding work in the private sector might therefore provide some evidence on the characteristics of university–industry linkages. In Norway, the "graduate survey" is carried out every second year among all spring graduate students in universities and colleges at the Master's

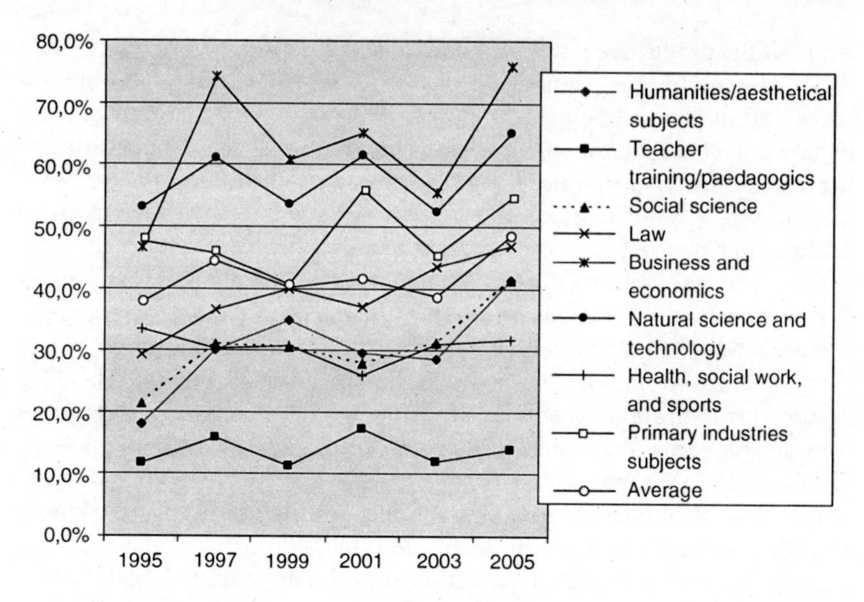

Figure 11.4. Share of graduates of different subjects leaving HEIs to work for the private sector, 1995–2005.

Source: NIFU STEP Graduate Survey.

degree level.[7] The share of students in different fields going to the private sector for their first postgraduate position during 1995–2005 is shown in Figure 11.4.

As expected, the highest share of private-sector postgraduate employment is found within business and economics studies, natural science and technology (including engineers) and primary industry subjects (agriculture, fisheries, etc.). There are large fluctuations during the decade in these shares, perhaps because of cyclical shifts in employer demand (private-sector employment shares for the leading fields of specialization tend to move together during the period). The figure shows that the share of private-sector postgraduate employment within all fields peaked in 2005; only 1997, a year near the peak of ".com era" boom, exhibits a comparably large share.

As was mentioned above, Norwegian industry representatives have expressed concern over an insufficient supply of domestic graduates, particularly in engineering and the natural sciences in the face of a rapidly aging population of employed scientists and engineers in Norway. Although employers argue that the "supply problem" is exacerbated by declines in the share of Norwegian university and college students electing these fields of specialization, the graduate survey shows that engineers and natural science graduates often face problems in their search for postgraduate employment. Figure 11.5

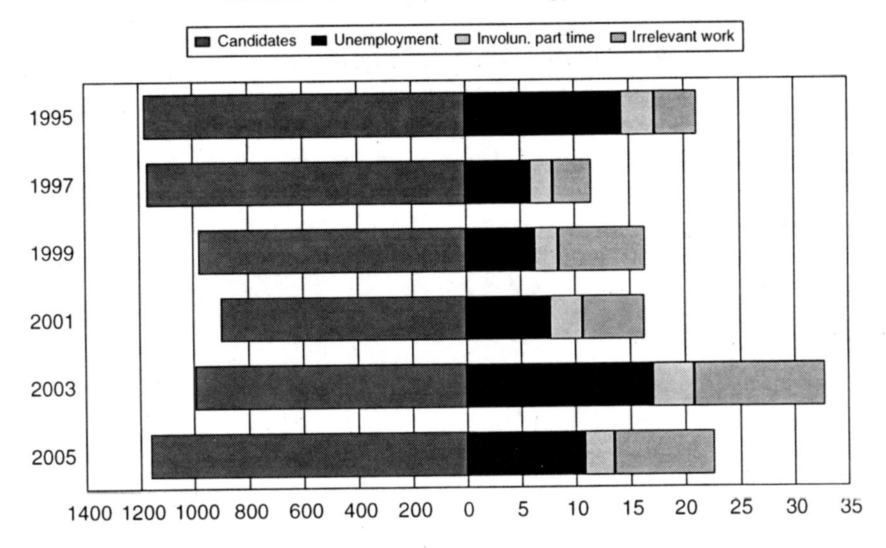

Figure 11.5. Labor market mismatch six months after graduation for Master degree graduates.

Source: NIFU STEP Graduate Survey. Number of graduates left, percentages right.

displays data on the number of Master's degree holders and employment "mismatches" for natural sciences ("cand.scient.") and technological ("sivilingeniør") fields within 6 months of graduation. An employment mismatch is defined by the survey as degree holder unemployment, involuntary part-time work and irrelevant work after graduation.

The figure shows that in 2003, for example, one-third of all Master degree candidates from natural science and technology experienced unemployment or some other type of employment mismatch. Even in the heated labour market of 2005, more than 20 percent of survey respondents reported problems in their transition from academia to work. Other data from the survey suggest that graduates from natural science and technology are more vulnerable to labour market fluctuations than graduates in other fields. Although industrial demand for graduates exhibits wide fluctuations, the public sector seems to play a role as a "buffer" in depressed economic cycles, for example, as the share of engineering graduates going to the public sector increases during economic downturns.

Postgraduate employment data for graduates with Bachelor degrees in engineering show patterns of fluctuation and mismatch. These engineers are trained in state colleges, many of which have developed specialized programs that meet the needs of regional industry, such as maritime electronics in Kongsberg, microelectronics in Horten and ICT in Halden. The state colleges also frequently play an informal role in training that does not appear in public statistics. For example, the South Trøndelag College in Trondheim has several hundred students working on projects for companies and other users each year (cf. Sotarauta *et al.* 2006). There are no systematic studies of student projects for industry, but preliminary investigations indicate that this is a substantial activity (cf. Brandt 2005).

PhDs in industry

Another important channel for university–industry interaction, as well as an important source of industrial "absorptive capacity," is the placement of PhD degree holders in industrial positions. Assembling data on industrial employment of PhD degree holders in Norway requires a merger of several databases. Figure 11.6 shows the number of PhDs—of any nationality but trained in Norway—employed in all Norwegian industries with more than 40 PhD degree holders during 1995–2005.[8]

Available figures and aggregates from Statistics Norway's online database (not as detailed as the one in Figure 11.6) show that employment went down in many traditional industries like power and water supply and chemicals around the turn of the millennium. Employment increased sharply in oil and gas extraction before 2000 but has not grown since (with 2006 as the last year in the data). Figure 11.6 thus serves to show both that employment of PhDs has gone steadily up in many industries—more than total employment in many of them although this cannot be shown directly—and that an increasingly higher share of the PhDs that are trained find work in industry.

Many industries, particularly computer services, chemicals and oil/gas, expanded their employment of PhDs significantly during this period. By 2005, the oil and gas sector alone employed nearly 500 PhD degree holders. Although PhD degree holders still accounted for only 1.6 percent of the employment in this sector (up from 1.3 percent a decade earlier), this is far above the national average of 0.45 percent.[9] Chemicals/pharmaceuticals, medical instruments and business consulting have about the same share of PhD holders as the oil and gas industry (Olsen 2007; the share is higher, as expected, in private R&D and educational services). Several industries have seen a sharp increase in the number of PhD degree holders within their workforce, e.g. the number of PhDs employed in the power/water supply industry quadrupled during 1995–2005 (with no growth in total employment in the industry in

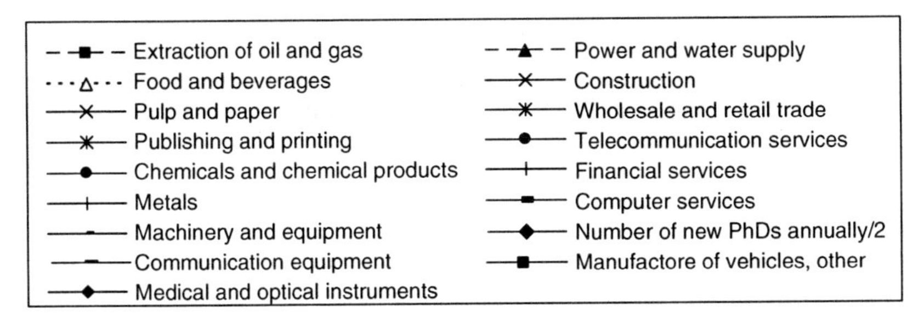

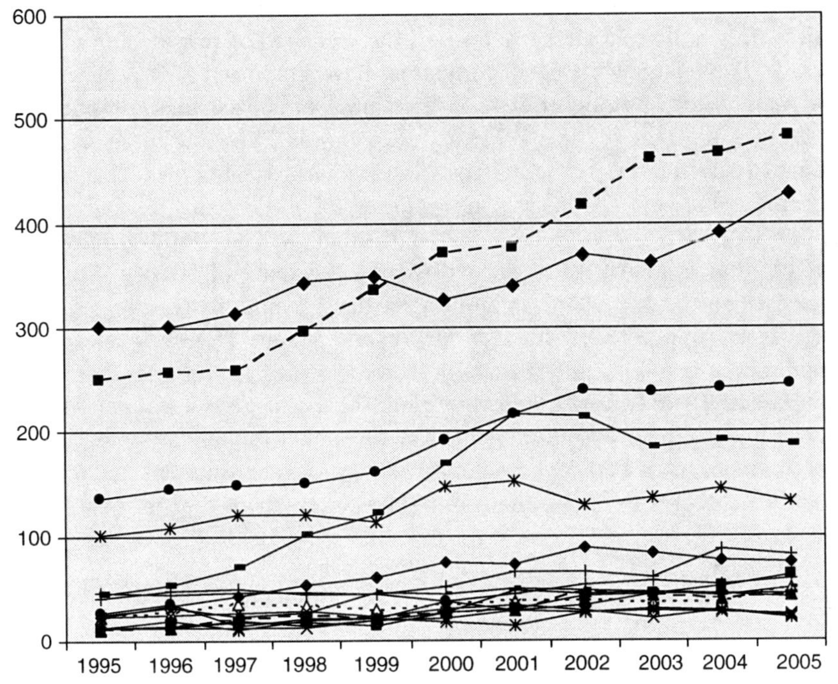

Figure 11.6. Number of PhDs in selected Norwegian industries.

Source: Gunnes *et al.* (2007).

this period), tripled in medical and optical instruments, and doubled in metals and foods. Altogether the number of PhDs employed by Norwegian industry increased by about 80 percent during the 1995–2005 decade. The share of PhD holders in the total private sector workforce has gone slowly up from a little below 0.4 percent at the turn of the millennium to close to 0.5 percent in 2007 (see Olsen 2007). In the mid 1970s 90 percent of the PhD holders found work in the public sector (including universities, hospitals, research institutes and public administration), but this share has stabilized at around 60 percent since the early 1990s (*ibid.*). In other words, even in a period of rapid expansion of PhD training, the share of PhD holders finding work in firms is historically high.

The employment within industry of PhDs from a particular university can strengthen its linkages with industry, as many PhDs keep in touch with their supervisors and may be more knowledgeable about university research activities and competences than Master's degree graduates. The data in Figure 11.6 (which as mentioned only contain degree holders from Norwegian universities) thus indicate a strengthening of the cross-sector relationship the last decade. Some large Norwegian companies have supported PhD programmes for many years, although different firms have organized their support differently. For example, Norsk Hydro often funds PhDs indirectly through financial support for research council programmes, Nycomed (the largest pharmaceutical firm in Norway, now part of GE Amersham) through grants at the University of Oslo, and Statoil through its own PhD programme, VISTA, that is managed by the Norwegian Academy of Science (see Gulbrandsen and Larsen 2000). In some specialized technical areas, large industrial firms support R&D less for the specific results than as a means for maintaining research on industrially relevant topics, training PhD students for potential employment, and observing these prospective employees in a research setting. In addition, industrial research support can assist in the development of an attractive teaching and learning environment that attracts qualified students to obtain the training that these firms seek in their future employees.

UNIVERSITY–INDUSTRY RELATIONS: A THIRD MISSION?

Although there are no clear definitions of a "third mission" for universities (Martin 2003), all Norwegian universities and colleges have published strategy documents stating that some mix of public outreach and economic development are missions that are as important as research and teaching. Many of these documents define the "third mission" as the direct transfer of knowledge to society, including the use of university research for more general reflection upon and critique of societal developments. In this section we take a closer look at one dimension of this emergent "third mission" in Norwegian universities, patenting and the creation of spinoff companies.

Patenting and other forms of commercialization

The number of articles on academic patenting and spin-off company formation is large and growing. Although issued patents are published by national governments, it is rarely easy to identify patents that result from university research. Only in countries where the higher education institutions themselves own the intellectual property rights to their employees' research results and/or apply consistently for patents in the institution's name, can researchers easily identify at least a subset of patents resulting from university research (recognizing that in some cases, university faculty may obtain patents that are not assigned to their employer institution).

Although the Bayh–Dole Act of 1980 requires that US universities be assigned the intellectual property resulting from publicly funded research performed by faculty, the situation in other industrial economies has been different until recently. Most European countries maintained a so-called "teacher exemption clause" or "professor's privilege," granting individual faculty members the right to commercial exploitation of their research results. Academic researchers applied for patents as private citizens, often using their home address or that of a partner, for example a company. Identifying patents resulting from university research under the "professor's privilege" policy thus is difficult. In Norway, legislation passed in 2003 removed the teacher exemption clause and allowed higher education institutions to gain ownership of the intellectual property rights to inventions resulting from research carried out at the institutions. At the same time, the universities and colleges were assigned responsibility to ensure that "practically relevant research results" come into use. Norwegian colleges and universities also received public funding to establish technology transfer offices (TTOs).

But as in other countries, commercialization of university inventions is not a new activity in Norway. In a survey in 2001, seven percent of all university researchers in Norway stated that their research had produced one or more patents (Gulbrandsen and Smeby 2005). As was noted earlier in this chapter, patenting, faculty consultancy activities, faculty involvement in product development and involvement with spin-off companies, all are strongly related to industrial funding and are most common in the technological disciplines (*ibid.*). Recent interviews of Norwegian professors indicates that faculty patenting is fairly common, although academic institutions have played a miniscule role in the process because of their lack of experience and expertise in managing the early stages of commercialization (patenting, licensing, etc.) of research results (Gulbrandsen 2005).

Quantitative studies of academic patenting in the United States (Henderson *et al.* 1998) showed that academic patents increased 15-fold between 1965 and

1988.[10] This increase in intensity is recognized to involve a set of interlinking changes in university policies (Gibbons *et al.* 1994; Webster 2003; Etzkowitz 1998; Etzkowitz *et al.* 2000), in technology (e.g. Zucker *et al.* 1998), and in US government patent policy (Jaffe and Lerner 2004). Mowery *et al.* (2004), highlighting the concentration of U.S. university patenting and licensing in the biomedical technologies since the 1970s, also argue that large federal investments in biomedical research in US universities have contributed to growth in US academic patenting.

A 2003 OECD report ("Turning Science into Business") summarized the results of a survey of TTOs and public research organizations to map the extent of patenting and spinoff firm formation. The study's key findings are summarized in Table 11.5 and interpreting these results must be done with care, because of the methodological difficulties associated with cross-national comparisons of results from surveys that employ varying data collection methods and sector definitions/borders. Nor are simple counts of the number of spin-off companies associated with a university or public laboratory reliable indicators of economic impact without greater cross-national consistency in the definition of a "spinoff company" and some effort to adjust the reported birth rates of new firms for the high mortality rates of such undertakings.[11]

Nevertheless, the data in Table 11.5 reveal that Norwegian universities and research institutes are active in commercialization of their research results. Norwegian research organizations (universities and research institutes) receive particularly high scores in formation of spinoff companies (Norway ranks second only to Switzerland in the countries in Table 11.5 for which data are available), which may have several explanations. First, support initiatives like the research council program FORNY has for years strongly emphasized this means of commercialization over patenting and licensing. Second, since much of the patenting of Norwegian universities and research institutes covers biomedical areas with little industrial activity in the country, commercialization of these patents may rely more heavily on formation of spinoff companies, funded by various public support programmes and a growing seed and venture capital sector. As a whole, however, there are still more spin-offs in the oil and gas industry. Unpublished Norwegian data from a survey of academic inventors show that almost one-third of the academic patents are utilized in start-up companies, a much higher share than in many other countries (see Meyer 2006). Many public research organizations, especially Norway's research institutes, have long traditions of spinoff formation.

In 2005, a database of Norwegian academic and research institute patenting was created through a merger of Norway's national patent database with the "researcher personnel register" at NIFU STEP, followed by a survey and a manual validation of matches. The database reveals a total of 569 researchers from

Table 11.5. Patent grants (one year), patent applications (same year) and spinoffs (same year) from HEIs and research institutes in various OECD countries

Country	Institution	Patent grants	Pat. applications	Spin-offs
Australia (2000)	Universities	219	586	32
	Institutes	279	248	15
	Total	**498**	**834**	**47**
Belgium (Flanders) (2001)	**Total (universities and institutes)**	57	121	15
Germany (2001)	Institutes only	747	1,058	37
Italy (2000)	Universities	34	102	27
	Institutes	30	88	9
	Total	**64**	**190**	**36**
Japan (2000)	**Total (universities and institutes)**	**163**	**567**	**6**
Korea (2000)	Universities	186	244	19
	Institutes	832	1448	37
	Total	**1,018**	**1,692**	**56**
Netherlands (2000)	Universities	64	111	27
	Institutes	103	101	10
	Total	**167**	**212**	**37**
Norway (2001)	Universities	20	40	16
	Institutes	28	43	51
	Total	**48**	**83**	**67**
Spain (2001)	**Total (universities and institutes)**	**64**	**133**	**11**
Switzerland (2001)	Universities	59	132	56
	Institutes	53	43	12
	Total	**112**	**175**	**68**
U.S. (2000)	Universities	3,617	6,135	390
	Institutes	1,486	2,159	n.a.
	Total	**5,103**	**8,294**	**n.a.**

Note: Universities includes all higher education institutions in most cases. Institutes includes all "public research organizations" in most cases. Numbers for patents for universities in Norway are estimates based on Iversen *et al.* (2007) of patents involving university scientists and applied for through the commercialization structure.

Source: OECD (2003).

Norwegian public research organizations (mainly universities and research institutes) who were involved in at least one patent application in 1998–2003. A total of 828 patents—or 10.6 percent of domestic patents—applied for during 1998–2003—involved at least one public sector researcher as an inventor.[12] 21 percent of all Norwegian patents in chemicals and pharmaceuticals applied for during this period in this sense "came from" universities and colleges, and a further 8.2 percent from research institutes. As in other industrial economies, the share of public R&D institutions in patenting is also high in instruments, where institutes and universities each contribute 11 percent of all Norwegian patents. International comparisons are difficult, but a comparison with Balconi *et al.* (2004), who followed a very similar approach for approximately the same period, suggests that university patenting may account for twice as large a share of Norwegian patents overall as is the case in Italy.[13]

The period following the 2003 passage of Norway's new law on academic patenting did not immediately produce an increase in university patenting; indeed, preliminary data suggest that the share of Norwegian patents accounted for by universities dropped from 2002 to 2003 (Iversen *et al.* 2007). The law created a period of uncertainty for some researchers about the nature of the changing division of labour in patenting and commercialization between researcher and institution. In addition, Norwegian universities only established TTOs to manage patenting and licensing in 2004. Further research is needed to see whether the legislative changes have any long-term impact on patenting with academic involvement.

CONCLUSION

There has been much attention to university–industry relations in recent decades, and mechanisms to encourage such interaction have been developed all over the world. Many of these policy initiatives are rooted in new beliefs and ideologies on innovation and knowledge production, as well as increased financial pressures on universities, growing policymaker concern over the pace of "technology-driven" economic growth, and changes in intellectual property rights regimes and scientific research advances. As the case of Norway shows, some of the mechanisms have been initiated by the academic community itself.

This chapter has presented data that indicate that recent growth in university–industry interactions in fact seems to be only the most recent phase of a long trend that was initiated decades ago, as part of broader efforts at technological upgrading of industry and increased openness to cross-sector collaboration. The 1980s, however, were a turning point, as formal research collaboration between firms and higher education institutions (HEIs) grew in Norway and throughout the OECD. These changes during the 1980s may reflect increasing knowledge needs in industry, but also were supported by change in academic culture after a decade or two of "radicalization" all over Europe. Increased industrial needs for academic knowledge stem from increased global competition and internationalization, more rapid technological advance, the "enrolment explosion" of the 1970s, which led to larger numbers of higher degree candidates in firms and thus an enhanced capacity to utilize new knowledge, and greater reliance by industry in Norway and elsewhere in Europe on external sources for R&D. Measures of formal university–industry research collaboration suggest that growth has stabilized or slowed since the mid 1990s, perhaps because the interaction has reached an "optimal level," where it is mutually beneficial to both parties. Norwegian data still show how the number of PhD graduates working in firms has increased with more than 80 percent from 1995 to 2005—and the share of PhD holders among private sector employees has gone up from 0.3 to close to 0.5 percent in this period. Norway's oil and gas industry has become very competence and technology intensive and now accounts for more than 10 percent of total industrial funding of Norwegian university research. In this industry, 1.6 percent of the employees hold a PhD degree.

This broad trend of gradually increasing interaction seems to affect the largest and most technologically advanced companies and industries in Norway, which are now supported by policies that seek to create "centres of excellence" through public-private R&D partnerships. Consistent with the political saliency of Norway's "small-scale" sector of enterprise (Chapter 2), policies to include small and medium-sized firms, and firms from throughout Norway's regions in these collaborations, also have been developed. Public agencies, state colleges, and the research institutes are important actors in these initiatives. Some Norwegian firms have a strong and expanding absorptive capacity, and special programmes for regions, clusters, SMEs, and so on, seek to build up absorptive capacity in other firms as well.

Although our analysis has included research, teaching, and commercialization data, we have not discussed other aspects much like consultancy, informal networking, and unpaid student work for companies which all may constitute an important part of the university–industry interface. It is natural, however, to assume that these activities are related to indicators of industry involvement in funding, networking and outputs. Industry surveys (CIS) show that the innovation activities in Norwegian firms are often oriented at collaboration, and that universities/colleges and research institutes are more frequent partners and sources of information for Norwegian firms than in most other European countries. Data on scientific co-authorships, and to some extent funding, partly confirm this picture, although countries tend to differ a bit depending on the indicators used.

Overall, this evidence of strong inter-institutional and cross-sector collaboration may contribute to an explanation of the "Norwegian paradox" of high industrial productivity and relatively low levels of industrial R&D investment (emphasized, e.g., in OECD country reports). The collaborative nature of the system ensures quick and efficient knowledge flows, and leading firms have qualified manpower to absorb and utilize new scientific and technological knowledge. R&D costs remain low because they are frequently shared among many different actors, perhaps enabling the exploitation of some efficiencies associated with a network organization. In this sense, it is natural to view universities not just as responsive to industrial needs but also an active part of professional communities oriented at transforming and utilizing the vast quantities of knowledge that are produced outside of the small country's borders. Innovation collaboration is partly a general small country phenomenon. But as Norway scores comparatively highly on many indicators of collaboration, there is probably a cultural, structural, and/or historical component here as well.

Public support for entrepreneurship and commercialization has a history of at least three decades in Norway. Patenting happens regularly and involves Norwegian public research organizations (at least) as frequently as in other countries. There seems to be a particular preference for creating spin-off companies in Norway. The 2003 legislations that removed the university "teacher exemption clause" and increased HEIs' responsibility for commercialization is too recent for an effective evaluation of its effects, particularly since the support structure of technology transfer offices was established only one year later. Thus far, academic "technology transfer" activities appear to remain very modest in scope by comparison with teaching and research. There is some worry that increased university willingness to take ownership to research results could impair other channels of university–industry interaction, but few data exist to support any such concern.

Finally, in Norway as in other industrial economies, different disciplines, industries, institutions, technological areas, and so on, exhibit very different patterns of interaction among research institutes, universities, and firms. The technology heavy university NTNU (and its neighbour institute SINTEF) dominates university–industry relations in Norway, regardless of what indicators one considers. Other HEIs also receive a significant share of their funding from industry, for example, the profession-oriented University of Life Sciences (UMB) and increasingly the comprehensive University of Bergen, and patenting and the creation of spin-off companies happen at all institutions. Student projects for companies are common many places, not least among the colleges, but are less visible in the statistics. It is essential that any assessment of Norway's public R&D infrastructure adopt a comprehensive perspective that covers all of its numerous components—many publications, patents and industry projects are carried out by HEIs in partnership with research institutes. The growing demands that HEIs collaborate with small businesses and more actively commercialize and transfer knowledge also mean that they may take on activities that traditionally have been the domain of the institutes. This development could influence inter-institutional relations in Norway's overall public-sector R&D system, as was discussed in Chapter 3.

NOTES

1. We mainly use "industry" in a broad sense referring to the business for-profit sector—otherwise we specify e.g. "manufacturing industry." In the term "university–industry relations" we also include colleges.
2. R&D operating costs excludes investments in large equipment and facilities; industrial funding is very rarely used for these purposes. Industrial funding is, however, a significant source of funding for equipment in the technological disciplines and specifically at NTNU.
3. 1 MNOK equals about 125,000 Euro.
4. The University of Stavanger is included in the state college sector where it belonged until the 2005 statistics.
5. With at least one author representing a Norwegian firm.
6. Innovation collaboration is defined as "active participation in joint R&D and other innovation projects with other organizations (either other enterprises or noncommercial institutions). It does not necessarily imply that both partners derive immediate commercial benefit from the venture. Pure contracting out of work, where there is no active collaboration, is not regarded as co-operation." Highly important information source is not defined further, although it is indicated that it refers to suggestions for new innovation projects or contributing to the implementation of existing projects (Eurostat 2004: 292–3).
7. Unfortunately there are no comparable data from other countries. A current EU project named REFLEX aims for a comparative perspective on the transition from higher education to work, but the data are not ready yet (April 2007).
8. "R&D services" is omitted which includes most of the research institute sector and has about 2,000 PhDs.
9. This is again based on data for PhD holders trained in Norway only, see Olsen (2007). An attempt at identifying PhD holders in industry with training from other countries will be carried out in 2008 by Olsen and colleagues. In universities, colleges and research institutes, the share of PhD holders with foreign degrees is 18.6 percent.
10. The explosion of university patents however has accompanied a peaking of this quality-measure during the mid 1980s, suggesting, "that the rate of increase of important patents from universities is much less than the overall rate of increase of university patenting in the period" (Henderson *et al.* 1998).
11. Some studies do find that academic spinoffs grow more slowly but tend to have a higher chance of surviving compared to spinoffs from private firms (Dahlstrand 1999).

12. The database consists of patent applications, some of which are later rejected or withdrawn. The study shows that applicants from universities and research institutes have a somewhat larger propensity of being granted a patent than applicants from the private sector.
13. Comparisons with Italian university patenting are very difficult, not least since the "professor's privilege" was introduced into Italian universities in 2003.

REFERENCES

Balconi, M., S. Breschi, and F. Lissoni (2004) "Networks of inventors and the role of academia: an exploration of Italian patent data," *Research Policy*, 33(1), 127–45.

Becker, Gary S. (1964) *Investment in Human Capital: A Theoretical and Empirical Analysis With Special Reference to Education*, NBER: New York.

Brandt, E. (2005) *Kartlegging av praksisbasert høyere utdanning.* Oslo: NIFU STEP, Skriftserie 8/2005.

Calvert, J. and P. Patel (2003) "University–industry research collaboration in the UK: bibliometric trends." *Science and Public Policy*, 30: 85.96.

Dahlstrand, Å. L. (1999) "Technology-based SMEs in the Göteborg region: their origin and interaction with universities and large firms." *Regional Studies*, 33: 379–89.

Etzkowitz, H. (1998) "The norms of entrepreneurial science: cognitive effects of the new university–industry linkages." *Research Policy*, 27: 823–33.

Etzkowitz, H., A. Webster, C. Gebhardt & B. R. C. Terra (2000) "The future of the university and the university of the future: evolution of ivory tower to entrepreneurial paradigm." *Research Policy*, 29(2), 313–30.

Eurostat (2004) Innovation in Europe: Results for the EU, Iceland, and Norway. Luxembourg: Eurostat.

Frölich, N. and A. Klitkou (2006) "Strategic management of higher education institutions: performance funding and research output," Conference on Science and Technology Indicators, Lugano 15–17 November 2006.

Gibbons, M., C. Limoges, H. Nowotny, S. Schwartzman, P. Scott, and M. Trow (1994) *The New Production of Knowledge. The Dynamics of Science and Research in Contemporary Societies.* London: Sage Publications.

Godin, B. (1998) "Writing Performative History: The New New Atlantis?" *Social Studies of Science*, 28, 465–83.

Gulbrandsen, M. (2005) " 'But Peter's in it for the money'—the liminality of entrepreneurial scientists," *VEST Journal for Science and Technology Studies*, 18: 49–76.

Gulbrandsen, M. and I. M. Larsen (2000) *Forholdet mellom næringslivet og UoH-sektoren—et krevende mangfold.* Oslo: NIFU, Rapport 7/00.

Gulbrandsen, M. and J.-C. Smeby (2005) "Industry Funding and University Professors' Research Performance," *Research Policy*, 34, 932–50.

Gunnes, H. *et al.* (2007) *Forskerrekruttering i Norge—status og komparative perspektiver.* Oslo: NIFU STEP, Rapport 2/2007.

Guston, D. H. and K. Kenniston (1994) *Introduction: The Social Contract for Science.* In Guston, D. H. and K. Kenniston (eds.), *The Fragile Contract.* Cambridge/London: MIT Press, pp. 1–41.

Henderson, R., A. B. Jaffe, and M. Trajtenberg (1998) "Universities as a Source of Commercial Technology: A Detailed Analysis of University Patenting, 1965–1988." *Review of Economics and Statistics*, 80(1), 119–27.

Hervik, A. and S. J. Waagø (1997) *Evavluering av brukerstyrt forskning på vegne av Nærings- og handelsdepartementet*, Trondheim: NTNU, Institutt for industriell økonomi og teknologiledelse.

Iversen, E., M. Gulbrandsen, and A. Klitkou (2007) "A baseline for the impact of academic patenting legislation in Norway," *Scientometrics*, 70: 393–414.

Jaffe, A. B. and J. Lerner (2004) *Innovation and its Discontents: How our Broken Patent System is Endangering Innovation and Progress, and what to do about it*, Princeton: Princeton University Press.

Larsen, M. T. (2007) *Academic Enterprise: A New Mission for Universities or a Contradiction in Terms?* Copenhagen: Copenhagen Business School: PhD Thesis.

Martin, B. R. (2003) "The changing social contract for science and the evolution of the university," In Geuna, A., A. J. Salter, and W. E. Steinmueller (eds.), *Science and Innovation. Rethinking the Rationales for Funding and Governance*, Edward Elgar, Cheltenham.

Martin, B. R. and H. Etzkowitz (2000) "The origin and evolution of the university species," *VEST Journal for Science and Technology Studies*, 13: 9–34.

Meyer, M. (2006) "Academic inventiveness and entrepreneurship: on the importance of start-up companies in commercializing academic patents," *Journal of Technology Transfer*, 31: 485–99.

Mowery, D. C., R. R. Nelson, B. N. Sampat, and A. A. Ziedonis (2004) *Ivory Tower and Industrial Innovation: University–Industry Technology Transfer before and after the Bayh–Dole Act.* Stanford, CA: Stanford University Press.

Nowotny, H., P. Scott, and M. Gibbons (2001) *Re-thinking Science: Knowledge and the Public in an Age of Uncertainty*, Cambridge: Polity Press.

OECD (2003) *Turning Science into Business: Patenting and Licensing at Public Research organizations*, Paris: OECD, Directorate for Science, Technology and Industry.

Olsen, Terje Bruen (2007) Doktorgrad—og hva så? Om doktorenes yrkeskarriere. Oslo: NIFU STEP, Report 20/2007.

Pechter, K. and S. Kakinuma (1999) "Coauthorship Linkages betwen University Research and Japanese Industry." In Branscomb, L.M., Kodama, F., R. Florida (eds.), *Industrializing Knowledge. University–Industry Linkages in Japan and the United States*, MIT Press, Cambridge, Massachusetts, pp. 102–27.

Rosenberg, N. and R. R. Nelson (1994) "American Universities and Technical Advance in Industry," *Research Policy*, 23: 323–348.

Smeby, J.-C. and S. L. Sundnes (2005) "Utgifter til forskning ved universitetene," in Gulbrandsen, M. & J.-C. Smeby (eds.), *Forskning ved universitetene: rammebetingelser, relevans og resultater*, Oslo: Cappelen akademisk forlag, pp. 27–45.

Sotarauta, M., Nauwelaers, C., Gulbrandsen, M., and Dubarle, P. (2006) *Supporting the Contribution of Higher Education Institutions to Regional Development: Peer Review Report:Trøndelag (Mid-Norwegian Region)*, Norway. OECD. Directorate for Education. Programme on Institutional Management in Higher Education (IMHE). Paris.

UFD (2004–5) *Vilje til forskning.* Oslo: Utdannings- og forskningsdepartementet (White Paper on Research from the Ministry for Education and Research).

Valentin, F. and R. L. Jensen (2003) "Discontinuities and Distributed Innovation: The Case of Biotechnology in Food Production," *Industry and Innovation*, 10(3), 275–310.

Van Looy, B., M. Ranga, J. Callaert, K. Debackere, and E. Zimmermann (2004) "Combining Entrepreneurial and Scientific Performance in Academia: Towards a Compounded and Reciprocal Matthew Effect?" *Research Policy*, 33, 425–41.

Webster, A. (2003) "Knowledge Translations: Beyond the Public/Private Divide?" *Journal of Education through Partnership*, 3(2), 7–22.

Zucker, L. B., M. R. Darby, and M. B. Brewer (1998) "Intellectual Human Capital and the Birth of US Biotechnology Enterprises," *The American Economic Review*, 88, 290–306.

12

The Technical-Industrial Research Institutes in the Norwegian Innovation System

Lars Nerdrum and Magnus Gulbrandsen

In developed countries publicly supported research is most often performed in two types of organizations; the higher education system and the public/non-profit research institutes that do not formally offer higher education or teaching activities. The place within national innovation systems of public research organizations with little or no teaching varies greatly among countries. Norway and Sweden represent contrasting models for public research. In Norway, research institutes are an integral, central part of the public research system, and interact closely both with universities and with firms and other actors (see Chapter 2). As Chapter 3 points out, many of these institutes (within meteorology, veterinary medicine, geology, and marine research, for instance) were established in the early part of the twentieth century, and the post Second World War period witnessed the emergence of many new institutes with a variety of users and orientations. Sweden, by contrast, has only a few small publicly funded research institutes outside of the higher education system.

Recent evaluations have criticized the large size of the Norwegian institute sector, suggesting that it may impede rather than support firm-level innovation (Arnold *et al.* 2001). How strong is the evidence underpinning this critical assessment? This chapter describes the research institutes in Norway and explores their relationship with other actors in Norway, focusing on the services supplied by the technical-industrial research institutes to firms and business in Norway and the importance of the institutes for Norwegian innovation.

CRITICISMS OF NORWAY'S RESEARCH INSTITUTES

Publicly funded industry- and mission-oriented research institutes have received little attention in contemporary theories of innovation and knowledge production. Many such laboratories have been reduced in size or closed in recent years in response to criticism of them as institutionalizations of the much criticized "linear" model of innovation (seeing innovations as spin-offs from basic research) or reflections of failed "national champions" policies, supporting individual firms in sectors of allegedly "strategic" importance. Nevertheless, in some countries, critical discussions of the role of research institutes have served to strengthen public laboratories, not least due to the need for independent expertise in society (Larédo and Mustar 2001). Thus, public laboratories or research institutes are still relevant actors in the national innovation system in many countries (Crow and Bozeman 1987; Nelson 1993; Nelson and Rosenberg 1993; Larédo and Mustar 2001; Larédo 2003; Ministry of Education and Research 2005), although their structure, function, and role within national innovation systems differ greatly among otherwise similar countries. In one of the few international comparisons, Slipersæter *et al.* (2003) find that the roles taken on by research organizations of this type differ greatly across countries, and it is difficult to compare them directly. This chapter's examination of Norway's research institutes uses unique data that cannot be readily compared with data from other countries or international organizations.

The critical evaluation of the Norwegian research institutes mentioned earlier asserted that there is "an enormous imbalance in the respective sizes of the institute and university sectors" in Norway (Arnold *et al.* 2001: 29), and identified three problems associated with the size of the research institute sector:

- The "absorptive capacity" (Cohen and Levinthal 1989) of Norwegian industry could be weakened by a tendency for trained researchers to work in research institutes rather than in industry.
- Their large size and strong links with Norwegian industry may prevent Norway's universities from modernizing their research practices and relationship with industry.
- The large commitment of Norwegian public funding to support for the research institutes prevents the growth of financial support for the nation's university system to level that would enable the system to reach a "critical mass" in more research areas.

None of these assertions was established empirically by the evaluators; rather they called for more research on the extent of the asserted problems associated with Norway's research institutes in the nation's innovation system. This chapter will provide some evidence to address these issues.

THE SIZE AND STRUCTURE OF THE NORWEGIAN INSTITUTE SECTOR

R&D statistics and the funding policies of the Research Council of Norway (RCN) may contribute to an impression that the institutes constitute a homogeneous sector, but the reality is much more complex. Research institutes in Norway are "... a highly heterogeneous collection of units with different tasks and target groups, different background, different financial basis, and different organizational and affiliation forms" (Ministry of Education and Research 2005: 169). Among the technical-industrial institutes, which are the central focus of this chapter, some are some clearly "mission-oriented," such as the Norwegian Defence Research Establishment and the Institute for Energy Technology (previously named the Institute for Atomic Energy). Others, such as SINTEF, are user-oriented serving a broad cross-section of Norwegian industry, and still others, such as many of the agricultural and environmental institutes, serve both purposes, supporting government missions and addressing the needs of industry. Some of the user-oriented institutes also collaborate closely with higher education institutions, as in the case of SINTEF and NTNU.

Measured by expenditure the research institutes perform approximately one-quarter of Norway's national R&D, only slightly less than the higher education system (Research Council of Norway 2006). Comparing the Norwegian research institutes with those of other countries, Slipersæter *et al.* (2003) found that although the share of total R&D expenditures spent in research institutes in Norway is relatively high (23 percent), public funding accounts for a comparatively small share of the institutes' overall R&D budget (61 percent). A recent examination of framework conditions for research institutes found that Norwegian institutes receive less public funding as a share of their budgets than institutes in Denmark, Sweden, the Netherlands, and Germany, but around the same as Finnish institutes (Brofoss and Slipersæter 2004). Public basic grants defined as lump-sum transfers account for only around 8 percent of the incomes of the so-called "technical industrial research institutes" (see below). In addition to basic grants, RCN supports "strategic institute programmes" that constitute around 13 percent of the institutes' income. The remainder of the income of Norway's research institutes is generated from projects and R&D services provided to national and foreign customers, of which many are public—although the private–public composition of the research portfolio varies among institutes. In order to obtain such projects, the institutes must compete with other institutes, universities or consultancy firms and are chosen for price, quality or relevance reasons like in any other client–supplier relationship.

An important policy document in the development of Norwegian policy in public financing of industrial R&D was the 1981 Green Paper of the Thulin Commission (NOU 1981; see also Chapter 2). According to Wiig and Mathisen (1994), this report originated the "double purpose policy" for Norway's technical-industrial research institutes, which has guided their operations since then. The double purpose policy has two facets. First, industrial institutes should constitute an R&D infrastructure to which firms can turn with their needs for competence, knowledge and R&D equipment that is economically impossible for them to acquire and maintain in-house. Second, institutes should serve as "intermediators" between firms and universities by interpreting and translating the technical needs of firms for university researchers in order to give them a sense of industrial needs and to inspire new research in industrially relevant areas. The Thulin commission based the "double purpose" policy on the premise that Norway has an unusually large number of small and medium-sized enterprises in industries that are not very R&D-intensive. Their needs for new knowledge, technology and problem solving are best met by publicly subsidized organizations with the mission of contributing to the growth and innovation of Norwegian industry. The research institutes were charged with responsibility for applied research and the development of linkages to the academic research frontier and graduate training, in order to reduce the innovation-related problems associated with the weak absorptive capacity of much of Norwegian industry.

This "double purpose policy" has guided policy toward Norway's technical-industrial research institutes since the Thulin Commission. More recently, however, the potential contributions of the research institutes to education of researchers and capacity building within industry have received greater attention. A recent White Paper (Ministry of Education and Research 2005: 172–3) argues that education of students, including doctoral training and supervision should be defined as an explicit mission for the institute sector. A new funding system for the institutes is currently being developed (Norges forskningsråd 2006) that will include incentives both for research "quality" (including doctoral training and publishing) and relevance (measured by share of customer-oriented activity), re-emphasizing the "dual role" for the sector.

Public data yield inconsistent estimates of the number of Norwegian firms purchasing R&D services from the institutes and of the revenues generated by these industrial contracts. One comparison (Brofoss *et al.* 2002) of different sources of research-institute revenues uses data from 1999 and 2000. According to the study, the Norwegian industrial R&D statistics show that fewer than 500 companies purchased R&D services from the institutes in 1999 for a total of 672 million NOK. These figures omit R&D contracts with companies employing fewer than 10 employees and report only a selection of the institute client firms that employ 10–50 people. Other data from 2000 on the industrial contracts of 41 institutes reveal that around 1,400 firms purchased services from these institutes, generating revenues of roughly 990 million NOK (Broch *et al.* 2001). These figures exclude projects less than 10,000 NOK (of which there are many) and projects where the company name could not be identified. Many of the projects covered by these contracts, however, may not constitute R&D as defined by the Frascati manual. The Norwegian Research Council's data showed that around 1,500 firms were involved in "user-controlled research programmes" involving institutes in 2000, and the institutes themselves stated in their annual reports that firms' purchases of services in 2000 generated approximately 1,450 million NOK in revenues.

More recent data based on Norwegian firms' R&D tax deductions through the SkatteFUNN programme (see www.skattefunn.no) show that 6,440 companies were involved in 13,046 projects eligible for R&D tax deductions in 2002–2006, and approximately 30 percent of these projects involved collaboration with public R&D organizations. The R&D tax deduction data reveal that 391 of the 1,569 active collaboration projects reported by firms in 2006 involved collaboration with the research institute SINTEF, while 106 projects linked firms with the technical university NTNU. The Norwegian institutes thus have a large number of industrial clients, but not all contracts can be defined strictly as R&D-related (which is probably one reason for the discrepancy between data sources).

The 110 organizations that belong to the Norwegian institute sector vary greatly in size, ownership, location, organizational status, areas of expertise (see Figure 12.1), and research orientation (Slipersæter *et al.* 2003). These organizations were established in different periods of Norway's economic development, but all were established to respond to needs for applied knowledge from government and industry (cf. Chapter 2; Skoie 1990 and 2005). For instance, some government ministries established research institutes to respond to their needs for knowledge and research in specific areas. The remainder of this chapter focuses on Norway's "technical-industrial research institutes," i.e. organizations that were established to collaborate with industry and to conduct contract research. This group accounts for the largest single share (41 percent) of total R&D expenditures by the Norwegian research institutes (Figure 12.1).

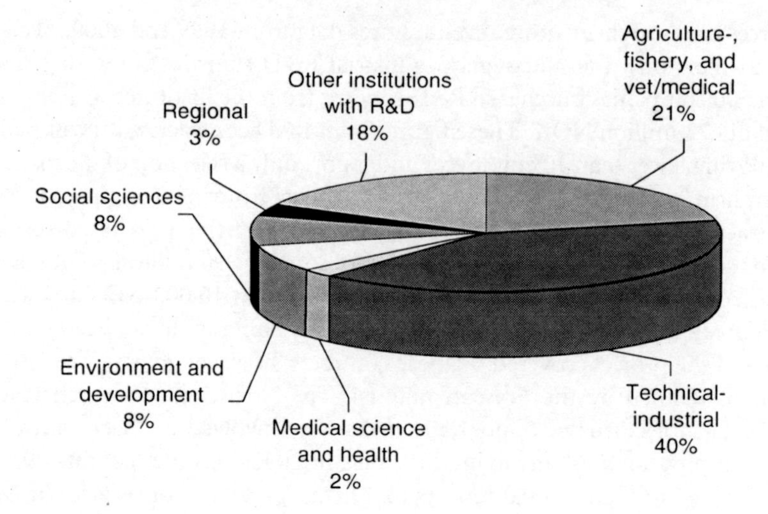

Figure 12.1. Total R&D expenditure in the institute sector in 2005 by field of research, 110 institutes with R&D activities in Norway (%).

Source: Research Council of Norway 2006.

SERVICES, ACTIVITIES, AND CUSTOMERS OF R&D INSTITUTES IN NORWAY

Our discussion of the effects of R&D in research institutes for industrial research draws on a survey of Norwegian firms that was conducted in 2002 (See Brofoss and Nerdrum 2002, for a more detailed description of the survey). Statistics Norway drew a sample of firms from the business register for the 30 most R&D-intensive two-digit NACE industries. Although firms of all sizes were selected, the sampling method favoured large firms that also tend to be more research-intensive than the average Norwegian firm.[1] After removing a large number of firms that were not relevant to the survey (lawyers, manpower firms, and other types of firms that are not involved in R&D), we ended up with a sample of 986 firms, of which 460 (47 percent) provided satisfactory responses.

Of the 460 firms responding to the survey, around one-third reported that they had purchased R&D from external suppliers in 2000. The firms constitute the smallest unit in the Business register. Many firms are owned by the same establishment, particularly is that very often the case for large establishments which are most often made out of many smaller firms, and intra-firm R&D purchases are consequently very common. Among the firms which reported in-house R&D, the share that had purchased R&D from external suppliers was 87 percent, indicating that in-house R&D and purchased R&D are complements, a finding largely consistent with other research and the framework of, for example, Cohen and Levinthal (1989). Around 22 percent of the firms responding to the survey had purchased R&D services from research institutes in Norway, while only around 7 percent of the firms had purchased R&D services from universities or colleges. The Norwegian research institutes thus are more important external sources of R&D for these firms, which are relatively R&D-intensive, than Norwegian universities. The comparison in Brofoss and Nerdrum (2002) of the survey results with data from Statistics Norway revealed that survey respondents reported larger expenditures on research institute services than appeared in public Norwegian R&D statistics. The discrepancy appears to reflect the stricter definition of R&D (Frascati definition) employed in the Statistics Norway R&D data than was used by respondent firms.

INDUSTRIAL PROJECTS IN RESEARCH INSTITUTES

The survey results in Brofoss and Nerdrum (2002) indicate that product development services were the most important type of services purchased by Norwegian firms from research institutes, followed by process development expertise. But testing and consultancy services also were important to many firms. Table 12.1 summarizes information on the number of projects, mean project size, and average firm-level expenditures at primary sector institutes, social sciences, "environment and development," and "science and technology" institutes, the taxonomy used by the Norwegian Research Council. The table uses data reported by Broch *et al.* (2001).

Table 12.1. Industrial customers and projects above €1,250 at 41 Norwegian research institutes in 2000.

Institute group (N)	No. ind. Projects	Mean proj.size (€)	€/customer
Primary (10)	466	39,826	55,804
Social Sciences (14)	240	24,080	40,497
Environment and dev. (5)	299	21,993	32,186
Science & Tech (12)	2,873	38,755	71,352
All (41)	3,878	31,961	52,248

Source: Data from Broch *et al.* (2001).

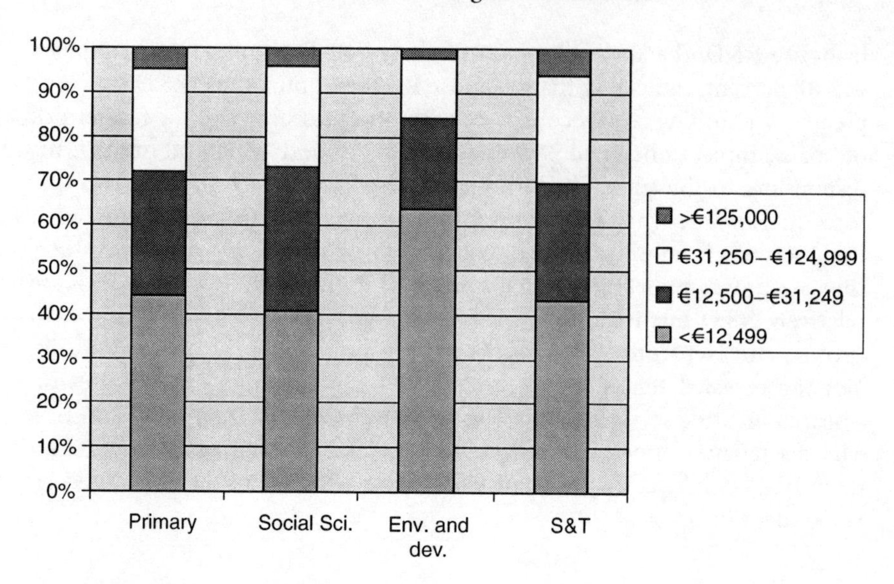

Figure 12.2. Distribution of industrial projects by size in 2000, only including projects above €1,250 at the 41 most industry-relevant research institutes. (Number of industrial projects in 2000.)

Source: Broch *et al.* (2001).

Table 12.1 excludes company-financed projects that are smaller than NOK 10,000 (€1,250) which in the institutes' records are frequently called "training", "upgrading", "teaching", or "seminar" projects, indicating the use by firms of these projects as vehicles for learning and keeping in touch with scientific and technological developments. Even without these small contracts, less than half of the remaining portfolio consists of projects greater than NOK 100,000 (EUR 12,500). The "environment and development" institutes have a particularly high share of small projects (see Figure 12.2), probably related to their provision of testing services (pollution, chemical composition of materials etc.). In other words, more than one-half of the research institutes' projects are relatively simple, short-term tasks that probably do not meet a strict definition of R&D. These data also do not reveal whether small projects may be related to larger, more complex projects that may be defined as R&D. The data in Table 12.1 also highlight the tremendous administrative challenges for many institutes resulting from the need to administer large numbers of projects that account for a small share of their turnover. Nevertheless, projects averaging more than €125,000 constitute about half of the turnover for the 41 institutes in Table 12.1, indicating the dependence of many institutes on a relatively small number of customers and projects.

Table 12.2. Reasons among firms for purchases of R&D services from Norwegian research institutes, factor analysis

Rotated Component Matrix	Component			
	R&D motivation	Accessible	Capacity	Lack of competence
Lack of own capacity generally	,007	,106	**,878**	,184
Lack of capacity in particularly busy periods	–,001	,367	**,740**	–,159
Lack of competencies in the firm	,064	–,012	,111	**,870**
Lack of equipment/testing facilities	,070	**,608**	,159	–,028
Strategy to increase the firm's R&D competences	,777	,092	–,081	–,064
Obtain participation in competence networks	**,839**	,222	–,084	,089
Access to public R&D funding	**,467**	–,154	,261	–,338
Increase the quality of in-house R&D	**,658**	–,158	,423	,275
Geographical proximity is important	–,021	**,630**	,270	,002
Only knowledgeable about Norwegian institutes	–,089	**,658**	–,095	,460
Personal contacts in the institute(s)	,462	**,629**	–,039	–,063
Percentage of variance explained	*19.9*	*16.8*	*15.5*	*11.2*
Cumulative percentage	*19.9*	*36.7*	*52.2*	*63.4*

Extraction Method: Principal Component Analysis. Rotation Method: Varimax with Kaiser Normalization. Rotation converged in 7 iterations.

WHY FIRMS PURCHASE R&D SERVICES FROM RESEARCH INSTITUTES

This section investigates firms' motives for purchasing R&D from research institutes, using data from the Brofoss and Nerdrum survey discussed earlier. Table 12.2 reports the results of our factor analysis of the responses to questions about why the firms purchase R&D services from Norwegian research institutes.[2] The analysis indicates that Norwegian firms use institutes for four different types of reasons. We call the first type *R&D motivation*, which refers to firms purchasing R&D services from research institutes as part of a strategy intended to increase their otherwise inadequate in-house knowledge and as a means of joining "competence networks," to use the term from the questionnaire. This cluster of reasons was emphasized by 58 percent of the respondents.

The second type of motive is labelled *Accessible*; these firms state that they lack their own equipment or test facilities, they stress geographical proximity and personal contacts, and they agree that they are familiar with only (some) Norwegian institutes. A closer look at the dataset reveals that half of the firms in and around the Trondheim area, where SINTEF and other important research institutes are located, emphasize geographical proximity. Only a handful of firms from other regions emphasize this aspect. Possible explanations to this geographically determined response pattern are that the firms in the Trondheim area have access to a more complete knowledge base through the local research infrastructure than firms in most other regions. Hence, the Trondheim-based firms rely more heavily on locally available external technological resources, primarily SINTEF and NTNU—yet other firms may not see the distance to Trondheim as a major obstacle to collaboration.

A third type of motive is labelled *Capacity*. Certain firms use the institutes due to a lack of in-house capacity. Other firms rely on the institutes for research services in particularly busy periods, using the institutes as a capacity buffer that is available when the employees of the firm itself are too busy with other tasks. Finally, the statement "lack of competencies in the firm" emerges as a separate factor in this analysis, and is characteristic of a small number of firms that otherwise appear to have little in common.

We found no significant differences between large and small firms in the answers to the above questions, but incomplete data on industry classifications reveal some significant differences among firms. Oil and gas companies score significantly higher than firms in other industries on the *R&D motivation* index, and "traditional manufacturing firms" score significantly higher on the *Accessible* index. Due to space limitations, we do not show these indexes here.

THE IMPORTANCE OF R&D IN RESEARCH INSTITUTES TO FIRMS

A factor analysis of survey respondents' assessments of the importance of R&D purchased from research institutes yields three different dimensions (Table 12.3). The first is called *Process*, referring to firms that emphasize the importance of research institutes for developing new or improved processes, for developing new work methods or tools, and for increasing quality and reliability in production.

Another group of firms emphasize *Market* considerations in their assessment of the importance of the R&D purchased from institutes. These firms agree that R&D from institutes is important for understanding the needs of customers and for helping to enter new markets. Inasmuch as most R&D services are purchased by Norwegian firms from institutes that specialize in technology and natural sciences, this emphasis by respondents on the role

Table 12.3. How important is R&D purchased from Norwegian research institutes for different dimensions of improvement development? Factor analysis

Rotated Component Matrix	Component		
	Process	Market	Product
Development of new or improved products	–,037	,384	,781
Development of new or improved processes	,815	,080	,230
Development of new materials	,274	–,180	,802
Development of new work methods/tools	,821	,223	,019
Improved quality and reliability in production	,798	,363	,071
Improved understanding of customer needs	,306	,811	,060
Obtaining/entering new markets	,189	,843	,052
Percentage of variance explained	*31.2*	*24.8*	*18.8*
Cumulative percentage	*31.2*	*56.0*	*74.8*

Extraction Method: Principal Component Analysis. Rotation Method: Varimax with Kaiser Normalization. Rotation converged in 6 iterations.

of purchased R&D in providing information on markets and user needs, is surprising. There is a high correlation between the market dimension and the "accessibility" factor discussed in the previous table, suggesting that the companies that know little about non-Norwegian research institutes are the ones that rely on the institutes for help with market development and understanding user needs. The last dimension in Table 12.3 services is called *Product*, highlighting the presence of a group of firms that emphasize the value of institute research services for developing new or improved products and new materials. There are no statistically significant differences among industries or between large and small companies in responses to this set of questions, although Norwegian food companies place greater emphasis on the importance of research institute services for process innovation and electric/electronics firms assign less importance to this aspect of the importance of purchased R&D services.

The survey data contain information on firms' motives for pursuing R&D collaboration in general, defining "collaboration" to include work with entities (other firms, universities) in addition to the Norwegian research institutes (Table 12.4). More than 65 percent of the respondent firms indicate that participation in networks of various types to obtain updated knowledge is "rather" or "very" important. A higher share of firms with fewer than one hundred employees indicate that networks are important, suggesting a greater reliance by such firms on outside sources of knowledge. Another important motive for collaboration is access to applied R&D results, which is stressed by more than 60 percent of the firms, especially firms with more than

Table 12.4. Firms' experiences with and opinions of the research institutes, factor analysis

Rotated Component Matrix	Component	
	Positive	Negative
The institutes' R&D services are fairly expensive	,212	,625
The institutes' R&D services are of high quality	,507	–,531
Personal acquaintances are important	,688	,130
Good former experience is important	,831	,142
The institutes' reputation is important	,799	–,076
The institutes have valuable test facilities and methods	,523	–,342
The firm needs its own core competencies and prefers to do R&D in-house	–,057	,786
Percentage of variance explained	34.0	*20.7*
Cumulative percentage	*34.0*	*54.7*

Extraction Method: Principal Component Analysis. Rotation Method: Varimax with Kaiser Normalization. Rotation converged in 3 iterations.

200 employees. Finally, "access to R&D skills" is an important for R&D collaboration of all types for 79 percent of the firms.

THE EXPERIENCES AND OPINIONS OF FIRMS REGARDING INSTITUTES

Table 12.4 summarizes the results of a factor analysis of survey responses to questions dealing with respondents' experiences with and opinions of Norwegian research institutes, based on their agreement or disagreement with statements characterizing the institutes and firms' experiences in collaboration with them. The data in Table 12.4 are based on responses from all firms answered the relevant questions, regardless of whether they used the institutes in the most recent year or not (there are no significant differences between users and non-users when it comes to size and industry affiliation). These opinions reflect a broader base of user experiences than those of institute clients in 2000—127 firms responded to the questions (see Table 12.5 for a summary), considerably more than the 99 firms from the survey that purchased services from research institutes in 2000.

There are two dimensions in this set of questions that distinguish between firms with *positive* experiences and perceptions and firms with *negative* or critical ones. The positive firms agree that the institutes' R&D services are

Table 12.5. To what extent do respondents agree with the following assertions about the research institutes and their activity? Frequencies of response (N = 127)

	Agree	Disagree	No opinion
Experience from earlier collaboration is important	88%	2%	11%
Rumours about the institutes are important	77%	5%	18%
Personal acquaintances are important	69%	12%	20%
The institutes have high quality	64%	8%	28%
The institutes possess valuable test sites and test methods	64%	9%	27%
The institutes are relatively expensive	56%	6%	39%
The firm needs own R&D and prefers to conduct R&D itself	40%	32%	29%

of high quality, and these firms also value the test facilities and methods of the institutes. The positive firms in particular emphasize the importance of several network-related factors, including personal acquaintances and former experience, in R&D collaboration. The negative or critical firms express a preference for building up their own core competences and performing R&D in-house. They claim that the institutes' services are fairly expensive (quite a few of the positive firms also express this view), and they disagree with the "high-quality" characterization of the institutes' services.

Less than 10 percent of the 127 firms who gave an opinion on the institutes (based on recent or prior experience) were clearly critical about the research institutes. Nearly 70 percent of respondents agree that personal acquaintances are important to their experiences (presumably, both positive and negative) with the institutes (Table 12.5). We found no systematic differences between large firms' and small firms' assessments of the research institutes. Although our industry information is, as mentioned, incomplete, it may be noted that 15 of the 19 electric/electronics firms express a preference for performing their R&D in-house. We are unable to determine whether this preference reflects negative experiences with the institutes, a perception by electric/electronics firms that the institutes lack sector-specific competences, or some other set of factors. More than half of the respondents (Table 12.5) also state that the research institutes' services are relatively expensive. As was mentioned earlier, institutional funding (public grants) for the research institutes is relatively low, which means that the institutes price their services at a level close to that of private consulting or technical-services firms. R&D directors of Norwegian multinationals emphasize in interview-based case studies that it is frequently cheaper to collaborate with institutes in other European countries than in Norway (Gulbrandsen and Godoe 2008).

In order to assess the validity of the critical statements concerning Norway's research institutes that were raised by the 2001 evaluation report that was discussed earlier, firms participating in the 2002 survey (Brofoss and Nerdrum, 2002) were asked to respond to the following assertions about Norwegian research institutes:

A Research institutes have unique competences
B Research institutes have good knowledge of R&D in industry
C There is little capacity in research institutes to conduct projects
D Secrecy needs prevent collaboration with research institutes
E Research institutes lack long-term funds and do not obtain fundamental research skills
F Research institutes absorb R&D competent labour that industry would need
G Research institutes prevent collaboration between firms and universities

Figure 12.3 summarizes the level of partial or total agreement with these assertions among firms that had reported having experience with research institutes in the survey and those that had not. Respondents expressing no opinion and those expressing disagreement form the remainder. Partial or total agreement with Assertions A and B—"institutes have unique competences" and "institutes have good knowledge of R&D in industry"—denotes a positive assessment of the research institutes by respondents. Agreement with the other statements, which express negative views concerning the institutes, denotes a more critical opinion of them. In general, respondents to the survey expressed positive views concerning the institutes, and the views of users were more strongly positive.

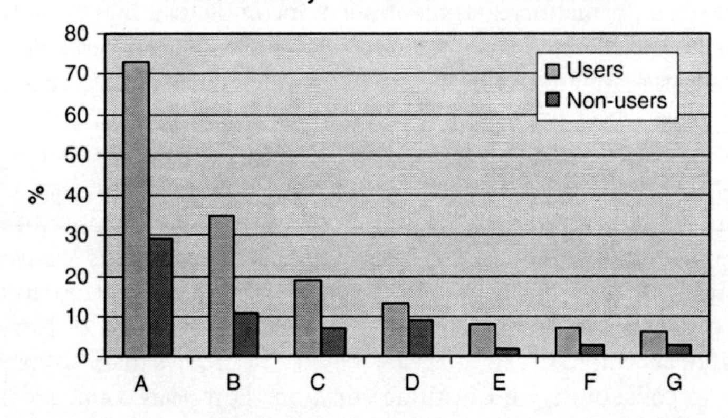

Figure 12.3. Share of users (N = 127) and non-users (N = 333) of research institutes who agree totally or partially with assertions (see list in text) about Norwegian research institutes.

Responses to all assertions from users of the research institutes differ significantly from those of nonusers (user and nonuser responses to assertion D differ significantly at the .05 confidence level, C at the .01 level and the rest at the .001 level). Neither users nor non-users provide strong support for the critical Assertion C concerning limited capacity in institutes, Assertion D, which states that firms' secrecy requirements prevent effective collaboration, or Assertion E, which claims that the institutes lack sufficient research skills due to lack of long-term funds.

The assertions F and G (respectively stating that the research institute absorb workers that otherwise could be employed by Norwegian firms in their in-house R&D operations; and that the institutes impede university-industry collaboration) directly test two of the critical statements from the 2001 evaluation concerning the size of the Norwegian institute sector (Arnold *et al.* 2001). Less than five percent of the 460 firms responding to the survey agreed with the assertion that institutes attract researchers and scientists that would be more usefully employed by Norwegian firms. Since the survey respondents include many R&D-intensive Norwegian firms that are most exposed to competition from the institutes for R&D workers, this lack of support for the 2001 evaluation's critical assessment is interesting.

The survey provides even less support for the argument advanced by Arnold *et al.* (2001) that research institutes prevent collaboration between firms and universities, or modernization in universities. Only 17 firms of 460 agreed with "Assertion G," which stated that the institutes impede firm-university co-operation. International R&D indicators provide some support for the views of survey respondents, since Norwegian firms' share of university R&D funding (in 2003–5 percent) is only slightly lower than the OECD average (6.1 percent in 2003; see Chapter 13 for additional discussion). Since Norwegian companies are less R&D-intensive and are concentrated in less R&D-intensive industries than companies in most other developed countries, industry-university collaboration may in fact be relatively high in Norway, given its industrial structure.

TEACHING AND MOBILITY ISSUES

Research institutes have a more complex role in the research and innovation system than simply supplying knowledge and services for industry and other users. Data from the Research Council of Norway show that almost 700 researchers in the research institutes (around 10 percent of the total workforce in the sector) supervised students at colleges and universities in 2002 (cf. Kaloudis and Koch 2004). During the same year, 450 graduate students who were working on a Master degree or a PhD were formally employed at an institute.

The boundary between Norway's research institutes and its higher education system thus is relatively porous. Institutes host and supervise PhD students and are involved in both teaching and basic research activities. Moreover, as Chapter 2 pointed out, a number of Norwegian universities and state colleges have expanded their ownership of geographically proximate research institutes since the turn of the millennium, in order to deepen their linkages with Norwegian industry and other sources of research funding. Indeed, some research institutes are difficult to distinguish from higher education organizations.

Researchers in Norwegian universities and research institutes also are linked through joint appointments. Researchers primarily employed in the higher education system worked 42 man-years (full time equivalents) in technical-industrial research institutes in 2004 (Norges forskningsråd 2005). Since most university researchers with research institute appointments hold part-time appointments in the institutes (roughly 20 percent FTE) a total contribution of 42 man-years translates into more than 200 university employees being jointly employed by research institutes and academic institutions. Research staff of the institutes contributed approximately 26 man-years of effort within higher education in 2004, suggesting that as many as 125 institute researchers also hold adjunct professorships.

These data also do not report on working relationships between universities and institute researchers that operate without such formal cross-appointments, and other evidence indicates that these informal relationships are extensive. According to figures from NTNU (the second largest university in Norway) and SINTEF (the largest research institute, of which a main body is located next to NTNU), around 500 researchers worked in both organizations in 2005 (source: www.ntnu.no). These data suggest that nearly 20 percent of the total combined scientific staff of SINTEF and NTNU are involved in collaborative research projects and teaching/supervision tasks.

Further evidence of the interaction between researchers in the two R&D sectors is given by co-authorship and project organization data. According to Kaloudis and Koch (2004), 57 percent of the scientific publications from the technical-industrial institutes in the period 1999–2002 had at least one co-author with an address from Norwegian universities and colleges. In 2004, technical-industrial research institutes supported "project-based collaborations with external organizations"—in practice, institute researchers with adjunct positions outside—involving almost 1,600 man-years; 26 percent of this total labour effort involved universities and colleges in Norway and abroad and 63 percent involved Norwegian firms (Norges forskningsråd 2005).[3] In other words, rather than being a barrier to university–industry interaction, considerable evidence suggests that Norway's research institutes may "lubricate" interinstitutional interaction in the innovation system through their strong linkages with both higher education and firms.

The research institutes generate other types of potentially beneficial spillovers of knowledge and personnel within the Norwegian innovation system. Nerdrum (1999) found that approximately 30 percent of the researchers who left SINTEF in the 15-year period 1974–1988 moved with a project they had worked on in SINTEF to a position in the firm or organization that had originally contracted for the project. An overwhelming majority of these achieved professional "success" in their subsequent careers, suggesting both that they are valuable and skilful workers and that their training at SINTEF contributed industry-relevant human capital. The technical-industrial research institutes provide a pool of skilled specialized labour to industry as a valuable by-product of the R&D services that industry purchases from the institutes.

Stenstadvold (1994) stated that the strategy and culture of SINTEF favoured transfer of researchers to industry, and characterized this personnel flow as part of the institute's implicit contract with Norway's government in exchange for basic funding. Broch *et al.* (2002) showed that roughly 7 percent of the institutes' research staff moved each year from the natural science and technology institutes to industrial employment. Knowledge transfers and spillovers through personnel movements from research institutes are another important contribution from the research institutes to the dynamic performance of Norway's geographical and sectoral innovation systems.

Stenstadvold (1994) used SINTEF data covering the 1965–90 period to argue that far more researchers moved from SINTEF to Norwegian industry than vice versa. Based on his more than 30 years as director of SINTEF, Stenstadvold argued that relatively few researchers from Norwegian industry moved to SINTEF, despite the institute's efforts to attract such individuals. This limited industry-institute flow of personnel may have reflected the relatively low wages in SINTEF or more attractive overall working conditions in Norwegian firms. But the broader issue of industry-institute personnel flows has not attracted extensive empirical attention, especially for more recent years.

Table 12.6 summarizes data on personnel movement from Norwegian industry to research institutes during 1997–2005. Although net flows to institutes from industry are generally positive, the incoming mobility to institutes from industry is far from negligible. Stenstadvold's observation that few

Table 12.6. Outgoing and incoming researcher mobility between technical-industrial research institutes and business sector. Absolute numbers, 1997–2005

	1997	1998	1999	2000	2001	2002	2003	2004	2005
Outgoing mobility	195	167	105	171	107	81	38	53	82
Incoming mobility	60	95	45	55	81	86	56	56	55
Net inflows	135	72	60	116	26	−5	−18	−3	27

Notes: FFI (The Defence Research Institute) is not included in the table due to missing and incomplete data. Shares do not always sum up due to rounding errors.
Source: Key figures on the institute sector, NIFU STEP.

industrial researchers were attracted to the institute sector may no longer be true, and this form of mobility was slightly higher during 2002–5 than during 1997–2000.

In Figure 12.4, incoming and outgoing researcher flows to and from the business sector are depicted as shares by researcher man-years in the technical-industrial research institutes during 1997–2005. As we noted earlier, personnel flows from technical-industrial research institutes to industry generally exceed the reverse flow, with the exception of the 2002–4 period. The trends in the Figure suggest that institute-industry mobility may be sensitive to demand for research personnel in industry, and from this perspective research institutes constitute a pool of skilled labor for firms. Not surprisingly, personnel flows from business to technical-industrial institutes appear to be less sensitive to the business cycle.

CONCLUDING REMARKS

This chapter has examined the technical-industrial research institutes, which form an important part of the research institute sector in Norway. Our empirical evidence suggests that these institutes provide valuable contributions to Norwegian innovation through a number of different channels. The institutes provide research results, research-based services, and valuable networks around and between firms, the higher education system, and other actors in the knowledge system. The institutes also serve the needs of smaller firms by supporting a large number of small projects and activities that range from courses to advising and connecting business representatives. These services are not normally classified as R&D, but can be considered as important

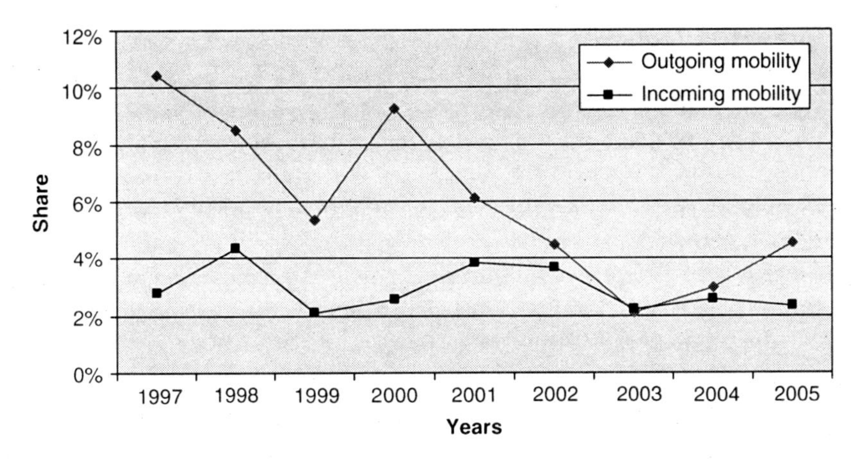

Figure 12.4. Outgoing and incoming mobility between technical-industrial research institutes to business sector. (Shares of researcher-years 1997–2000)

Source: Key figures on research institutes, NIFU STEP.

by-products and complementary services to the R&D conducted by the institutes.

Our survey of "users" and "non-users" of the services provided by the technical-industrial institutes suggests that Norwegian firms view the institutes as possessing "unique" competencies and providing high-quality R&D and related services. Most firms use research institutes because they lack the skills, the R&D capacity, equipment or understanding of research methodology to conduct R&D projects in-house. Research institutes therefore provide important contributions to firm-level innovation in Norway, although Norwegian firms generally evaluate the institute services as expensive. The high prices for institute services may reflect the relatively modest contribution of public funds to the research institutes (see Kaloudis and Koch 2004). In addition, however, many of the Norwegian institutes have considerable market power and few competitors (especially in some regions), which may enable them to charge high prices for their services.

The data presented in this chapter highlight three roles for the technical-industrial research institutes in Norway's innovation system. First, the institutes represent a *learning partner* for Norwegian firms. Such learning happens primarily through collaborative projects—sometimes with the institutes in charge within public programmes, in other cases (probably most often) with the firm(s) in charge in contract-based projects. Some projects may include the use of institute personnel as instructors for colleagues in industry or as part-time workers in firms, universities and other organizations.

Second, the institutes contribute to *increasing absorptive capacity* within Norwegian firms and/or to *overcoming problems of low absorptive capacity*. These contributions to firms' absorptive capacity operate through two mechanisms. The first mechanism relies on the numerous projects and personal contacts established between industry and institute staff, which seem to be quite stable over time. The other mechanism is the substantial flow of institute staff leaving to work in industry (or elsewhere), often following their projects and bringing their expertise into the user organization. On average, 7 percent of institute staff moved to industry per year during 1997–2005, substantially higher than staff mobility from Norwegian universities to other sectors, which averaged 1 percent per year during 1991–2001 (Nerdrum and Sarpebakken, 2006). Both of these mechanisms help firms adopt new knowledge and technologies and/or to define problems and seek solutions.

Third, the research institutes have a *lubrication* or *intermediation* role in the research and innovation system, facilitating interaction among institutions and firms. They have close relations to higher education, including personnel in adjunct and ad-hoc positions, joint laboratories and joint projects with and without industry involvement. In some respects, the institutes may act as a buffer zone between universities and industry that supports rather than discourages university–industry interaction. Recent developments at the research frontier are quickly communicated to the firms by their regular partners in institutes in a language closer to their day-to-day operations. Universities for their part learn about technological developments and new ideas that may inform the basic research agenda, but at the same time are shielded by the research institute from the most mundane industrial requests for assistance. In this manner, institutes act as a flexible complement to the universities without playing a strict intermediation role in a linear relationship. Norwegian companies also use institutes to increase their capacity for R&D, meaning that the institutes may enhance the array of options and flexibility for industry as well. In this fashion, the research institutes act as a *flexible repository* in the national innovation system. These issues deserve further investigation, as research institutes in general are much less studied than research units in higher education.

NOTES

1. The sample was drawn from firms from the 30 most R&D intensive industries (at NACE 2-digit level). All firms with more than 100 employees, 29 percent of the firms with between 40–99 employees, and 2 percent of the firms between 10–49 employees were selected, making a first sample size of around 2,000 firms.

2. We used principal components as extraction mechanism, eigenvalue above one as selection method and Varimax rotation with Kaiser normalization. The analysis was undertaken using SPSS 15.
3. This data source does not allow a distinction between universities/colleges in Norway and abroad. The questionnaire furthermore just specifies "Norwegian companies," some of which may of course be foreign-owned but located in Norway.

REFERENCES

Arnold, E., S. Kuhlmann and B. van der Meulen (2001) *A Singular Council. Evaluation of the Research Council of Norway*, Technopolis, Brigthon.

Broch, M., M. Gulbrandsen, L. Nerdrum and M. Staude (2001) *Forskningsinstituttene og næringslivet. Delrapport 1: aktørene i samspillet.* Oslo: NIFU, Skriftserie 30/2001.

—— M. Staude, and S. O. Nås (2002) *Hvem er forskningsinstituttenes næringslivskunder?*, Delrapport 2, STEP-rapport R-08, oktober.

Brofoss, K. E. and L. Nerdrum (2002) "Forskningsinstituttene og næringslivet. Delrapport 3: Bedriftenes kjøp av FoU fra instituttene," NIFU Skriftserie 23/2002.

—— M. Gulbrandsen, L. Nerdrum, and S. O. Nås (2002) "Forskningsinstituttenes betydning for FoU i næringslivet—Syntese og utfordringer", NIFU Skriftserie 26/2002.

—— and S. Slipersæter (2004) *Forskningsinstituttenes rammebetingelser for internasjonal konkurranse*, NIFU STEP skriftserie nr. 19/2004, Oslo.

Cohen, W. M. and D. A. Levinthal (1989) "Innovation and Learning: The Two Faces of R&D," *Economic Journal*, vol.99, no. 397, September, pp. 569–96.

Crow, M. and B. Bozeman (1987), "R&D Laboratory Classification and Public Policy: the Effects of Environmental Context on Laboratory Behavior," *Research Policy*, vol. 16, pp. 229–58.

Gulbrandsen, M. and H. Godoe (2008), " 'We really don't want to move, but...' Identity and Strategy in the Internationalisation of Industrial R&D," forthcoming in *Journal of Technology Transfer*.

Kaloudis, A. and P. Koch (2004) *De næringsrettede forskningsinstituttenes rolle i det fremtidige innovasjonssystemet*, NIFU STEP Rapport 4/2004, Oslo.

Larédo, P. (2003) "Six major challenges facing public intervention in higher education, science, technology and innovation," *Science and Public Policy*, vol. 30, pp. 4–12.

—— and P. Mustar (eds.) (2001) *Research and Innovation Policies in the New Global Economy. An International Comparative Analysis*, Cheltenham: Edward Elgar.

Ministry of Education and Research (2005) *Vilje til forskning*, St.meld.nr. 20, 2004–2005, White Paper, 18th March.

Nelson, R. R. (1993) "A retrospective." In Nelson, Richard R. (ed.), 1993, *National Innovation Systems: A Comparative Analysis.* New York: Oxford University Press, pp. 505–23.

——and N. Rosenberg (1993) "Technical innovation and national systems." In Nelson, R. R. (ed.), 1993, *National Innovation Systems: A Comparative Analysis.* New York: Oxford University Press, pp. 3–21.

Nerdrum, L. (1999) *The Economics of Human Capital: A Theoretical Analysis Illustrated Empirically by Norwegian Data*, Scandinavian University Press, Oslo.

——and Bo Sarpebakken (2006) "Mobility of foreign researchers in Norway", *Science and Public Policy*, vol. 33 (3), April, pp. 217–29.

Norges forskningsråd (2005) *Årsrapport 2004. Forskningsinstituttene. Delrapport for teknisk-industrielle institutter*, Oslo, juni.

——(2006) *Nytt basisfinansieringssystem for instituttsektoren. Forslag fra Norges forskningsråd, oktober 2006.*

NOU (1981) 30A, "Forskning, teknisk utvikling og industriell innovasjon" (Thulin Commission).

Research Council of Norway (2006) *Report on Science and Technology Indicators for Norway 2005.*

Skoie, H. (1990) "Økt markedsstyring eller mer langsiktighet?" ("Increased market stearing or more long-term policy?"), in Hans Skoie and Einar Ødegård (eds.), *De teknisk-industrielle forskningsinstitutter i 1990-årene*, NAVFs utredningsinstitutt, Rapport 5/90, pp. 63–73.

——(2005) *Norsk forskningspolitikk i etterkrigstiden.* Oslo: Cappelen akademisk forlag.

Slipersæter, S., K. Wendt, and B. Sarpbakken (2003) "Instituttsektoren i et internasjonalt perspektiv belyst ved FoU-statistiske data", NIFU skriftserie nr. 30/2003.

Stenstadvold, K. (1994) *Hvor ble det av forskerne etterpå? Og hva mener de i dag? En mobilitetsstudie*, SINTEF rapport STF01A 94010, oktober.

Wiig, O. and W. C. Mathisen (1994) *Instituttsektoren—mangfold og utvikling. En studie med særlig vekt på perioden 1983–91* ("*The institute sector—heterogeneity and development. A study with particular emphasis on the period 1983–91*"), Utredningsinstituttet, Rapport 5/94.

13

Industrial R&D Policy in Norway: Who Gets the Funding and What are the Effects?

Tommy H. Clausen

The public support system for R&D is a defining feature of most nations' innovation systems (Lundvall and Borrás 2005), and postwar Norway is no exception. Chapter 4 of this volume describes the evolution of industrial R&D policies in Norway from the early postwar years through the early 1990s, emphasizing the rise and decline of Norwegian policies that sought to support the development of industries characterized by high R&D intensity.

Norway's low level of R&D spending was a source of great concern for Norwegian policymakers, industry experts and academics throughout this period. Indeed, increasing industry R&D performance was a primary motive for the "national champion" policies discussed in Chapter 4. Although the *content* of this type of policy has shifted over the years, in the sense that different types of industries and technologies have been selected for support, the *objective* has remained the same: Industrial R&D policy in Norway has encouraged national champions do to more R&D.

Although Chapter 4 argues that the national champion strategy was largely discarded during the 1990s, this chapter presents data from contemporary industrial R&D policy in Norway that suggests that national champions still benefit disproportionately from R&D subsidies. This evidence suggests that at least some manifestations of the national champion strategies of earlier postwar decades persist in Norwegian R&D policy. In many respects, such persistence is a good example of the path dependency that characterizes institutional and policy development in most democratic industrial societies, as the Introduction to this volume points out.

A second objective of this chapter is using firm-level data to assess the effectiveness of contemporary technology programs in achieving three key objectives: (1) supporting the development of national champions,

(2) correcting (financial) market failures, and (3) upgrading the technological capabilities of existing industries (Blanes and Busom 2004). This chapter accordingly examines the characteristics of the firms that receive funding from the most important technology programs in Norway, and whether R&D subsidies have stimulated firms to increase their self-financed R&D efforts.

This last issue is important against the backdrop of the "Lisbon Agenda" in the European Union, adopted in 2000 for the decade ahead. Among the central goals of the Lisbon Agenda are facilitating growth, R&D and innovation in the private sector, leading to an increase in national expenditures on R&D in all EU member states to at least 3 percent of GDP.[1] The Norwegian government has largely endorsed these policy goals (NHD 2006). To what extent are its current policies consistent with these objectives?

R&D SUPPORT TO FIRMS IN NORWAY

As was noted above, Norway's public support system for industrial R&D has for much of the post-1945 period supported national champions in order to enhance the R&D intensity of large scale enterprises. Up to the 1980s policymakers implemented a "research driven" industrialization strategy where it was important to fund the research activities of large enterprises and to support emerging technologies. A main aim was the development of a more diversified industrial base that would lessen Norway's dependence on resource-based industries, such as oil and gas, energy and metals. But by the 1990's, these "national champion" policies had lost much of their support and were replaced by new policies that supported development projects in the private sector in order to solve user-specific and practical problems (Chapter 4). This chapter updates Chapter 4's analysis and examines the effects of this policy reorientation on patterns of Norwegian industrial R&D support.

Rather than focusing on the stated goals of recent Norwegian R&D policies, this chapter seeks to examine patterns of resource allocation within current technology programs. Our analysis draws on a classification by Blanes and Busom (2004)—updated in this chapter for the Norwegian case—in distinguishing among the following policy objectives for Norwegian technology programs: (1) correcting market failures, (2) developing national champions, and (3) upgrading industry technological capabilities. This taxonomy of program goals will aid our assessment of a potential mismatch between policy documents and programmatic design and practice.

Policy objectives pursued by technology programs

The objectives of R&D programs summarized above should determine the allocation of program funds to specific programs, the distribution of money across industries, and the screening rules used to select firms and projects for funding. These screening rules and policy objectives will determine firm "participation" in a technology program, where "participation" is defined as meaning that a firm is subsidized by a technology program. We discuss these policy objectives in more detail immediately below, in order to motivate our empirical analysis.

Correcting market failures

R&D policy interventions have traditionally been justified by the potentially large gap between the private and the social rate of return to R&D investments. Large gaps between the private and the social rate of return to R&D arise due to incomplete appropriability and spillovers between firms (Arrow 1962). Firms will not invest their own funds in R&D because the knowledge and information gained from these investments will "spill over" to competitors. Classical market failures are thus something that applies to all R&D active firms—or potentially R&D active firms—regardless of size, age, industry affiliation, etc. By subsidizing innovative activity in the private sector, especially large research projects close to basic science, policymakers can make novel innovation projects viable (Nelson 1959; Arrow 1962). Norwegian technology programs that seek to correct these market failures should thus allocate funds to more uncertain projects close to basic science in the private sector. We analyze program data below in this Chapter to determine whether the types of activities supported by Norwegian industrial R&D subsidy programs are consistent with a strategy of correcting market failures.

Another type of market failure results from capital market imperfections that create a "wedge" between the rate of return required by a firm investing its own funds in a R&D project and the return required by external investors (see Hall 2002 for a review). This type of market failure means that some innovations will fail to be developed because the cost of external capital is too high, especially for small, young and cash-constrained firms (Hall 2002). This type of market failure argument differs from the Nelson-Arrow one, and suggests that policies to address the effects of financial market imperfections on R&D investment should target small and young firms. Technology programs whose main aim is to correct financial market failures should also be less inclined to fund firms with established innovation capabilities, instead seeking to support innovation within firms that in absence of the subsidy would not spend money on innovation. Our examination of the characteristics of firms that receive Norwegian R&D subsidies will shed light on the importance of this objective within contemporary R&D policy.

Promoting national champions

A second and contrasting objective in R&D and innovation policy is fostering national champions. Historically, this objective has been an important influence on industrial R&D policy in Norway. Wicken's (2000) historical analysis of Norwegian R&D policy concluded that economics of scale and scope were important characteristics of firms supported by Norwegian industrial R&D policies. Large, diversified and financially strong industrial R&D performers, many with a group membership,[2] were targeted by these Norwegian R&D subsidies through much of the postwar period. Norwegian R&D subsidies that seek to promote national champions thus should support larger, diversified domestic firms, many of which have innovation capabilities and are exporters.

Technological upgrading

A third goal of industrial R&D policy is technological upgrading of technologically lagging firms in traditional industries, which tend to be older and employ a large fraction of the workforce. R&D subsidy programs that promote such technological upgrading should support larger and older firms according to the classification developed by Blanes and Busom (2004). It is on the other hand questionable whether technological lagging firms in traditional industries in Norway can be described as large. According Wicken's analysis in Chapter 2 (this volume) the Norwegian economy consists of a "small-scale decentralized" layer dominated by small and family owned business enterprises in traditional industries. At least for our Norwegian case, there are as such strong reasons to believe that technology programs pursuing "technological upgrading" will support smaller and older firms. We depart from the classification developed by Blanes and Busom (2004) in this regard. Diversified firms and companies that is a part of an enterprise group will be less likely to obtain subsidies from "technology upgrading" programs because such firms have access to resources and knowledge at the group level. Fast-growing firms and successful exporters, both of which are less likely to be technological laggards, also are less likely to receive subsidies from programs aimed at technological upgrading within lagging firms in traditional industries.

Drawing on Blanes and Busom (2004), Table 13.1 summarizes the expected relationships between firm characteristics and the selection rules and policy

Table 13.1. R&D program policy objective and their expected selection rules

	Financial market failures	National champions	Technological upgrading
Prior innovation activity	–	+	–
Firm age	–	?	+
Firm size	–	+	–
Foreign ownership	?	–	?
Group	–	+	–
Diversified	–	+	–
Export activity	–	+	–
Growth	?	+	–

objectives of R&D programs. The technology program objectives listed in Table 13.1 are linked to firm characteristics, rather than the types of R&D being supported by subsidies. As such, only the "financial market failure," as opposed to market failures linked to spillovers and nonappropriability, is listed in the table as a program objective that can be linked to recipient firm characteristics.

In Table 13.1 the question marks (?) indicate that the relationship between specific firm characteristics and R&D subsidy objectives is indeterminate. A "+" sign signals that larger values for the variable in question will enhance the probability of being subsidized. Firm size will be positively (and significantly) related to receiving subsidies if technology programs are primarily focused on supporting national champions. A negative sign in Table 13.1 indicates that larger values for the variable in question will decrease the probability of being subsidized, e.g. the coefficient of firm size will be significantly negative if technology programs aim to correct (financial) market failures, since these are hypothesized to be more severe among small and start-up firms.

In the section below we describe and discuss the five most important technology programs in Norway. We also provide some descriptive statistics that illustrate the importance of industrial R&D policies and public funding for different types of firms and industries in the Norwegian context. We are especially interested in the distribution of subsidies among industries and firm size classes. The programs are the following:

1. SND: the State Industrial and Development Fund, known since 2004 as "Innovation Norway." SND was established in 1993 by merging the "Industry fund" (Industrifondet, established in 1979), the "SME fund" (Småbedriftsfondet, founded in 1965) the "Development fund" (Distriktenes Ut byggingsfond, established in 1961) and the Industry

Bank (Industribanken, established in 1936. The SND supports innovation and growth in industry.

2. NRC: the Norwegian Research Council. The NRC was established in 1993 in a merger of 5 different research councils. The NRC provides R&D subsidies to firms in the form of research grants with few restrictions on use. Most NRC subsidies are allocated among firms on a competitive basis, based on reviews by industry and scientific experts.
3. Ministries: Firms also may receive support directly from Governmental Ministries. Little is known about the actual role of Ministries in the public support system for R&D and innovation in Norway. Unfortunately, we have no information on the distribution among specific ministries, the policy objectives of which may differ, of R&D subsidies.
4. European Union (EU): Norwegian firms can receive subsidies from the European Union, especially through the Framework program. Little is known about the role of EU subsidies in the Norwegian innovation system.
5. FUNN: The predecessor to the current R&D tax credit policy, SkatteFUNN. FUNN, which was established in 2001, was operated as a subsidy scheme by the NRC, which stated that the program's goal was to provide firms with an "easy and un-bureaucratic" access to public R&D financing. The program was also especially directed towards encouraging university–industry interactions.

Rye (2002) provided a historical review of twelve important evaluation studies of Norway's public R&D subsidy, including the NRC and SND (and their predecessors). In these evaluation studies, funding from the SND and NRC occupy a key role. Rye (2002) argues that Norwegian R&D subsidy programs fall into two broad categories, and subsidize different types of projects. The first type of technology program, primarily the NRC and its predecessors, supports projects within industry that are uncertain, have a high research component, and are "far from the market," i.e. not likely to be commercialized immediately. A second type of R&D program, primarily those administered by SND, supports less uncertain projects within firms that have a high development component—these are "closer to the market" and commercialization phase.

This distinction between "research" and "development" subsidies links R&D support to the characteristics of the R&D, rather than the characteristics of the recipient firm. Research subsidies for private firms can be characterized as policies designed to correct the Arrow-Nelson *classical market failures* associated with knowledge spillovers and nonappropriability. The existence of these *classical market failures* may not be linked to measurable firm characteristics. For example, a Norwegian R&D subsidy program might support the development of national champions and seek to correct classical market failures at the same time.

Rye's (2002) review focused primarily on SND and NRC programs, and predated a large-scale reorganization of Norway's R&D subsidy programs in the 1990s. In addition, a number of technology programs not covered by Rye, such as FUNN and EU, have come to play an important role as sources of R&D subsidies for Norwegian firms. Rye's analysis also did not discuss the role of R&D subsidies from Norwegian Ministries. This chapter's analysis includes all of these programs not examined by Rye, but classifies them as supporting R&D activities that either belong to the "research" or "development" categories, i.e. far from or near to commercialization. My classification of these R&D subsidy programs relies on several sources of information and analysis. Discussions with Norwegian technology historians and policymakers supports a categorization of Ministry R&D subsidy programs as targeting "close to the market" activities, similarly to the SND subsidies examined by Rye. These information sources led me to classify EU and FUNN R&D subsidies as primarily supporting "research" activities that are "far from the market," similar to those supported by the NRC. This interview information was corroborated by the results of a factor analysis that is presented in the appendix to this chapter. In summary, then, the analysis below treats R&D subsidies from NRC, FUNN, and the EU as directed primarily to supporting "research" projects, while the R&D subsidies from Ministries and SND are classified as supporting "development" projects in industry.

Who gets the funding?

Official statistics do not reveal the distribution of R&D subsidies among firms from the technology programs discussed above, but Table 13.2 provides estimates of the size of the R&D subsidies provided by each of them. The calculations are based on a R&D survey conducted at the enterprise level in Norway for a representative sample of firms with 10 employees or more. This survey, which is discussed in more detail in section 3.1, is the source of the data used in the official Norwegian R&D statistics.

Table 13.2 shows that public R&D subsidies for Norwegian industry amounted to 944 million NOK in 2001 (about 118 million Euros). Since total internal R&D spending from public and private sources in the private sector in 2001 was 12,614 million NOK, the data in Table 13.2 suggest that R&D

Table 13.2. R&D and subsidy statistics in million NOK for the private sector in Norway[3]

	Sum in Mill. NOK.
Subsidies from EU	101
Subsidies from Ministries	514
Subsidies from SND	85
Subsidies from FUNN	42
Subsidies from NRC	202
Total public funding	944
Total internal R&D in private sector	12,614
N = 3899/11832 (weighted)	

Source: Author's calculation based upon weighted R&D survey data for 2001 (data discussed below).

subsidies financed slightly more than 7 percent of internal R&D activity in Norwegian industry. The amount of R&D funding provided by the five programs summarized earlier differs greatly. Norwegian government Ministries were by far the largest source of direct public R&D funding for Norwegian firms in 2001, accounting for well over 50 percent of total R&D funding, followed by the NRC, EU, SND and FUNN.

Table 13.3 presents data on the distribution of R&D expenditures and public R&D funding among Norwegian industries.[4] The table also contains data on industry-level employment the 2.digit NACE level (as well as each industry's share of total Norwegian employment), in order to provide some sense of the relative economic importance of the industries receiving public R&D funding.[5]

The figures for total R&D in Table 13.3 are based on the responses by Norwegian firms in the R&D survey, aggregated to the 2-digit NACE level. A few industrial sectors are responsible for most of the internal R&D spending in Norway, most notably NACE 72, 32, 24, and 29 (respectively, computers, radio equipment, chemicals, and machinery), which account for about 45 percent of total R&D and about 11 percent of employment. These four sectors also received approximately 63 percent of total public R&D subsidies from the 5 programs described above. Firms in the machinery industry, NACE 29, were responsible for 7.4 percent of total internal R&D, received 37 percent of public R&D subsidies, and accounted for only 3 percent of employment. Public R&D subsidies for Norwegian industry are concentrated in a few sectors that account for a relatively small share of total Norwegian employment.

Table 13.3. Distribution of R&D, public R&D funding, and employment at the industry level

	Internal R&D in Mill. NOK.	% of total internal R&D	Public R&D in Mill. NOK	% of total public R&D	Number of employees	% of total employment
Fishing & fish farming (5)	272.5	2.2	15.1	1.6	3,799	0.5
Mining of coal and lignite (10)	0	0.0	0	0.0	273	0.0
Petroleum and natural gas (11)	754.5	6.0	22	2.3	30,204	4.3
Mining of metal ores (13)	5.5	0.0	0.3	0.0	360	0.1
Other mining (14)	12.8	0.1	4.6	0.5	2,441	0.4
Food prod. and beverages (15)	369	2.9	23.2	2.5	52,653	7.6
Tobacco (16)	0	0.0	0	0.0	480	0.1
Textiles (17)	25.5	0.2	0.5	0.1	3,961	0.6
Apparel & dying (18)	35.5	0.3	1.7	0.2	1,090	0.2
Leather products (19)	6.6	0.1	0.6	0.1	359	0.1
Wood products (20)	56.6	0.4	7.5	0.8	13,178	1.9
Pulp & paper (21)	159.9	1.3	0.9	0.1	8,503	1.2
Publishing & printing (22)	32.1	0.3	0.5	0.1	28,641	4.1
Chemicals (24)	1057.8	8.4	24.8	2.6	23,182	3.3
Rubber & plastics (25)	57.9	0.5	7.7	0.8	5,783	0.8
Non-metallic minerals (26)	66.8	0.5	3.9	0.4	9,001	1.3
Basic metals (27)	405.7	3.2	2.2	0.2	12,981	1.9
Fabricated metal prod. (28)	140.9	1.1	12.6	1.3	16,648	2.4
Machinery & equip. N.E.C (29)	934.3	7.4	351.6	37.3	20,761	3.0
Office machinery & comp. (30)	67	0.5	2.5	0.3	367	0.1
Electric machinery & app. (31)	392	3.1	11.6	1.2	7,585	1.1
Radio, television, com. (32)	1742	13.8	35.2	3.7	6,161	0.9
Medical & optical instr. (33)	511.3	4.1	32.7	3.5	5,652	0.8
Motor vehicles (34)	417.2	3.3	6.6	0.7	5,555	0.8
Other transportation equip. (35)	274	2.2	16.2	1.7	30,823	4.4
Furniture (36)	128.3	1.0	5	0.5	10,541	1.5
Recycling (37)	28.3	0.2	2	0.2	1,001	0.1
Electricity, gas and water (40)	89.7	0.7	6	0.6	16,420	2.4
Coll. and distrib. of water (41)	0.5	0.0	0	0.0	247	0.0
Construction (45)	278.5	2.2	14.2	1.5	104,526	15.0
Wholesale trade (51)	345.8	2.7	35.7	3.8	74,061	10.6
Land transport (60)	12.9	0.1	7.3	0.8	32,180	4.6
Water transport (61)	40.4	0.3	1.6	0.2	23,663	3.4
Air transport (62)	4.2	0.0	0	0.0	14,330	2.1
Auxiliary transport act. (63)	27.4	0.2	0.6	0.1	22,863	3.3
Post & telecommunication (64)	693.8	5.5	20.6	2.1	12,145	1.7
Financial intermediation (65)	262.2	2.1	7.5	0.8	31,459	4.5
Insurance & pension (66)	93.8	0.7	0.2	0.0	8,533	1.2
Act. Auxiliary to nace 65 (67)	98.9	0.8	0	0.0	3,331	0.5
Computers and related (72)	1885.6	14.9	177.6	18.8	29,158	4.2
Research and development (73)	113	0.9	20	2.1	166	0.0
Other business activities (74)	712.9	5.7	61	6.5	20,564	3.0
Total	12613.6	100.0	943.8	100.0	695,629	100

Source: Own calculation based upon weighted R&D survey data for 2001 (data discussed below).

Table 13.4. Actual and relative distribution of total R&D and employment according to size classes

	Total R&D in Mill. NOK	% of total R&D	Public R&D in Mill. NOK.	% of public R&D	Number of employees	% of total employment
10–30 emp.	1,920	15.2	140.2	14.9	133,953	19.3
31–50 emp.	1,103	8.7	70	7.4	60,420	8.7
51–100 emp.	1,033	8.2	60.6	6.4	76,629	11.0
101–300 emp.	2,510	19.9	209.3	22.2	117,364	16.9
301–500 emp.	745.9	5.9	31.4	3.3	62,244	8.9
501 > emp.	5,302	42.0	431.9	45.8	245,019	35.2
SUM	12613.9	100	943.4	100	695,629	100

Source: Own calculation based upon weighted R&D survey data for 2001 (data discussed below).

Data on the distribution of Norwegian public R&D subsidies among firms of different sizes are presented in Table 13.4. Most of the firm-size classes exhibit a high correlation between their share of overall R&D, their share of public R&D subsidies, and their share of employment. For instance, firms with between ten and thirty employees performed 15 percent of total R&D in the private sector, received 14.9 percent of public funding, and accounted for about 19 percent of employment. This characterization is slightly less accurate when applied to the small number of very large (greater than 500 employees) firms in Norway, which perform 42 percent of total industrial R&D, received 45.8 percent of total R&D subsidies, and accounted for slightly more than 35 percent of total employment. The weaker correlation between R&D and employment shares among the very largest firms is not surprising, since these firms tend to operate in highly capital-intensive industries.

What are the effects?

How if at all do Norwegian R&D subsidies influence the firms receiving funding? A large commercial research institute sector in Norway has evaluated the "effects" of R&D subsidies on firm performance, concluding that NRC subsidies to private firms have a high private and social rate of return (Hervik 2004; Hervik *et al.* 2004, 2005). Evaluations of the other major technology program in Norway, SND, reached similar conclusions (Madsen and Brastad 2005; 2006; Hatling *et al.* 2000; Hauknes *et al.* 2000). No systematic evaluation has been published, however, for the largest single set of R&D subsidy programs within Norway, those overseen by Norwegian government ministries.

In her review of the most important evaluation studies in Norway, Rye (2002) found that both "research subsidies" and "development subsides" were associated with a high additionality rating,[6] although the first type of R&D support was more highly valued by respondents. The overall conclusion that seems to emerge from Rye's work is that all types of R&D subsidies stimulate additional self-financed R&D investment from Norwegian firms.

Most of these evaluation studies, however, are based on self-assessments in which managers of subsidized firms are asked to rate whether the subsidized project would have been initiated without a subsidy. This source of data may be biased, since respondents are likely to report a high payoff in order to make the R&D program look successful, and thus, enhance their prospects for further funding in the future (Klette *et al.* 2000). Few studies have used more advanced evaluation designs that compare subsidized firms with a control group of unsubsidized firms.

Klette and Møen (1998) represents an exception to this characterization of the Norwegian evaluation literature. They used a panel database of high-tech business units and a quasi-experimental control group design in order to evaluate whether subsidies stimulated or substituted for private R&D spending. This study concluded that subsidies neither stimulated nor substituted for private R&D investment in Norwegian industry. Although Klette and Møen did examine different types of technology programs, including subsidies from ministries, they found no evidence that the effects of these different technology programs on firms' R&D investment differed. Since the Klette and Møen analysis was carried out, the public support system for R&D and innovation in Norway has been reorganized, as was noted above. The effects of Norway's restructured R&D support programs merit a new analysis, as do the effects of ministry R&D subsidies, which have received much less attention.

This analysis examines the effects of Norwegian R&D subsidies with an econometric quasi-experimental design that compares the behavior of subsidized firms with that of a control group of unsubsidized firms. The study employs objective performance measures and a sample of firms that is representative of most Norwegian industries. In order to minimize bias in the statistical analysis, we use instrumental-variable (IV) regression techniques.[7] Most R&D policy evaluations in Norway have not "corrected" for the possibility that firms either self-select or are non-randomly picked by policymakers to participate in technology programs. As Klette *et al.* (2000) point out, these characteristics of R&D subsidy programs can lead to biased conclusions about the effectiveness of the subsidy being evaluated unless statistical techniques are used to control for selection effects.

METHOD, DATA, AND VARIABLES

The first part of the analysis uses logistic regression analysis in order to estimate firms' propensity to receive a subsidy. In this regard we are only looking at the existence of a subsidy, and not the total value or size of the subsidy, which can introduce some "noise" if a small number of firms each receive very large subsidies. This skewed distribution may well characterize the distribution of Ministry R&D subsidies. Our analysis is done separately for each of the five technology programs discussed earlier, enabling us to estimate differences in the objectives and the effects on firms' behavior of different Norwegian R&D subsidy programs. This part of the analysis also examines whether technology programs in Norway aim to (1) correct financial market failures, (2) develop national champions, or (3) support technological upgrading of existing industry.

The second part of the analysis estimates the effects of R&D subsidies on recipient firms' R&D, focusing on firms with self-financed R&D activities. This analysis probes the extent of "additionality" in the effects of R&D subsidies within Norwegian industry, a widely employed measure of program success. This part of the analysis incorporates the value of individual firms' subsidies in the econometric analysis.

Data

The statistical analysis draws on a Norwegian survey, conducted at the enterprise level, that combined the third version of the Community Innovation Survey (CIS 3) and an R&D survey. Although the survey was implemented in 2002 by Statistics Norway, the questions refer mainly to the 1999–2001 time period, while the data on R&D subsidies refer to 2001. This combined survey contains information about firms' innovation activities (based on the CIS survey) and questions about the financing of their R&D activities. The questionnaire was sent to a representative sample of Norwegian firms with 10 employees or more, and returned by 3899 firms, yielding a response rate of 93 percent. The variables we use in the analysis are defined in Table 13.5 below. We include the "firm characteristics" variables in this analysis in order to allow us to test the importance of the different policy objectives that I discussed earlier within Norwegian R&D subsidy programs and summarized in Table 13.1. Industrial sector dummy variables (based on 2.digit NACE) are also included in the analysis, although results are not reported due to space limitations.

Table 13.5. Definition of the variables used in the analysis

Variable	Definition
Size	Number of employees in 1999. Used as log of (employees +1).
Age	Number of years from when the firm was established. Used as log of (age +1).
Group	Binary variable where value 1 indicates that a firm belongs to a group through the following survey question: Is the enterprise a part of a group?
Foreign ownership	Binary variable where value 1 indicates that a firm is foreign owned.
Diversification	Binary variable where value 1 indicates that a firm is diversified.
Export intensity	Exports in 1999/turnover in 1999.
Patent	Binary variable where "1" indicates presence of one or more valid patents in 2001.
Research subsidy	Amount of subsidies a firm receives from NRC, EU and FUNN in 2001. Used in logarithmic form defined as the log of (subsidy +1).
Development subsidy	Amount of subsidies a firm receives from SND and Ministries in 2001. Used in logarithmic form defined as the log of (subsidy +1).
Private internal R&D	Private expenditure on internal R&D in 2001 (total internal R&D—subsidy). Used in logarithmic form defined as the log of (private internal R&D).
Research expenditure	Amount of funds allocated to research activities (applied & basic research) in 2001. Used in logarithmic form defined as log of (research expenditure +1).
Development expenditure	Amount of development expenditure in 2001 as provided by the firm manager. Used in logarithmic form defined as log of (development expenditure +1).
Growth	Growth in the number of employees from 1999 to 2001.

ANALYSIS

The first part of this section's econometric analysis utilizes a series of logistic regressions to estimate the firm-level probability of receiving a subsidy from a specific program. The second part of the analysis examines the "effects" on firms' self-financed R&D spending of the subsidies.

Detecting innovation policy practice

This part of the analysis estimates the influence of firm characteristics on the probability of receiving R&D subsidies from each of our five technology programs. Among other things, this analysis will enable an assessment of the importance of the three key objectives that were discussed earlier for each of the five most important technology programs in Norway: (1) correcting financial market failures, (2) supporting national champions, or (3) encouraging technological upgrading in existing industries (Table 13.1).

Table 13.6. Determinants of receipt of a research subsidy from NRC, EU, or FUNN

Variables	NRC technology program		EU technology program		FUNN Technology program	
	Beta	Odds ratio	Beta	Odds ratio	Beta	Odds ratio
Size	0.46***	1.58	0.65***	1.91	0.57***	1.77
Age	−0.12	0.89	−0.42**	0.66	0.05	1.05
Group	−0.07	0.93	−0.51	0.60	−0.48	0.62
Patent	1.56***	4.78	1.77***	5.90	0.72***	2.05
Export activity	0.56*	1.75	1.04**	2.84	1.28***	3.61
Diversification	−0.25	0.78	−0.30	0.74	−0.15	0.86
Foreign ownership	−0.29	0.75	0.51	1.66	−0.06	0.94
Growth	−0.31	0.74	−0.01	0.99	0.002	1.00
Constant	−5.05	0.001	−6.02	0.002	−6.77	0.001
	$R^2 = 0.21/N = 3{,}633$		$R^2 = 0.28/N = 3{,}633$		$R^2 = 0.22/N = 3{,}633$	

***sig. at the 0.01 level, **sig. at the 0.05 level, and *sig. at the 0.1 level.

We first estimate the firm-level probability of receiving a subsidy from technology programs that support "research projects" in the private sector. As previously discussed, NRC, EU, and FUNN technology programs may be the primary sources of this type of funding in the Norwegian innovation system. The results in Table 13.6 indicate that the largest firms in Norway are the most likely to receive R&D subsidies from the NRC, EU, and FUNN technology programs. Export-oriented firms and companies with established innovation capabilities (measured by the patent variable) are also significantly more likely to attract a subsidy from these programs. Based on the classification of policy objectives in Table 13.1, it seems that NRC, EU, and FUNN follow a policy objective to support the development of national champions, as firm size, patenting and export activity are positively associated with receiving subsidies from these technology programs.

EU funding is however a possible exception as we find that younger companies are significantly more likely to receive subsidies from EU programs. Although one should not overstate this finding, it is one of the only pieces of evidence of a "financial market-failure" objective for any of the five major Norwegian R&D subsidy programs. It is interesting that EU R&D subsidies, a policy area over which Norwegian policymakers have little control, at least partly seems to correct such market failures in the Norwegian innovation system. Nevertheless, as was noted above, firm size, export and innovation activity also are positive associated with the receipt by Norwegian firms of EU R&D subsidies, and all of these firm-level characteristics are associated with "national champion" objectives in R&D subsidy programs.

Table 13.7. Determinants of getting access to a development subsidy from SND or ministries

Variables	SND technology program		Ministry technology program	
	Beta	Odds ratio	Beta	Odds ratio
Size	−0.16	0.85	0.38***	1.47
Age	0.10	1.10	−0.28	0.76
Group	0.47*	1.60	−0.19	0.83
Patent	1.06***	2.88	0.45	1.56
Export activity	0.70**	2.02	1.05**	2.87
Diversification	−0.39	0.68	−0.20	0.82
Foreign ownership	−0.90**	0.41	−1.91***	0.15
Growth	−0.03	0.97	−0.08	0.92
Constant	−3.37	0.02	−4.45	0.01
	$R^2 = 0.11/N = 3,633$		$R^2 = 0.13/N = 3,633$	

*** sig. at the 0.01 level, ** sig. at the 0.05 level, and * sig. at the 0.1 level.

The results displayed in Table 13.6 suggest that the EU, NRC, and FUNN programs, aim to support the development of national champions by subsidizing research projects in industry. Hence, there seems to be a rather strong "correlation" between the objective to support national champions and the ambition to correct classical market failures among these technology programs. Although this "correlation" may at first sight seem odd, it can arguably be explained by the argument that mainly large firms have the complementary resources to launch large scale basic R&D projects (Rosenberg 1996). Large-scale projects close to basic science are also put forth as a type of inventive activity that suffers from classical market failures (Nelson 1959). An important complementary resource may in this case be possession of developed innovation capabilities.

Table 13.7 presents the results of regression specifications that estimate the firm-level propensity to get access to a "development subsidy" from SND or ministry programs. Firms that are members of enterprise groups, firms with some history of patenting, and companies that export more intensively are significantly more likely to receive subsidies from the SND program, in contrast to companies with a foreign ownership. The SND program, which is widely characterized as subsidizing "development" activities within recipient firms, thus seems to focus on the development of national champions. It is nevertheless interesting that firm size is not a significant predictor within the SND specification, suggesting that smaller as well as larger Norwegian firms do receive R&D subsidies from SND.

The results in Table 13.7 on the firm-level factors that result in R&D subsidies from Norwegian government ministries indicate that foreign-owned firms and smaller enterprises are less likely to receive subsidies from this source. Export-intensive companies, however, are more likely to receive Ministry R&D subsidies. These econometric results are consistent with the presence of "national champion" objectives within the R&D subsidy programs of Norwegian ministries. Other than firm size, foreign ownership, and export activity, however, few variables are significant in predicting the receipt by a given firm of a ministry R&D subsidy. As such, it is difficult to classify the objectives of Ministry R&D subsidy programs according to the policy objectives summarized in Table 13.1. This result is hardly surprising, in view of the heterogeneity in goals, political constituencies, and industry "clients" among the ministries responsible for these R&D subsidies. The role of R&D support from Ministries to firms in Norway continues to pose a research challenge.

What are the effects?

As was noted earlier, IV regressions were used to estimate the effects of "research" and "development" subsidies on private R&D spending. The results are reported in Table 13.8 below. Econometric details, such as overidentification tests and discussion of instruments, are available in a working paper (Clausen 2007).[8]

The pattern in Table 13.8 is clear: "Research subsidies" have significant and positive impacts on the self-financed R&D spending of firms, while "development subsides" have a negative and significant influence on R&D spending. The results in the rightmost columns of Table 13.8 indicate that "research subsidies" stimulate privately financed research spending. By comparison, "development subsidies" appear to "crowd out" privately funded spending on development, although these subsidies do not appear to significantly affect privately funded "research" spending (the coefficient is negative but not statistically significant).

Summary and discussion of the results

This chapter has shown that the most important Norwegian R&D subsidy programs support the development of national champions. There are few if any signs that these subsidy programs in Norway aim to correct financial market failures among small and young firms. A possible exception to this

Table 13.8. The effect of R&D subsidies on private R&D spending

	Private R&D exp. (log)		Development expenditure (log)		Research expenditure (log)	
	Coefficient	Std. err	Coefficient	Std. Err	Coefficient	Std. Err
Development subsidy (log)	−.662540***	.24788	−.667775**	.28923	−.246018	.497780
Research subsidy (log)	.364787*	.198279	.145737	.231352	1.34681***	.398163
Size (log)	.421912***	.117426	.50032***	.137012	.031835	.235802
Age (log)	−.052154	.077373	−.01680	.090278	−.089231	.155372
Group	.291447*	.16708	.278321	.194949	.234967	.335512
Foreign ownership	−.201416	.21513	−.452987*	.251015	.142110	.432003
Diversified	−.188462	.172999	−.157434	.201856	.028455	.347398
Patent	.538703***	.205399	.700167***	.23966	.084816	.412462
Export intensity	.893588***	.229159	.98459***	.267383	.557149	.460172
Growth	.008167***	.002431	.008737***	.002836	−.001214	.004881
Constant	5.96974	.374364	5.52215	.43681	.970285	.751759
R^2	–		–		–	
N	1,019		1,019		1,019	

*sig. at the 0.1 level, **sig. at the 0.05 level, and ***sig. at the 0.01 level.

characterization is EU financing, which appears to support younger firms to a greater extent.

A closer look at the firm characteristics that predict the receipt by firms of R&D subsidies from the most important technology programs in Norway reveals that export-intensive companies are significantly more likely to receive R&D subsidies from all of our five technology programs. Subsidizing R&D activity in the export active segment of Norwegian industry thus could be portrayed as a continuation of traditional policies of supporting Norwegian firms in international markets.

Firms with some level of innovation capabilities, based on their previous patenting activity, were significantly more likely to access public R&D funding from four out of five technology programs. This suggests that Norway's public support system for R&D tend to be directed toward more innovative firms, excluding firms without previous innovation activity. Are non-innovative Norwegian firms in fact denied access to public R&D funding, or do such firms not apply for public R&D funding because they are not innovative? The first scenario would in fact suggest the existence of a self-reinforcing mechanism in the Norwegian public support system for R&D where innovative firms receive subsidies and are able to initiate more projects, raising their chances of receiving subsidies in subsequent years. The second scenario implies the existence of a large business segment in Norway where firms devote few resources to innovation and as such do not apply for public R&D funding. These are important topics for further research.

The largest firms in Norway are more likely to receive R&D subsidies. With the possible exception for the SND program, large firms are more likely to receive R&D subsidies from the most important technology programs in Norway compared to smaller firms. Larger firms seem to be especially successful in gaining access to such subsidies, keeping in mind that the analysis does not measure the influence of firm size on the scale of the subsidies that they receive. The overall results suggest that "barriers to entry" may limit the participation by smaller and younger firms in the R&D subsidy programs of the Norwegian Innovation System. The limited presence of these firms in Norwegian (as opposed to EU) R&D subsidy programs may reflect the fact that technology programs in Norway have often sought to support the development of national champions. Another "barrier to entry" can also exist, however, that has little to do with the goals of Norwegian R&D subsidy programs. If most young and small companies simply do not innovate, then their inability to launch new R&D projects may limit their access to R&D subsidies.

One recent innovation policy initiative that is designed to improve support for small-firm R&D is SkatteFUNN, a tax credit scheme that granted tax credits worth roughly 500 million NOK to finance R&D projects in the private sector (SSB, 2008) in 2006. Approximately 60 percent of these tax credits were allocated to companies with less than fifty employees. The focus of SkatteFUNN on small firms thus appears to differ from that of most other Norwegian R&D subsidy programs (the EU funding being a possible exception, because of its support for smaller firms). A recent evaluation show that SkatteFUNN stimulates industrial enterprises to do more R&D, and that the program is especially beneficial to SMEs without prior experience in doing R&D (Cappelen *et al.* 2008; Alsos *et al.* 2008).

Does the use by the Norwegian government of five separate R&D subsidy programs (keeping in mind that we treat as a single program the R&D subsidies of Norwegian ministries, which in fact includes many heterogeneous programs) reflect diversity in the goals or targets of these programs? In fact, this chapter's analysis suggests that this array of R&D subsidy programs, which support both "development" and "research" projects within firms, focus on a relatively homogeneous group of firms. Export-oriented, innovative and large firms are significantly more likely to receive R&D subsidies from all five technology programs, with the exception of the EU programs (as I noted above, SkatteFUNN, which is not included in the econometric analysis, also may now provide additional support for smaller firms). Norway's "portfolio" of R&D

subsidy programs thus is relatively undiversified, and seems largely focused on developing national champions. Moreover, the tendency for "national champion" programs to subsidize development (rather than research) expenditures within firms may result in the substitution of private for public funds within industry, rather than achieving "additionality."

This evidence suggests that Norwegian policy departs from some recommendations in the Lisbon Agenda. With the exception of EU subsidies, young firms tend to receive less support from Norwegian R&D subsidy programs. Firms lacking established innovative capabilities and companies without an "(international) growth potential" also receive little support from these technology programs. As such, these Norwegian R&D subsidy programs have limited potential to transform non-innovative companies into "innovating" and "high performing" firms.

Norway's R&D subsidy programs generally do not seem to assign a high priority to correcting financial market failures, since these programs provide limited support to small and young firms. Nevertheless, the Norwegian R&D subsidy programs that support "research" rather than "development" activities, such as the NRC and EU programs, stimulate additional firm-financed R&D investment. The evidence and analysis of this chapter suggests that Norwegian R&D subsidy programs primarily support large, successful firms, and thus may support R&D and innovation within the industries in which Norwegian firms have long excelled. At the same time, however, the relative lack of subsidies for smaller, younger firms may reduce the diversity and "radicalness" of technological alternatives being pursued by Norwegian enterprises.

This analysis is limited to firms with ten or more employees as of 2001. Most firms in Norway have fewer than ten employees, and we therefore lack information on the effects of R&D subsidies on firms in this size class within Norwegian industry. Another promising area for future research involves the analysis of this "very small firm" segment of Norwegian industry to better understand the effects of R&D subsidies on their R&D and innovation-related activities.

CONCLUSION

This chapter has examined the characteristics of firms that receive R&D subsidies from the five most important technology programs in Norway, as well as the effects of these subsidies on firm-financed R&D investment. The analysis showed that Norwegian R&D subsidies primarily support large companies with a strong export orientation and innovation capabilities. Firms within this segment of the firm population receive both "development" and "research" subsidies from programs that appear to be aimed primarily at supporting the development of national champions.

Our analysis suggests that this area of Norwegian innovation policy is somewhat inconsistent with the recommendations of the Lisbon Agenda. Rather than focusing almost exclusively on relatively large "national champions," Norwegian innovation policy should give greater consideration to supporting innovation in firms without established innovative capabilities. Transforming such companies into innovative firms could stimulate growth and innovation in the Norwegian Innovation System by increasing the diversity of innovation strategies within Norwegian industry.

The results also suggest that subsidies for "research" within firms increase firm-financed R&D spending. Subsidizing research activity in areas characterized by a large gap between the private and social rate of return to investments in innovation stimulates companies to spend more on R&D, especially research (Nelson 1959; Arrow 1962). Recipients of research subsidies are large, export-oriented companies with prior experience in innovation, and therefore may possess the necessary complementary resources that enable the creation and implementation of large-scale R&D projects close to basic science (Rosenberg 1996). One implication of this analysis is that Norwegian R&D subsidy programs should increase subsidies for research, as opposed to development, within such firms, in order to increase R&D spending in Norwegian industry. As public resources allocated to development projects within large successful companies appear to substitute for private R&D spending, such resources might be used more efficiently by supporting R&D within small and young firms. Such Industrial R&D policy might increase the number of small and young firms with an approach to innovation in the Norwegian economy, thereby stimulating technological diversity and, conceivably, working against the systemic "lock-in" that Narula (2002) claims has been a characteristic of the Norwegian innovation system.

APPENDIX

We have conducted a principal components factor analysis with varimax rotation in order to analyze whether there are some similarities between R&D projects subsidized by technology programs in Norway. The analysis is motivated by Rye's (2002) review. In this review R&D subsidies from SND were deemed to be allocated to firms with projects "close to the market" (or development projects) while subsidies from

Table 13.A1. Factor loadings among binary R&D subsidies (rotated solution)

Variables	Factor loadings	Factor loadings
NRC subsidy	0.73	0.22
EU subsidy	0.75	−0.01
Ministry subsidy	0.15	0.66
SND subsidy	−0.01	0.81
FUNN subsidy	0.64	0.04
Cumulative proportion of variance explained	33%	53%
N = 3899	First factor	Second factor

NRC were distributed to companies with projects "far from the market" (or research projects). In Table 13.A1, results from the factor analysis are reported.

According to the reported results, two underlying latent factors are identified. The subsidies allocated by FUNN, NRC, and EU all load high on the first factor. NRC is the reference indicator in this regard. Because the NRC program has been found to support firm projects "far from the market", other technology programs grouped together with NRC share this characteristic. The SND and Ministry subsidy variables load high on the second factor. This finding suggests that subsidies from these two technology programs are allocated to firm projects "close to the market."

NOTES

I would like to thank Bronwyn Hall, Jarle Møen, Dirk Czarnitzki, David Mowery, Bart Verspagen, Jan Fagerberg, Martin Srholec, Svein Olav Nås, Magnus Gulbrandsen, Pierre Mohnen, Ådne Cappelen, Olav Wicken, and Joost Heijs for helpful comments and suggestions on previous versions of this chapter. Funding from the Norwegian Research Council, P. M. Røwde's Foundation and the Ruhrgas Foundation is greatly acknowledged.

1. The Lisbon Agenda also advocates the development of a supportive environment for Small and Medium sized companies (SMEs) in order to facilitate and strengthen the growth and innovative potential of these firms. Improved access to public R&D funds, more efficient use of subsidies to leverage private R&D spending, and improving innovation support services, are other key features of the Lisbon Agenda (NHD 2006).
2. Wicken uses the Norwegian term "Konsern."
3. The statistics for NRC, which operates the FUNN program, do not include the FUNN program. EU funding is also included in total public funding, which is not done in official R&D data from Statistics Norway.

4. These statistics can deviate some from those produced by Statistics Norway. We calculate these statistics using the enterprise level, while Statistics Norway uses the business unit level.
5. We have used the number of employees as provided by the firm managers in the R&D survey as discussed below. These employment statistics will deviate from those produced by Statistics Norway because we use survey data in the calculation of the employment figures.
6. Additionality can be defined as the change in industry-financed R&D spending, company behavior or performance that would not have occurred without the public program. In other words, does a public R&D subsidy result in higher or lower levels of firm-financed R&D investment?
7. Methodological discussions and details are available in a working paper (Clausen 2007).
8. In the IV analysis I included four variables in the first stage regression that were excluded from the second stage regression: (1) Amount of funding at the industry level from SND and Ministries, (2) amount of funding at the industry level from NRC, EU, and FUNN, (3) firms' distance to the nearest regional headquarter of SND, (3) firms' distance to the national (and only) headquarter of NRC.

REFERENCES

Alsos, G., Clausen, T., Ljunggren, E., Madsen, E. L. (2008) Evaluering av SkatteFUNNs adferdsaddisjonalitet. I hvilken grad har SkatteFUNN ført til endret FoU adferd i bedriftene? NF rapport nr.13.

Arrow, K. J (1962) "Economic Welfare and the Allocation of Resources for Invention," in R. Nelson (ed.), *The Rate and Direction of Inventive Activity*, Princeton University Press: Princeton, pp. 609–25.

Blanes, J. W., Busom, I. (2004) "Who Participates in R&D Subsidy Programs? The Case of Spanish Manufacturing Firms," *Research Policy*, 33, 1459–76.

Cappelen, Å., Fjærli, E., Foyn, F., Hægeland, T., Møen, J., Raknerud, A., Rybalka, M. (2008) Evaluering av SkatteFUNN—Sluttrapport. SSB rapport 2008/2.

Clausen, T. H. (2007) "Do Subsidies have Positive Impacts upon R&D and Innovation Activities at the Firm Level?" *TIK Working Paper on Innovation Studies No. 20070615.*

Hall, B. H. (2002) "The Financing of Research and Development," *Oxford Review of Economic Policy*, 18, 35–51.

Hatling, L., Herstad, S., Isaksen, A. (2000) "SND og distriktsutvikling—rolle, virkemidler og effekter," STEP report 05/00.

Hauknes, J., Broch, M., Smith, K. (2000) "SND og bedriftsutvikling—rolle, virkemidler og effecter," STEP report 04/00.

Hervik, A. (2004) "Kunnskapsstatus—Samfunnsøkonomisk avkastning fra forskning," Rapport 0406, Møreforskning.

Hervik, A., Bræin. L., Bergem, B. G. (2004) "Resultatmåling av brukerstyrt forskning, Anslag til samfunnsøkonomiske nytte/kostnadsanalyser av brukerstyrt forskning," Rapport 0407, Møreforskning.

Hervik, A., Bræin. L., Bergem, B. G. (2005) "Resultatmåling av brukerstyrt forskning," Rapport 0509, Møreforskning.

Klette, T. J., Møen, J. (1998) "R&D Investment Responses to R&D Subsidies: A Theoretical Analysis and Microeconometric Study," presented at NBER summer institute 1998.

Klette, T. J., Møen, J., Griliches, Z. (2000) "Do Subsidies to Commercial R&D Reduce Market Failures? Microeconometric Evaluation Studies". *Research Policy*, 29, 471–95.

Lundvall, B. A., Borrás, S. (2005) "Science Technology and Innovation Policy," in J. Fagerberg, D. Mowery, R. Nelson (eds.), *The Oxford Handbook of Innovation*, Oxford University Press.

Madsen, E. L., Brastad, B. (2005) "Kundeundersøkelse av bedriftsrettede virkemidler fra Innovasjon Norge. Etter undersøkelse i 2006 blant bedrifter som fikk finansiert bedriftsutviklingsprosjekter i 2002," NF Rapport nr. 6/2005. Nordlandsforskning.

—— —— (2006) "Kundeundersøkelse av bedriftsrettede virkemidler fra Innovasjon Norge. Etter undersøkelse i 2006 blant bedrifter som fikk finansiert bedriftsutviklingsprosjekter i 2002," NF Rapport nr. 9/2006. Nordlandsforskning.

Narula, R. (2002) "Innovation Systems and 'Inertia' in R&D Location: Norwegian Firms and the Role of Systemic Lock-in," *Research Policy*, 31, pp. 795–816.

Nelson, R. R. (1959) "The simple economics of basic scientific research," *Journal of Political Economy*, 67, 297–306.

NHD (2006) *The EU Lisbon strategy—A Norwegian Perspective. A policy document from the Ministry of Trade and Industry.* Retrieve from: http://odin.dep.no/filarkiv/271140/Lisboa06.pdf

Rosenberg, N. (1996) "Why do Firms do Basic Research (with their own Money)?" *Research Policy*, 19, 165–74.

Rye, M. (2002) "Evaluating the Impact of Public Support on Commercial Research and Development Projects. Are Verbal Reports of Additionality Reliable?" *Evaluation*, 8, 227–48.

SSB (2008) www.ssb.no

Wicken, O. (2000) "Næringslivets FoU og politiske tiltak for å regulere denne." *TIK Working Paper*, Centre for Technology, Innovation and Culture, University of Oslo.

Index

Figures, tables, boxes and appendices are indexed in bold.